HARNESSING DIVIDENDS FROM DRYLANDS

Innovative Scaling up with Soil Nutrients

Harnessing Dividends from Drylands

Innovative Scaling up with Soil Nutrients

Edited by

K.V. Raju and Suhas P. Wani
International Crops Research Institute for the Semi-Arid Tropics (ICRISAT), Patancheru, Telangana, India

CABI

CABI is a trading name of CAB International

CABI
Nosworthy Way
Wallingford
Oxfordshire OX10 8DE
UK

Tel: +44 (0)1491 832111
Fax: +44 (0)1491 833508
E-mail: info@cabi.org
Website: www.cabi.org

CABI
745 Atlantic Avenue
8th Floor
Boston, MA 02111
USA

Tel: +1 (617)682-9015
E-mail: cabi-nao@cabi.org

A catalogue record for this book is available from the British Library, London, UK.

Library of Congress Cataloging-in-Publication Data

Names: Raju, K. V., editor. | Wani, S. P. (Suhas Pralhad), 1952– editor.
Title: Harnessing dividends from drylands : innovative scaling up with soil nutrients / edited by Dr. K.V. Raju and Dr. S.P. Wani.
Description: Boston, MA : CAB International, [2016] | Includes bibliographical references and index.
Identifiers: LCCN 2016019493 | ISBN 9781780648156 (hbk) | ISBN 9781780648170 (epub)
Subjects: LCSH: Dry farming–India–Case studies. | Crop yields–India–Case studies. | Soils and nutrition–India–Case studies. | Agricultural productivity–India–Case studies.
Classification: LCC SB110 .H38 2016 | DDC 631.5/860954–dc23 LC record available at https://lccn.loc.gov/2016019493

ISBN-13: 978 1 78064 815 6

Commissioning editor: Ward Cooper
Editorial assistant: Emma McCann
Production editor: Lauren Povey

Typeset by SPi, Pondicherry, India.
Printed and bound in the UK by CPI Group (UK) Ltd, Croydon, CR0 4YY.

Contents

Contributors

A.V.R. Kesava Rao, Scientist (Agroclimatology), ICRISAT Development Center (IDC), Regional Program – Asia. E-mail: k.rao@cgiar.org

G. Pardhasaradhi, Manager – Soil Analytical Laboratory, ICRISAT Development Center (IDC), Regional Program – Asia. E-mail: g.pardhasaradhi@cgiar.org

Gajanan Sawargaonkar, Senior Scientist (Agronomy), ICRISAT Development Center (IDC), Regional Program – Asia. E-mail: g.sawargaonkar@cgiar.org

Girish Chander, Senior Scientist (NRM), ICRISAT Development Center (IDC), Regional Program – Asia. E-mail: g.chander@cgiar.org

K. Krishnappa, Resident Project Scientist – Karnataka, ICRISAT Development Center (IDC), Regional Program – Asia. E-mail: k.kamma@cgiar.org

K. Srinivas, Lead Scientific Officer, ICRISAT Development Center (IDC), Regional Program – Asia. E-mail: k.srinivas@cgiar.org

K.H. Anantha, Senior Scientist (NRM), ICRISAT Development Center (IDC), Regional Program – Asia. E-mail: k.anantha@cgiar.org

K.L. Sahrawat, Formerly of ICRISAT Development Center (IDC), Regional Program – Asia.

K.V. Raju, Theme Leader, Policy & Impact, Regional Program – Asia. E-mail: kv.raju@cgiar.org

Kaushal K. Garg, Senior Scientist (NRM), ICRISAT Development Center (IDC), Regional Program – Asia. E-mail: k.garg@cgiar.org

Mukund D. Patil, Senior Scientist (Soil Physics), ICRISAT Development Center (IDC), Regional Program – Asia. E-mail: m.patil@cgiar.org

Sudi Raghavendra Rao, Senior Manager (Watersheds), ICRISAT Development Center (IDC), Regional Program – Asia. E-mail: s.r.rao@cgiar.org
Suhas P. Wani, Regional Director – Asia. E-mail: s.wani@cgiar.org
V. Nageswara Rao, Lead Scientific Officer, Resilient Dryland Systems (RDS), Innovation Systems for the Dryland Program. E-mail: v.nageswararao@cgiar.org

All the contributors are from International Crops Research Institute for the Semi-Arid Tropics (ICRISAT), Patancheru 502324, Telangana, India.

Foreword

Currently, crop yields of smallholder farmers are two to five times lower than attainable yields. This provides a great opportunity for harnessing the potential of dryland agriculture, particularly in Asia and Africa. Farmers are not able to harness the benefits of new technologies, products, institutions and policies due to multiple constraints. As a result, their current productivity levels are very low, resulting in poor economic returns. There is an urgent need for innovations in the science of delivery rather than technology development per se to scale up science-led technologies for sustainable development in Asia and Africa. Scientists who undertake the discovery and proof of concept stages often assume dissemination and uptake will happen on their own. However, this is not so; unless focused attention is given to uptake and delivery, results from research will not reach smallholder farmers and not be adopted on a large scale. This long-felt gap has been bridged by the innovative mission mode programme 'Bhoochetana', which is implemented by the Department of Agriculture, Government of Karnataka with technical support from the International Crops Research Institute for the Semi-Arid Tropics (ICRISAT).

The Bhoochetana Mission Program began in 2009 and this book *Harnessing Dividends from Drylands: Innovative Scaling up with Soil Nutrients* brings together the results of a novel scaling-up initiative implemented during 2009–2012. The second phase of Bhoochetana covers all 30 districts for both irrigated and dryland crops with strong leadership by the Government of Karnataka with technical support from ICRISAT. This book covers the micro to macro perspective, and describes in detail how the Bhoochetana Mission Program was conceived and has evolved to reach more than 5 million farmers in 4 years. The chapters describe soil nutrient mapping as a knowledge-based entry point for building a rapport with the community and ensuring good community participation. New technologies such as geographical information system (GIS), information and communications technology (ICT), knowledge dissemination, monitoring and evaluation as well as analytical tools in addition to innovations in the areas of institutions, policies and delivery mechanisms have also contributed immensely to achieving the desired impact, and all these aspects have been covered in a number of chapters. Climate change is disrupting the livelihoods of smallholder farmers in the drylands. Over the 4 years of the project, large weather variability has resulted in 2 years of severe drought in the state of Karnataka. This provided a good opportunity for assessing the robustness of the interventions to minimize the impacts of weather variability on agricultural productivity and increase farmers' incomes. The benefits, such as increased water productivity and incomes along with social and economic benefits, were largely due to adoption of a holistic systems approach, starting with assessing the institutions to detailed monitoring and evaluation of programme implementation and impact assessment.

This book is unique in its detailed coverage of understanding the process of scaling up technology in developing countries which has benefited more than 5 million smallholder farms in Karnataka. The things learnt that are documented in this book will serve as a guide to researchers, development agents, development investors, policy makers and all who are interested in the science of delivery and adoption of improved technologies that empower farmers to become commercially successful.

Dr David Bergvinson

Director General
ICRISAT

Preface

The biggest challenge of the 21st century for humankind is to achieve sustainable development and ensure food and nutritional security for the ever-increasing human population, which has already reached 7.3 billion and is estimated to exceed 9 billion by 2050. The task of achieving food and nutritional security and improving the livelihoods of smallholder farmers is becoming challenging with declining per capita availability of land and water, particularly in the tropical countries of Asia and Africa. There exists a huge potential of drylands to be harnessed in Asia and Africa as current farmer yields are two- to fivefold lower than the achievable potential yields. For the past two to three decades, the crop yield gap, particularly in drylands, clearly indicates the failure of uptake of science-led development by smallholder farmers, as several technologies/products that have been developed by scientists remain on the shelf. The missing link between the pilot stage and impact/output stage has been overlooked by the biophysical and social scientists over the years as they continue to work in their own areas of expertise in the reductionist manner. The science of delivery, which calls for integration and building partnerships with different stakeholders and linking smallholder farmers to the markets to ensure tangible benefits for the farmers, has been demonstrated as a proof of concept in Bhoochetana, a mission programme of the Government of Karnataka technically supported by ICRISAT.

Another important aspect which contributed to the success of the Bhoochetana Mission Program is the use of demand-driven technologies rather than supply-driven technologies through which farmers' participation was ensured. Tangible economic benefits due to adoption of new interventions such as soil health mapping, used as an entry point activity for improving the livelihoods, proved the effectiveness of a science-led approach. This was followed up with improved cultivars and soil, crop, water and pest management intervention along with an innovative knowledge dissemination approach and institutional arrangements for regular monitoring and evaluation. The Department of Agriculture, Government of Karnataka adopted weekly video conferencing for effective monitoring and evaluation, which served not only the purpose of monitoring but also as a learning process for all the participants.

This book, *Harnessing Dividends from Drylands: Innovative Scaling up with Soil Nutrients*, is an assembly and synthesis of things that have been learned by a large number of stakeholders which have been presented by the scientists working in their areas of specialization. This book will be a very useful resource for the development investors, policy makers, researchers and development agents to stress the need to undertake a demand-driven, science-led, integrated approach for rural development rather than a compartmental approach, which has been promoted over the years by the reductionist way of looking at science and development separately.

K.V. Raju and Suhas P. Wani
ICRISAT

Acknowledgements

We gratefully acknowledge the Government of Karnataka (GoK) for providing the opportunity to scale up the productivity enhancement initiative through a mission-mode programme 'Bhoochetana' in all the districts of Karnataka state in India. In particular we acknowledge the support provided by the Chief Ministers and Agricultural Ministers of the GoK for their support for implementation of Bhoochetana in the state since 2009. We sincerely thank Chief Secretaries of GoK, mainly S.V. Ranganath, Kaushik Mukherjee and Arvind Jadhav; also Additional Chief Secretary cum Development Commissioners, mainly Shanta Kumari, Subhir Hari Singh, V. Umesh, G.V. Krishna Rau, Meera Saxena, Abijit Das Gupta, Kaushik Mukherjee and Latha Krishna Rao for their guidance and support for executing the mission programme efficiently. We also thank all the Principal Secretaries of Agriculture, Commissioners and Directors along with Joint Directors of Agriculture, Assistant Directors of Agriculture and other officers whose unstinting support and active participation ensured the success of the programme. We sincerely acknowledge the support of four Universities of Agricultural Sciences (Bengaluru, Raichur, Dharwad and Shivamogga) and their vice-chancellors and the State Level Coordination Committee members for their support and technical help for university scientists and consortium partners – the Department of Agriculture, Department of Horticulture, Department of Water Resources, Department of Finance, Department of Economics and Statistics, Watershed Development Department, Karnataka State Natural Disaster Management Cell, Karnataka State Seed Corporation, community-based organizations, *Krishi Vigyan Kendras*, Watershed Committees, user groups and Watershed

Associations. We highly appreciate the efforts of farm facilitators and lead farmers and millions of farmers in Karnataka who adopted the improved crop management practices and made the Bhoochetana initiative successful. This publication is based on an output from a project funded by the Government of Karnataka, India. The views expressed are not necessarily those of GoK. We sincerely thank the former ICRISAT Director General, Dr William D. Dar, and the present Director General, Dr David Bergvinson, who supported the Bhoochetana mission partnership between ICRISAT and GoK. We are thankful to Sheila Vijayakumar for editorial assistance, N. Srilakshmi for meticulous secretarial assistance and M. Chaitanya for enabling the production of figures.

1

Drylands for Food Security: A Macro Perspective

K.V. Raju* and Suhas P. Wani

International Crops Research Institute for the Semi-Arid Tropics (ICRISAT), Patancheru, India

1.1 Introduction

Across the planet Earth, livelihoods are dependent on farming in drylands and such farmers have to face the challenges this presents[1] particularly in the tropics, which are densely populated. Harsh climatic conditions laced with low crop productivity increase vulnerability. Over the decades, the world has realized that using drylands for food production is a challenging task. This is in spite of enormous research nurtured on drylands by leading organizations and implemented by public and private agencies across several countries. Some 6.5 million km^2 in over 55 countries are classified as dryland tropics (Fig. 1.1). More than 2 billion people currently live in the drylands, with 600 million considered to be poor.[2] In practice, reducing land degradation and improving dryland resilience is based on the premise that land is the resource base that sustains the livelihoods of most poor dryland inhabitants; any variation in weather, climate and economic circumstances strongly affects their well-being.

1.2 A Changing World

An overview of recent developments and research publications has indicated that the tropical drylands are changing rapidly, and the challenges

*kv.raju@cgiar.org

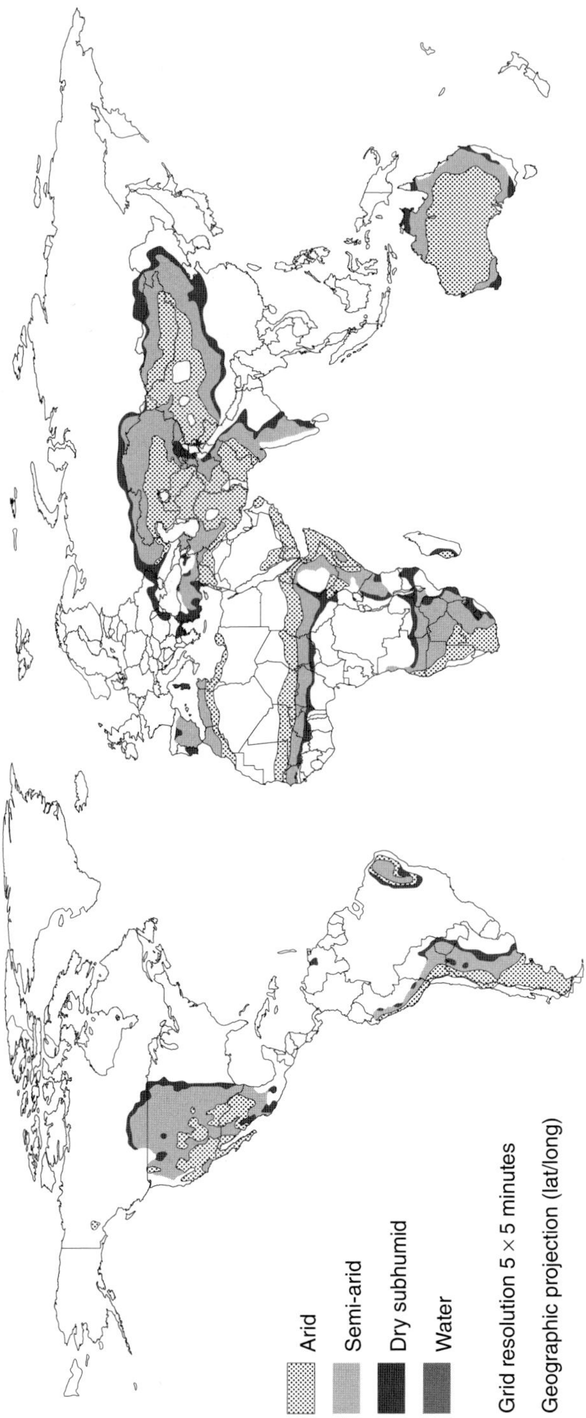

Fig. 1.1. Distribution of drylands in the world. (From FAO, 2002.)

are mounting largely because of the increased vulnerability of drylands to the impacts of climate change. The globalization and liberalization of food markets is increasingly affecting the demand for and supply of dryland crops. Accelerating environmental threats such as climate change, land degradation and biodiversity loss add new pressures to long-standing challenges of poverty, food and nutrition insecurity, drought, increasing population, and rising costs of food and inputs in the dry tropics. A 'perfect storm' of these converging pressures is looming over the drylands, and it will punish the poor most severely. In recent decades, great advances in scientific methodologies and technologies are increasing humankind's ability to address these challenges, especially in such areas as biotechnology, geospatial sciences, modelling and information and communication technology.

These advances, if harnessed for the good of the poor people, hold the potential to increase enormously the power and effectiveness of research in a range of areas. They will be particularly effective: (i) in improving the use of genetic resources (genetic material of plants that is of value for present and future generations of people) and genetic information for crop improvement; (ii) in the analysis and integration of complex environmental information; (iii) in the management of unprecedented quantities of data emerging from advanced research; and (iv) in improving exchange of knowledge with partners.

Despite scientific advances, however, a crisis continues. The drylands are home to the deepest pockets of poverty on Earth. The incidence of poverty is highest in rural areas, where agriculture is the main occupation for 50–100% of the poor in Asia and Africa; the 2008 World Development Report presented compelling empirical evidence from a wide range of countries that supports this finding (World Bank, 2008). Nearly every year, drought and hunger hit hard somewhere in the drylands. The external image of the tropical drylands is one of perennial crisis and suffering, a zone that threatens to remain dependent on emergency relief aid. Table 1.1 shows the global and continent-wise rainfed area and percentage of total arable land that is rainfed. The projected tropical drylands in 2050, based on climate change models, in Africa and Asia are presented in Fig. 1.2. The maps in Figs 1.1 and 1.2 were produced starting with year 2000 data from the CRU-TS 3.0 Climate Database and using the HADCM 3 output of the A2a scenario to determine the year 2050 areas.

Yet there are signs of hope. The incidence of absolute poverty (purchasing power parity adjusted income less than US$1/day) has declined substantially in dryland Asia, from 54.7% in 1990 to 20.7% in 2010, benefiting 745 million Asians (ADB, 2014). However, absolute poverty remains stagnant at a high level in Africa (47%), and the African poor lie further below the dollar-a-day poverty line than the Asian poor. Asia's higher total population indicates there are still twice as many poor in the drylands as

Table 1.1. Rainfed areas in the world. (From FAO, 2010a, b.)

Continent/region	Total arable land (million ha)	Rainfed area (million ha)	Rainfed area (%)
World	1551.0	1250.0	80.6
Africa	247.0	234.0	94.5
Northern Africa	28.0	21.5	77.1
Sub-Saharan Africa	218.0	211.0	96.7
Americas	391.0	342.0	87.5
Northern America	253.5	218.0	86.0
Central America and Caribbean	15.0	13.5	87.7
Southern America	126.0	114.0	90.8
Asia	574.0	362.0	63.1
Middle East	64.0	41.0	63.4
Central Asia	40.0	25.5	63.5
Southern and Eastern Asia	502.0	328.0	65.4
Europe	295.0	272.0	92.3
Western and Central Europe	125.0	107.5	85.8
Eastern Europe	169.0	164.0	97.1
Oceania	46.5	42.5	91.4
Australia and New Zealand	46.0	42.0	91.3
Other Pacific Islands	0.57	0.56	99.3

there are in dryland Africa (185 versus 95 million), although the gap is decreasing because of poverty incidence trends plus higher population growth rates in Africa than in Asia (approximately 3% versus 2%, both expected to decline by a percentage point by 2020) (United Nations, 2014).

Despite Asia's declining poverty rate, childhood malnutrition remains at a totally unacceptable rate (42% of dryland children are malnourished), even higher than in Africa (27%). Childhood malnutrition rates are higher in the tropical drylands than in any other agroecological zone (Walker, 2010). Battling these scourges will be even more challenging in view of the perfect storm looming over the dryland poor.

1.3 Dryland Development Dynamics: Global Level

The Walker report (2010) analysed dryland trends and implications flowing from two major recent World Development Reports (World Bank, 2008, 2009) on agricultural development and the geography of poverty. The analysis concluded that national economies can be broadly described in terms of three states of agricultural development: (i) subsistence agriculture; (ii) transitional agriculture; and (iii) urban market-oriented agriculture. The emergence from subsistence to market-oriented agriculture reduces poverty because markets stimulate demand for a wider diversity of

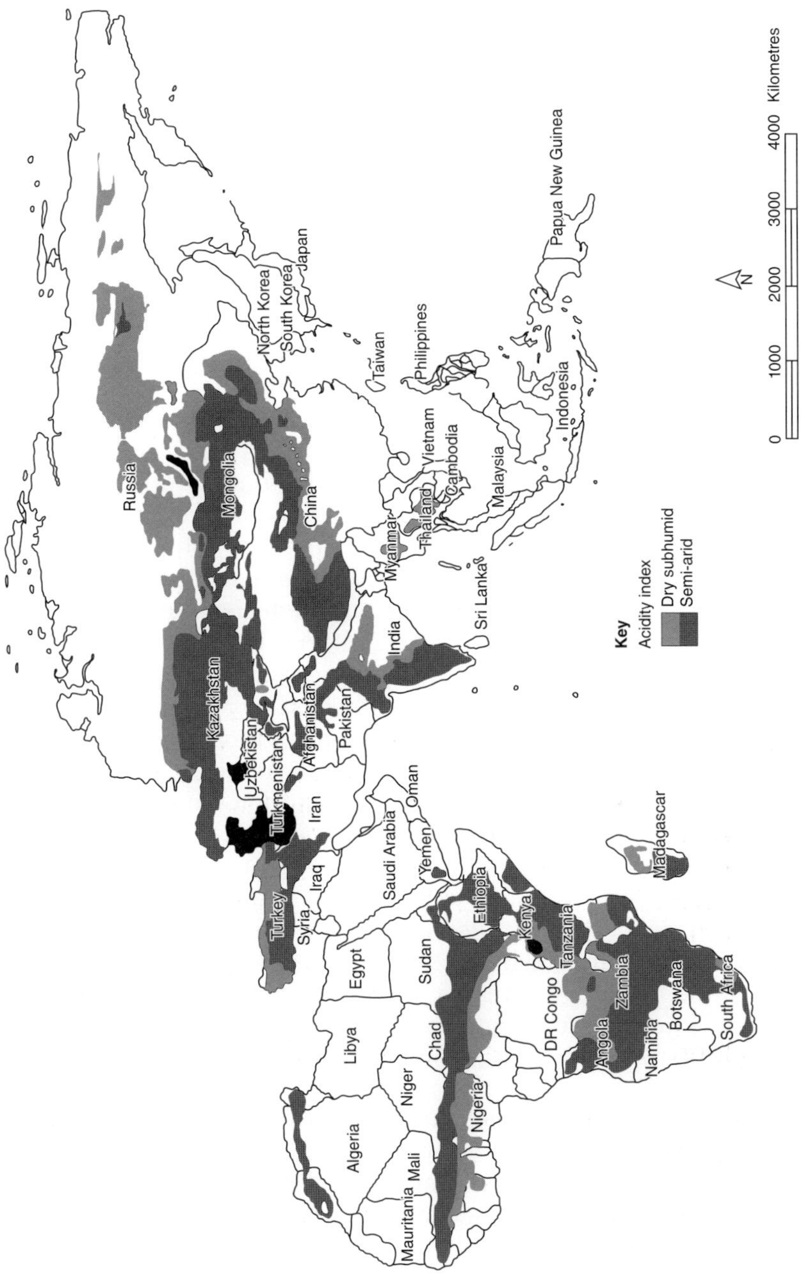

Fig. 1.2. Projected tropical drylands of Africa and Asia in 2050 based on climate change models.

higher-value foodstuffs and agro-industrial products that raise rural incomes (as well as creating opportunities beyond agriculture). Even the unique, long-term Village Level Studies of International Crops Research Institute for the Semi-Arid Tropics (ICRISAT) provide ground-level details to complement the World Bank's macro-scale picture (Uttam *et al.*, 2014).

These Village Level Studies confirm that where poverty has declined it is because of diversification beyond staple food crops into market-connected activities as well as off-farm employment. ICRISAT staff concurred with this perspective in their regional consultations (ICRISAT, 2010); helping the poor emerge from subsistence agriculture to sustainable market-oriented agriculture emerged as a consensus priority. Sub-Saharan Africa's drylands today are largely subsistence based, while dryland Asia finds itself in transition towards a market-oriented agricultural economy. There are, of course, local variations within each of these large regions, especially depending on distances to urban markets, ease of transportation and communications' infrastructure. Also, individual smallholder farm households carry out differing blends of subsistence versus market-oriented activities, and this mix changes over years in response to market forces. Nevertheless, subsistence versus commercial orientation explains much about rural dryland poverty and insecurity, and about livelihood strategies and decision making.

As Asian farmers have diversified into cash crops, the monetary value of ICRISAT's five mandate crops (sorghum (*Sorghum bicolor*), pearl millet (*Pennisetum glaucum*), chickpea (*Cicer arietinum*), pigeon pea (*Cajanus cajan*) and groundnut (*Arachis hypogaea*)) as a percentage of the total value of all crops grown in the drylands fell from 50% in 1950 to 20% in 2005/06 in India. A major factor on the supply side has been increasing access to irrigation which has displaced rainfed crops, including cereals (generally referred as coarse cereals) traditionally consumed by the very poor, in favour of government price supported crops, such as rice (*Oryza sativa*) and wheat (*Triticum aestivum*). On the demand side, an important factor has been a shift in consumer preference towards these two cereals with growing incomes. However, the ascendance of markets for a range of other commodities has also been a major factor (maize (*Zea mays*), soybean (*Glycine max*), rape (*Brassica napus*), castor (*Ricinus communis*), potato (*Solanum tuberosum*) and horticultural and floricultural crops, along with dairy, poultry and eggs). Such market-oriented diversification opens the door to new contributions through research to dryland farming systems. Smallholder farm households that are beginning to emerge out of poverty, hunger and environmental degradation are still vulnerable. The Walker (2010) analysis noted that when the dollar-a-day threshold that defines absolute poverty is relaxed to US$2/day, poverty rates in dryland Asia and Africa become similar (about 80% of the dryland population in both the regions lies below the US$2/day poverty line). This suggests that even though market-oriented development has begun to increase incomes, especially in the Asian drylands, escape from poverty is still in the early stages. Institutions and policies must change to enhance the adoption of new technologies if the pace of commercialization, and consequently of poverty reduction, is to accelerate. This scenario makes it explicit to innovate the scaling-up models quickly to reduce eradication of poverty as per the proposed sustainable development goals.

Livestock are as important as crops in value share of production in the drylands, though with strong variations across locations and the regions. They diversify and hedge risks: for example, cattle can move from one grazing area to another to escape drought (although restrictions on land access are increasingly constraining this option in many areas). Cattle diseases are also less frequent in dry areas. Livestock are a strategic investment when cash is available and a cash source when times are hard (sometimes referred to as 'walking wealth' or 'mobile bank'). They also generate ecological synergies with crops: for example, farmers allow herders to graze their animals on crop stubble so that subsequent crops can benefit from soil-enriching manure deposits.

A demand-driven livestock revolution is underway in South Asia and Africa (as seen in the growing demand for milk and meat), offering an opportunity for smallholder farm households to increase their incomes as well as capture the farming benefits highlighted above. Leading member organizations of the Consultative Group on International Agricultural Research (CGIAR) have been working on these relevant issues: ICRISAT, the International Center for Agricultural Research in the Dry Areas (ICARDA) and the International Institute of Tropical Agriculture (IITA) are focused on diversified farming systems and the International Livestock Research Institute (ILRI) is focused on livestock issues, to exploit crop–livestock synergies. The same perspective is maintained regarding agroforestry issues, which are the core topic of the World Agroforestry Centre (ICRAF). The World Vegetable Center (AVRDC) works on diversifying farming systems to integrate high-value, nutrition-enhancing vegetable crops into smallholder dryland farming.

1.4 Food Insecurity in Drylands

The agricultural productivity has seen a rapid growth in the past 50 years due to improved crop cultivars, fertilizer use and expansion in irrigated agriculture. The world food production outstripped the population growth. However, there are regions of food insecurity. Of the 7.3 billion population in 2013, about 805 million people face food insecurity. About 60% of them live in South Asia and sub-Saharan Africa. Food and crop demand is estimated to double in the next 50 years.

According to a global comprehensive assessment of water for food and health, it is possible to produce the required food to feed the growing population – but it is probable that today's food production and environmental trends, if continued, will lead to crises in many parts of the world (Molden *et al.*, 2007).

The importance of rainfed agriculture varies regionally, but produces most food for poor communities in developing countries. In sub-Saharan Africa more than 95% of the farmland is rainfed, while the corresponding figure for Latin America is almost 90%, for South Asia about 60%, for East Asia 65% and for Near East and North Africa 75% (FAOStat, 2005). Most countries in the world depend primarily on rainfed agriculture for food grains. Despite large strides made in improving productivity and environmental conditions in many developing countries, a great number of poor families in Africa and Asia still face poverty, hunger, food insecurity and malnutrition, where rainfed agriculture is the main agricultural activity. These problems are exacerbated by adverse biophysical growing conditions and the poor socio-economic infrastructure in many areas in the semi-arid tropics (SAT). This region is home to 38% of the poor in developing countries, of which 75% live in rural areas. Over 45% of the world's hungry and more than 70% of its malnourished children live in the SAT.

The importance of rainfed sources of food weighs disproportionately on women, given that approximately 70% of the world's poor are women (WHO, 2005). Agriculture plays a key role for economic development (World Bank, 2005) and poverty reduction (Irz and Roe, 2000), with evidence indicating that every 1% increase in agricultural yields translates to a 0.6–1.2% decrease in the percentage of absolute poor (Thirtle *et al.*, 2002). On average for sub-Saharan Africa, agriculture accounts for 35% of gross domestic product (GDP) and employs 70% of the population (World Bank, 2000) while more than 95% of the agricultural area is rainfed (FAOStat, 2005).

1.5 Crop Yields in Drylands

In the past 40 years, agricultural land use has expanded by 20–25%, which has approximately contributed to 30% of the overall grain production growth during the period (FAO, 2002; Ramankutty *et al.*, 2002). The remaining yield outputs originated from intensification through yield increases per unit land area. However, the regional variation is large, as is the difference between irrigated and rainfed agriculture. In developing countries rainfed grain yields are on average 1.5 t/ha, compared with 3.1 t/ha for irrigated yields (Rosegrant *et al.*, 2002), and an increase in production from rainfed agriculture has mainly originated from land expansion.

Trends are clearly different for different regions. With 99% of rainfed production of main cereals such as maize, millet and sorghum, the cultivated cereal area in sub-Saharan Africa has doubled since 1960 while the yield per unit land has nearly been stagnant for these staple crops (FAOStat, 2005). In South Asia, there has been a major shift away from more drought-tolerant low-yielding crops such as sorghum and millet while wheat and maize have approximately doubled in area since 1961 (FAOStat, 2005). During the same period, the yield per unit land for maize and wheat has more than doubled per unit land (Fig. 1.3). For predominantly rainfed systems, maize crops per unit land have nearly tripled and wheat more than doubled during the same time period.

1.6 Untapped Potential: Rainfed Crops

In several regions of the world rainfed agriculture generates among the world's highest yields. These are predominantly temperate regions, with relatively reliable rainfall and inherently productive soils. Even in tropical regions, particularly in the subhumid and humid zones, agricultural yields in commercial rainfed agriculture exceed 5–6 t/ha (Rockström and Falkenmark, 2000; Wani *et al.*, 2003a, b). At the same time, the dry subhumid and semi-arid regions have experienced the lowest yields and the weakest yield improvements per unit of land (Table 1.2). Here, yields fluctuate between 0.5 t/ha and 2 t/ha for cereal crops, with an average of 1 t/ha in sub-Saharan Africa, and 1–1.5 t/ha in South Asia, Central Asia and West Asia and North Africa (CWANA) for rainfed agriculture (Rockström and Falkenmark, 2000; Wani *et al.*, 2003a, b).

Yield gap analysis carried out by Comprehensive Assessment, for major rainfed crops in semi-arid regions in Asia and Africa and rainfed wheat in West Asia and North Africa, revealed large yield gaps with farmers' yields being a factor of two to four times lower than achievable yields for major rainfed crops (Wani *et al.*, 2011). Singh *et al.* (2009) discussed detailed yield gap analysis for major rainfed crops in different parts of the world. Figure 1.4 illustrates examples of observed yield gaps in various countries in Africa, Asia and the Middle East. In countries in eastern and southern Africa the yield gap is very large. Similarly, in many countries in West Asia, farmers' yields are less than 30% of achievable yields, while in some Asian countries the yields are about 50%. Historic trends present a growing yield gap between farmers' practices and farming systems that benefit from management advances (Wani *et al.*, 2003b).

Evidence from a long-term experiment at ICRISAT, Patancheru, India, since 1976, demonstrated the virtuous cycle of persistent yield increase through improved land, water and nutrient management in

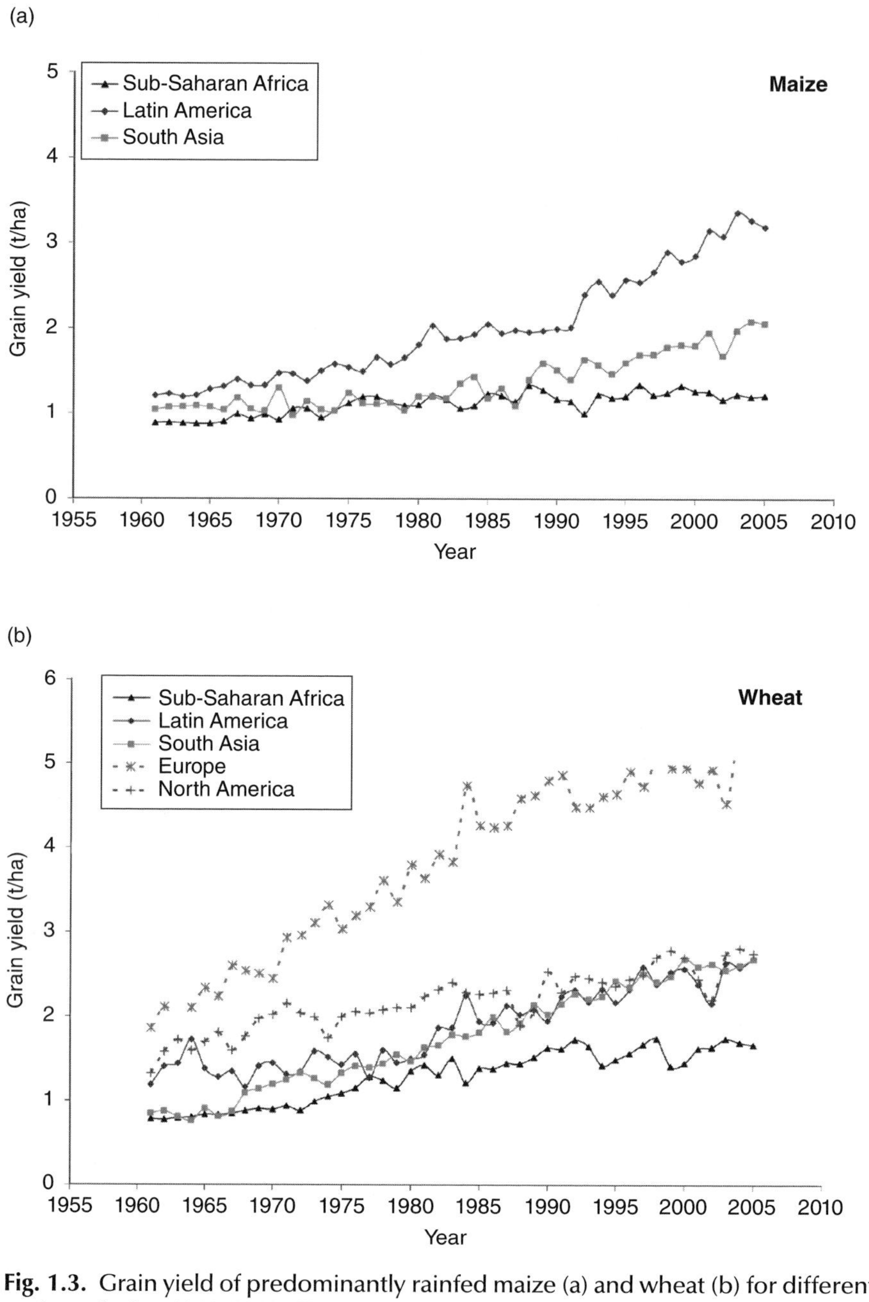

Fig. 1.3. Grain yield of predominantly rainfed maize (a) and wheat (b) for different regions during 1961–2010. (From FAOStat, 2005.)

rainfed agriculture. Improved systems of sorghum/pigeon pea intercrops produced higher mean grain yields (5.1 t/ha) compared with 1.1 t/ha average yield of sole sorghum in the traditional (farmers') post-rainy system where crops are grown on stored soil moisture (Fig. 1.5).

Table 1.2. Types of water stress and underlying causes in semi-arid and dry subhumid tropical environments. (From Falkenmark and Rockström, 2004.)

Types of water stress	Dry spell	Drought
Meteorological	Occurrence: 2 out of 3 years Impact: Yield reduction Cause: Rainfall deficit of 2–5 weeks during crop growth	Occurrence: 1 year out of 10 years Impact: Complete crop failure Cause: Seasonal rainfall below minimum seasonal plant water requirement
Agricultural (human induced)	Occurrence: > 2 out of 3 years Impact: Yield reduction or complete crop failure Cause: Low plant water availability and poor plant water uptake capacity	Occurrence: > 1 out of 10 years Impact: Complete crop failure Cause: Poor rainfall partitioning leads to seasonal soil moisture deficit to produce harvest

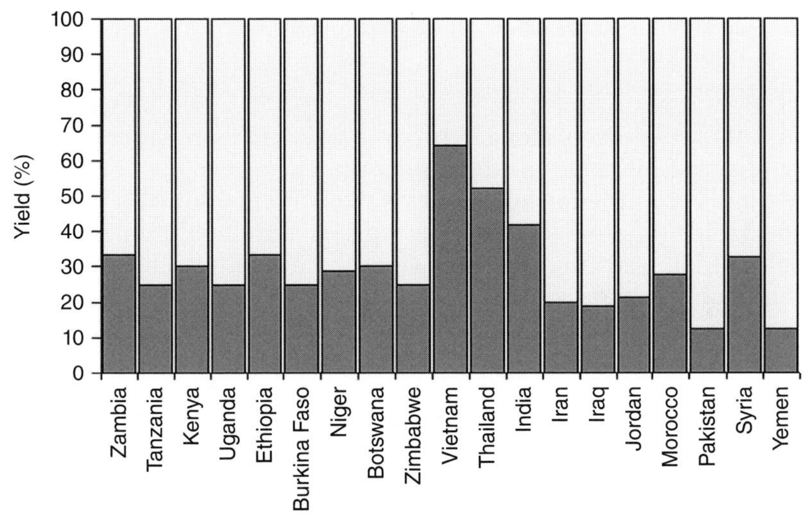

Fig. 1.4. Examples of observed yield gap (for major grains) between farmers' yields and achievable yields (100% denotes achievable yield level, and columns indicate the actual observed yield levels). (From Rockström *et al.*, 2007.)

The annual gain in grain yield in the improved system was 82 kg/ha/year compared with 23 kg/ha/year in the traditional system. The large yield gap between attainable yield and farmers' practice, as well as between the attainable yield of 5.1 t/ha and potential yield of 7 t/ha, shows that a large potential of rainfed agriculture remains to be tapped. Moreover, the improved management system is still continuing to provide an increase in productivity as well as improving soil quality (physical, chemical and

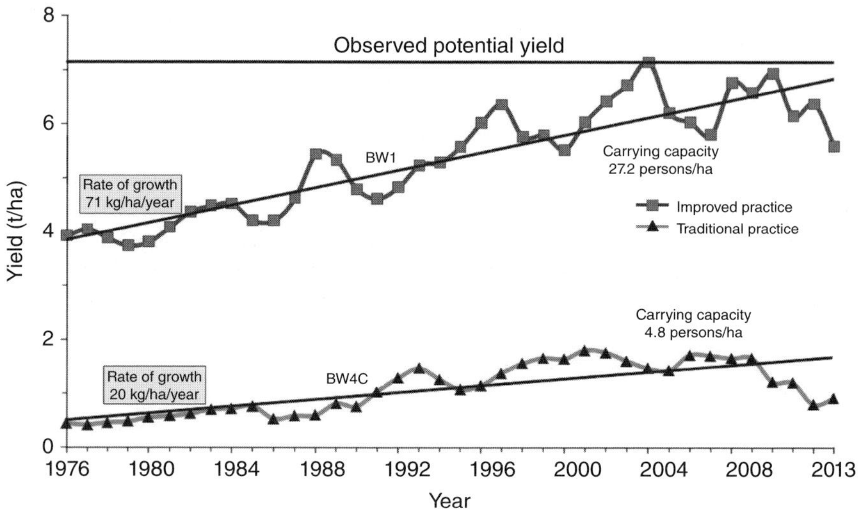

Fig. 1.5. Three-year moving average grain yield of sorghum/pigeon pea intercrop under improved and traditional management in a deep Vertisol catchment at ICRISAT, Patancheru, India.

biological parameters) along with increased carbon sequestration of 330 kg/ha/year (Wani *et al.*, 2003a).

Farmers' yields continue to be very low compared with the experimental yields (attainable yields) as well as simulated crop yields (potential yields), resulting in a very significant yield gap between actual and attainable rainfed yields. The difference is largely explained by inappropriate soil, water and crop management options used at the farm level, combined with persistent land degradation.

The vast potential of rainfed agriculture needs to be unlocked through knowledge-based management of natural resources for increasing the productivity and income to achieve food security in the developing world. Soil and water management plays a very critical role in increasing agricultural productivity in rainfed areas in the fragile semi-arid tropical systems.

1.7 Rainfed Agriculture: Key Constraints

An insight into the inventories of natural resources in rainfed regions shows a grim picture of water scarcity, fragile environments, drought and land degradation due to soil erosion by wind and water, low rainwater use efficiency (35–45%), high population pressure, poverty, low investments in water use efficiency measures, poor infrastructure and inappropriate policies (Wani *et al.*, 2003b, c; Rockström *et al.*, 2007). Drought and land degradation are interlinked in a cause and effect relationship, and the two combined are the main causes of poverty in the farm households. These rainfed areas are also prone to severe land degradation.[3] A global assessment of the extent and form of land degradation showed that 57% of the total area of drylands occurring in two major Asian countries, namely China (178.9 million ha) and India (108.6 million ha), are degraded (UNEP, 1997).

Current rainfed agriculture cannot sustain the economic growth and needed food security. There is an urgent need to develop a new paradigm for soil and water management. ICRISAT's studies in Africa and Asia have identified several key constraints to more widespread technology adoption (Ryan and Spencer, 2001). Other institutes have independently reached similar conclusions for other agroecosystems. So there is general agreement on the key challenges before us. These are:

- Lack of a market-oriented smallholder production system where research is market-led, demand-driven and follows the commodity chain approach to address limiting constraints along the value chain. For example, ICRISAT's work on community watersheds for improving livelihoods in Asia and developing groundnut markets in Malawi aims to address this issue.
- Poor research–extension–farmer linkages, which limit transfer and adoption of technology. For example, ICRISAT's work on farmer field schools in Africa and the consortium approach to integrated management of community watersheds in Asia aims to strengthen these linkages.
- Need for policies and strategies on soil, water and biodiversity to offset the high rate of natural resource degradation. These issues are central to ICRISAT's consortium approach to integrated community watershed management.
- Need to focus research on soil fertility improvement, soil and water management, development of irrigation, promotion of integrated livestock–wildlife–crop systems and development of drought mitigation strategies. These issues are addressed by several ICRISAT programmes, for example: (i) low-input soil fertility approaches in Africa; (ii) micronutrient research in Asia; and (iii) the Sahelian Eco-Farm.
- Need to strengthen capacities of institutions and farmers' organizations to support input and output marketing and agricultural production systems. Such capacity building is a primary goal of Soil Water Management Network (SWMnet) of ASARECA (Association for Strengthening Agricultural Research in Eastern and Central Africa) and Eastern and Central Africa Regional Sorghum and Millet Network (ECARSAM) in Eastern and Central Africa, and of seed systems/germplasm improvement networks globally.
- Poor information flow and lack of communication on rural development issues. These are being addressed by ICRISAT's VASAT

Consortium (Virtual Academy for the Semi-Arid Tropics) globally and specifically ICRISAT's Bio-economic Decision Support work with partners in West Africa.
- Need to integrate a gender perspective in agricultural research and training as seen in ICRISAT's work on human immunodeficiency virus (HIV)/acquired immune deficiency syndrome (AIDS) amelioration in India and southern Africa.

A major research challenge faced in integrated natural resource management (INRM) is to combine the various 'information bits' derived from different stakeholders, and distil these into decision rules that they can use (Snapp and Heong, 2003). ICRISAT's participatory research in southern Africa demonstrated that with micro-dosing alone or in combination with available animal manures, farmers could increase their yields by 30–100% by applying as little as 10 kg of nitrogen/ha (Dimes *et al.*, 2005; Ncube *et al.*, 2006; Rusike *et al.*, 2006).

In much of agricultural research, the multidisciplinary team approach has often run into difficulties in achieving impact because of the perceived disciplinary hierarchy. The integrated genetic and natural resource management (IGNRM) approach in the Community Watershed Consortium pursues integration of the knowledge and products of the various research disciplines into useful extension messages for development workers that can sustain increased yields for a range of climatic and edaphic conditions.

In Asia, the integrated community watershed management approach that aims to promote income-generating and sustainable crop and livestock production options as an important component of improved management of watershed landscapes is a live example of how IGNRM led to significant benefits in a poor area and this holistic participatory approach is transforming the lives of resource-poor small and marginal farmers in the dryland areas of Asia (Wani *et al.*, 2006a).

ICRISAT and the national agricultural research systems in Asia have developed in partnership an innovative and upscalable consortium model for managing watersheds holistically. In this approach, rainwater management is used as an entry point activity starting with *in situ* conservation of rainwater and converging the benefits of stored rainwater into increased productivity by using improved crops, cultivars, suitable nutrient and pest management practices, and land and water management practices. The IGNRM approach has enabled communities not only to harness the benefits of watershed management, but also achieve much of the potential from improved varieties from a wider range of crops. The household incomes and overall productivity have increased considerably throughout selected benchmark sites in Asia (Table 1.3). The benefits not only accrue to landholding households, but also to the landless marginalized groups through the creation of greater employment opportunities. The greater resilience of crop income in the watershed villages during the drought year in 2002 is particularly noteworthy. While the share of crops in household income declined from 44% to 12.2% in the non-project villages, crop income remained largely unchanged from 36% to 37% in the watershed village. The loss in household income in the non-project villages was largely compensated by migration and non-farm income, which increased from 49% in an average year to 75% during the drought year in 2002. Much of this gain originates from improved soil fertility management and increased availability of irrigation water and integration of improved cultivars and cropping patterns into the watershed systems.

Table 1.3. Effect of integrated watershed interventions on alternative sources of household income.[a]

Year	Village group[a]	Statistics	Crop income	Livestock income	Off-farm income	Household income
2001 (average year)	Non-project	Mean income (₹ '000)	12.7	1.9	14.3	28.9
		Share of total income (%)	44.0	6.6	49.5	100.0
	Watershed project	Mean income (₹ '000)	15.4	4.4	22.7	42.5
		Share of total income (%)	36.2	10.4	53.4	100.0
2002 (drought year)	Non-project	Mean income (₹ '000)	2.5	2.7	15.0	20.2
		Share of total income (%)	12.2	13.3	74.5	100.0
	Watershed project	Mean income (₹ '000)	10.1	4.0	13.4	27.6
		Share of total income (%)	36.7	14.6	48.7	100.0

[a]Sample size is n = 60 smallholders in each group (ICRISAT data).

1.7.1 Soil health: a key driver

Soil health is severely affected due to land degradation and soil fertility is the limiting factor to increased yields in rainfed agriculture (Stoorvogel and Smaling, 1990) and is in need of urgent attention. ICRISAT's on-farm diagnostic work in different community watersheds in different states of India as well as in China, Vietnam and Thailand showed severe mining of soils for essential plant nutrients. Exhaustive analysis showed that 80–100% of farmers' fields are deficient not only in total nitrogen but also micronutrients such as zinc, boron and secondary nutrients such as sulfur (Table 1.4). In addition, soil organic matter, an important driving force for supporting biological activity in soil, is very much in short supply particularly in tropical countries. Strategic long-term catchment research at ICRISAT has shown that legume-based systems particularly with pigeon pea could sequester 330 kg carbon up to 150 cm depth in Vertisols at Patancheru, India, under rainfed conditions (Wani *et al.*, 2003a). Under the National Agricultural Technology Project (NATP), ICRISAT, the National Bureau of Soil Survey and Land Use Planning (NBSS&LUP), the Central Research Institute for Dryland Agriculture (CRIDA) and the Indian Institute of Soil Science (IISS) have identified carbon sequestering systems for Alfisols and Vertisols in India (ICRISAT, 2005). In sub-Saharan Africa soil nutrient mining is particularly severe. It is estimated that approximately 85% of African farmland in 2002–2004 experienced a loss of more than 30 kg/ha of nutrients/year (IFDC, 2006).

Table 1.4. Percentage of farmers' fields deficient in soil nutrients in different states of India. (From Rego *et al.*, 2007; Sahrawat *et al.*, 2007.)

State	No. of farmers' fields	Soil nutrients[a]					
		OC (%)	AvP (ppm)	K (ppm)	S (ppm)	B (ppm)	Zn (ppm)
Andhra Pradesh	1927	84	39	12	87	88	81
Karnataka	1260	58	49	18	85	76	72
Madhya Pradesh	73	9	86	1	96	65	93
Rajasthan	179	22	40	9	64	43	24
Gujarat	82	12	60	10	46	100	82
Tamil Nadu	119	57	51	24	71	89	61
Kerala	28	11	21	7	96	100	18

[a]OC, Organic carbon; AvP, available phosphorus; K, potassium; S, sulfur; B, boron; Zn, zinc.

In India, farmers' participatory watershed management trials in more than 300 villages demonstrated that the current subsistence farming has depleted soils not only of macronutrients but also of micronutrients such as zinc and boron and secondary nutrients such as sulfur beyond the critical limits. A substantial increase in crop yields was experienced after micronutrient amendments, and a further increase by 70–120% when both micronutrients and adequate nitrogen and phosphorus were applied, for a number of rainfed crops (maize, sorghum, mung bean (*Vigna radiata*; green gram), pigeon pea, chickpea, castor and groundnut) in farmers' fields (Rego *et al.*, 2005).

1.8 Silent Revolution in Drylands: Meta-analysis

Detailed meta-analysis of 311 watershed case studies in India revealed that watershed programmes are silently revolutionizing rainfed areas with positive impacts (benefit–cost ratio of 1:2.14, internal rate of return of 22%, increased cropping intensity by 64%, increased irrigated area by 34%, reduced runoff by 13% and increased employment by 182 person-days/ha/year) (Joshi *et al.*, 2005). However, 65% of the watersheds were performing below average as they lacked community participation, programmes were supply driven, equity and sustainability issues were elusive and a compartmental approach was adopted (Joshi *et al.*, 2005) (Table 1.5).

Through the use of new science tools (i.e. remote sensing, geographical information system (GIS) and simulation modelling) along with an understanding of the entire food production–utilization system (i.e. food quality and market) and genuine involvement of stakeholders, ICRISAT-led watersheds effected remarkable impacts on resource-poor farm households in the SAT (Wani *et al.*, 2003a). Some of them are discussed below.

Reducing rural poverty in the watershed communities is evident in the transformation of their economies. The ICRISAT model ensured improved productivity with the adoption of cost-efficient water harvesting structures (WHS) as an entry point for improving livelihoods. Crop intensification and diversification with high-value crops is one leading example that allowed households to achieve production of basic staples and surplus for modest incomes. The model has provision for improving the capacity of farm households through training and networking and for alleviating livelihood-enhanced participation especially of the most vulnerable groups such as women and the landless. For example, the self-help groups common in the watershed villages of India and an improved initiative in China provide income and empowerment of women. The environmental clubs whose conceptualization is traced from the Bundi watershed of Rajasthan, India, inculcated environmental protection, sanitation and hygiene among the children.

Building on social capital made a huge difference in addressing rural poverty of watershed communities. This is evident in the case of Kothapally watershed in Telangana state (erstwhile Andhra Pradesh), India. Today, it is a prosperous village on the path of long-term sustainability and has become a beacon for science-led rural development. In 2001, the average village income from agriculture, livestock and non-farming sources was US$945 compared with the neighbouring non-watershed village income of US$613. The villagers proudly professed: 'We did not face any difficulty for water even during the drought year of 2002. When surrounding villages had no drinking water, our wells had sufficient water.'

Table 1.5. Benefits of watersheds – summary of meta-analysis. (From Joshi *et al.*, 2005.)

Indicator	Particulars	No. of studies	Mean	Mode	Median	Min	Max	*t* value
Efficiency	Benefit–cost ratio	128	2.14	1.70	1.81	0.82	7.06	21.25
	Internal rate of return (%)	40	22.04	19.00	16.90	1.68	94.00	6.54
Equity	Employment (person-days/ha/year)	39	181.50	75.00	127.00	11.00	900.00	6.74
Sustainability (%)	Irrigated area (%)	97	33.56	52.00	26.00	1.37	156.03	11.77
	Cropping intensity (%)	115	63.51	80.00	41.00	10.00	200.00	12.65
	Rate of runoff (%)	36	−13.00	−33.00	−11.00	−1.30	−50.00	6.78
	Soil loss (t/ha/year)	51	−0.82	−0.91	−0.88	−0.11	−0.99	39.29

To date, the village prides itself with households owning five tractors, seven lorries and 30 auto-rickshaws. People from surrounding villages come to Kothapally for on-farm employment. With more training on livelihood and enterprise development, migration is bound to cease. Between 2000 and 2003, investments in new livelihood enterprises such as a seed oil mill, a tree nursery and worm composting increased the average income by 77% in Powerguda, a tribal village in erstwhile Andhra Pradesh.

Crop–livestock integration is another facet harnessed for poverty reduction. The Lucheba watershed, Guizhou province of southern China has transformed its economy through modest injection of capital-allied contributions of labour and finance, to create basic infrastructures such as access to roads and drinking water supply. With technical support from the consortium, the farming system was intensified from rice and tending livestock (pig raising) and growing horticultural crops (fruit trees like *Ziziphus* sp.; vegetables like beans, peas (*Pisum sativum*) and sweet potato (*Ipomoea batatas*)) and groundnut. In forage production, wild buckwheat was specifically important as an alley crop as it was a good forage grass for pigs. This cropping technology was also effective in controlling erosion and increasing farm income in sloping lands (Marothia, 2013). This holds true in many watersheds of India where the improvement in fodder production has intensified livestock activities such as breed improvement (artificial insemination and natural means) and livestock centre/health camp establishment (Wani *et al.*, 2006b).

In Tad Fa and Wang Chai watersheds in Northeast Thailand, there was a 45% increase in farm income within 3 years. Farmers earned an average net income of US$1195 per cropping season. A complete turnaround in livelihood system of farm households was inevitable in ICRISAT-led watersheds. The watersheds were also highly effective in reducing annual runoff by 59% (131 mm versus 320 mm) and soil loss by 82% (6.1 t/ha versus 34.2 t/ha) thereby controlling land degradation, which is one of the major problems in Northeast Thailand (Wangkahart *et al.*, 2012).

Increasing crop productivity is a common objective in all the watershed programmes, and enhanced crop productivity is achieved after the implementation of soil and water conservation practices along with appropriate crop and nutrient management. For example, the implementation of improved crop management technology in the benchmark watersheds of Andhra Pradesh increased maize yield 2.5 times and sorghum yield threefold (Wani *et al.*, 2006a). Overall, in the 65 community watersheds (each measuring approximately 500 ha), implementing best-bet practices resulted in significant yield advantages in sorghum (35–270%), maize (30–174%), pearl millet (72–242%), groundnut (28–179%), sole pigeon pea (97–204%) and intercropped pigeon pea (40–110%). In Thanh Ha watershed of Vietnam, yields of soybean, groundnut and mung bean increased by three- to fourfold (2.8–3.5 t/ha) as compared with baseline yields (0.5–1.0 t/ha), reducing the yield gap between potential and farmers' yields. A reduction in nitrogen fertilizer (90–120 kg urea/ha) by 38% increased maize yield by 18%. In Tad Fa watershed of Northeast Thailand, maize yield increased by 27–34% with improved crop management.

Improving water availability in the watersheds was attributed to efficient management of rainwater and *in situ* conservation, establishment of WHS and improved groundwater levels. Findings in most of the watershed sites reveal that open wells located near WHS have significantly higher water levels compared with those away from the WHS. Even after the rainy season, the water level in wells nearer to WHS sustained good groundwater yield. In the various watersheds of India such as Lalatora (in Madhya Pradesh), the treated area registered a groundwater level rise by 7.3 m. At Bundi, Rajasthan, the average rise was 5.7 m and the irrigated area increased from 207 ha to 343 ha. In Kothapally watershed in Telangana state (erstwhile Andhra Pradesh), the groundwater level rise was 4.2 m in open wells. The various WHS resulted in an additional groundwater recharge per year of approximately 428,000 m^3 on average. With this improvement in groundwater availability, the supply of clean drinking water was guaranteed. In Lucheba watershed in China, a drinking water project, which constitutes a water storage tank and pipelines to farm households, was a joint effort of the community and the watershed project. This solved the drinking water problem for 62 households and more than 300 livestock. Earlier every farmer's household used to spend 2–3 h a day fetching drinking water (Marothia, 2013). This was the main motivation for the excellent farmers' participation in the project. On the other hand, collective pumping out of well water established an efficient water distribution system and enabled a farmers' group to earn more income by growing watermelon (*Citrullus lanatus*) with reduced drudgery as women had to carry water on their heads from a long distance. Pumping of water from the river as a means to irrigate watermelon has provided maximum income for households in Thanh Ha watershed in Vietnam (Wani *et al.*, 2006b).

Supplemental irrigation can play a very important role in reducing the risk of crop failures and in optimizing the productivity in the SAT. In these regions, there is good potential for delivering excess rainwater to storage structures or groundwater because even under improved systems there is loss of 12–30% of the rainfall as runoff. Striking results were recorded from supplemental irrigation on crop yields in ICRISAT benchmark watersheds in Madhya Pradesh. On-farm studies made during 2000–2003 post-rainy seasons showed that chickpea yield (1.25 t/ha) increased by 127% over the control yield (0.55 t/ha), and groundnut pod yield (1.3 t/ha) increased by 59% over the control yield (0.82 t/ha) by application of two supplemental irrigations of 40 mm. Similar yield responses in mung bean and chickpea crops were obtained from supplemental irrigation at the ICRISAT Development Center in Patancheru. Our results showed that crops on light-textured soils such as Alfisols respond better with supplemental irrigation. Clearly, there is potential to enhance productivity and reduce the risks of crop failures through application of harvested water through supplemental irrigation at critical stage(s) of the crop (Pathak *et al.*, 2009).

1.9 Indian Agriculture

1.9.1 An overview

Over the last decade Indian agriculture has become more robust, with record production of food grains and oilseeds, particularly in drylands; 60% of 142 million ha of arable land is rainfed (i.e. about 85 million ha). Increased procurement, consequently, has added huge stocks of food grains in the granaries. India is one of the world's top producers of rice, wheat, milk, fruits and vegetables. However, given that India is still home to a quarter of all undernourished people in the world and since on average almost half the total expenditure of about half the households is on food, increasing the efficiency of the farm-to-fork value chain is crucial for eliminating poverty and malnutrition.

As a concomitant of growth, the share of agriculture and allied sectors in GDP declined to 15.2% during the Eleventh Five Year Plan (Government of India, 2014) and further to 13.9% in 2013/14 (provisional estimates). While it still accounts for about 54.6% of total employment (Census 2011), there has been a decline in the absolute number of cultivators, which is unprecedented, from 127.3 million (Census 2001) to 118.7 million (Census 2011). This is indicative of a shift from farm to non-farm employment, causing real farm wages to rise by over 7% annually in recent years (GoI, 2014).

The resilience of Indian agriculture is evident in that this sector last posted negative growth in 2002/03 and has registered a remarkable average growth rate of 4.1% during the Eleventh Five Year Plan (2007/08 to 2011/12). As per the provisional estimates for 2013/14, the growth rate of agriculture GDP was 1.4% and 4.7%, respectively, during the first 2 years of the Twelfth Five Year Plan (Government of India, 2014) period (see Table 1.6).

In addition, a structural change in the composition of agriculture, showing diversification into horticulture, livestock and fisheries, is noticeable. The horticulture sector contributed 30.4% of agriculture GDP, while the livestock sector contributed over 4.1% of the total GDP in 2012/13 (GoI, 2014). A resilient agriculture with increasing contributions from horticulture and livestock is evident. Agriculture being a state subject, the primary responsibility for increasing agricultural production and productivity, exploiting untapped potential and enhancing

Table 1.6. Agriculture sector: key indicators (percentage at 2004/05 prices). (From Central Statistics Office (CSO) and Directorate General of Commercial Intelligence and Statistics (DGCI&S), as quoted in GoI, 2014.)

Item[a]	2009/10	2010/11	2011/12	2012/13	2013/14
Growth in agriculture GDP	0.8	8.6	5	1.4	4.7[b]
Share in total GDP	14.6	14.6	14.4	13.9	13.9[b]
Of which agriculture	12.3	12.4	12.3	11.8	NA
Share in total GCF	7.3	6.3	7.0	7.1	NA
Of which agriculture	6.7	5.8	6.5	6.5	NA
GCF as percentage of agriculture GDP	20.1	18.5	20.8	21.2	NA
Of which private sector	16.7	15.7	18.0	18.1	NA
Agriculture exports (includes marine products) as percentage of total exports	8.2	8.0	10.1	11.8	11.9 (P)

[a]GDP, Gross domestic product; GCF, gross capital formation; P, provisional; NA, not available.
[b]Quarterly estimates of GDP as of 30 May 2014.

incomes of the farming community rests with state governments. Their efforts are supplemented by many centrally sponsored and central sector schemes.

Area, production and productivity

Substantial progress in acreage and production are recorded for 2013/14. As per the third Advance Estimates, the acreage under food grains has increased to about 126.2 million ha, and to 28.2 million ha under oilseeds. Record production of food grains at 264.4 million t and oilseeds at 32.4 million t is estimated (Tables 1.7 and 1.8).

Production boost

After achieving the goal of increasing food grains production by 20 million t, new targets have been set under the National Food Security Mission of India, to produce an additional 25 million t of food grains by 2016/17: 10 million t of rice, 8 million t of wheat, 4 million t of pulses and 3 million t of coarse cereals. The focus is on cropping systems and on small and marginal farmers through development of farmer producer organizations and creating a value chain and providing market linkages. Funding under Rashtriya Krishi Vikas Yojana (RKVY) during the Twelfth Plan will be through production growth (35%), infrastructure and assets (35%), sub-schemes (20%) and 10% flexi-fund (GoI, 2014).

Given the limitations in expanding agricultural land, improvements in yield levels hold the key for long-term output growth. Groundnut has shown the largest jump in yield, while productivity increases are

Table 1.7. Area, production and yield of major crops in India in 2013/14. The data are third Advance Estimates. Figures in parenthesis indicate the percentage change over 2012/13. (From Directorate of Economics and Statistics, Department of Agriculture and Co-operation, as quoted in GoI, 2014.)

Group/commodity	Area (million ha)	Production (million t)	Yield (kg/ha)
Food grains[a]	126.2 (4.47)	264.4 (2.88)	2095 (−1.55)
Rice	43.9 (2.57)	106.3 (1.05)	2419 (−1.75)
Wheat	31.3 (4.33)	95.8 (2.46)	3059 (−1.86)
Coarse cereals	25.5 (2.98)	42.7 (6.64)	1672 (2.83)
Maize	9.3 (6.90)	24.2 (8.52)	2602 (1.40)
Pearl millet	7.9 (8.22)	9.2 (5.75)	1161 (−3.09)
Pulses	25.4 (9.01)	19.6 (7.10)	770 (−2.41)
Chickpea	10.2 (20.00)	9.9 (12.50)	974 (−5.98)
Pigeon pea	3.9 (0.00)	3.4 (13.33)	857 (10.44)
Oilseeds	28.2 (6.42)	32.4 (4.85)	1149 (−1.63)
Groundnut	5.5 (17.02)	9.5 (102.10)	1723 (73.17)
Rapeseed and mustard	6.5 (1.56)	7.8 (−2.50)	1208 (−4.28)
Cotton[b]	11.7 (−2.50)	36.5 (6.73)	529 (8.85)
Sugarcane	5.0 (0.00)	348 (2.11)	70 (0.00)

[a]Includes cereals, coarse cereals and pulses.
[b]Million bales of 170 kg each.

Table 1.8. Compound growth rates of area (A), production (P) and yield (Y) of principal crops in India during 1980/81 to 1989/90, 1990/91 to 1999/2000 and for 2000/01 to 2013/14. (From Department of Agriculture and Co-operation, as quoted in GoI, 2014.)

	Growth rate (% per annum)								
	1980/81 to 1989/90[a]			1990/91 to 1999/2000[a]			2000/01 to 2013/14[b]		
Crop	A	P	Y	A	P	Y	A	P	Y
Rice	0.41	3.62	3.19	0.68	2.02	1.34	0.00	1.82	1.82
Wheat	0.46	3.57	3.10	1.72	3.57	1.83	1.35	2.65	1.29
Coarse cereals	−1.34	0.40	1.62	−2.12	−0.02	1.82	0.25	2.96	2.70
Pulses	−0.09	1.52	1.61	−0.60	0.59	0.93	1.59	3.72	2.10
Sugarcane	1.44	2.70	1.24	−0.07	2.73	1.05	1.34	2.10	0.75
Nine oilseeds[c]	1.51	5.20	2.43	0.86	1.63	1.15	2.35	4.71	2.31
Cotton	−1.25	2.80	4.10	2.71	2.29	−0.41	3.22	13.53	9.99

[a]Base: Triennium ending (TE) 1981/82 = 100.
[b]Base: TE 1993/94 = 100; data as per second Advance Estimates.
[c]Groundnut, castor, sesame, rapeseed/mustard, linseed, niger, safflower, sunflower and soybean.

significant in the case of cotton (*Gossypium* sp.) and pigeon pea, as they have been achieved against declining/stagnant acreage. The compound growth rate of area, production and productivity during 2000/01 to 2013/14 has been higher than in the previous two decades for coarse cereals, pulses, oilseeds and cotton, while it has largely declined for rice and wheat.

1.9.2 Price policy supports dryland crops

The Indian government's price policy for major agricultural commodities has twin objectives: (i) to ensure remunerative prices to growers for their produce to encourage higher investment and production; and (ii) to safeguard the interests of consumers by ensuring supplies at reasonable prices. Towards these ends, the Commission for Agricultural Costs and Prices recommends minimum support price (MSP) based on certain economic criteria. Subsequently, the centre announces MSPs for 24 major agricultural commodities, including sugarcane (*Saccharum officinarum*), before each season, taking into account the views of state governments and the ministries/departments concerned. There have been substantial increases in the MSPs in the last few years, especially for pulses, oilseeds and cotton (Table 1.9).

1.9.3 Challenges

Productivity levels

It is heartening that India ranks first in the world for productivity of grapes (*Vitis* spp.), banana (*Musa* spp.), cassava (*Manihot esculenta*), peas and papaya (*Carica papaya*). However, despite efforts, the productivity levels of Indian agriculture are still way below global standards (Table 1.10). Without new technology and quality inputs, growth acceleration will be difficult to achieve at these productivity levels.

Net availability and per capita availability

The net availability of food grains has increased in 2013 at 229.1 million t, showing a 15% increase over the previous year. The per capita net availability of food grains spurted to 186.4 kg/year from 164.3 kg/year, and the net availability of edible oils also increased from 12.7 kg/year to 15.8 kg/year over the same period. These performances gain significance as the agriculture sector is the source of livelihood and food security for a vast majority of low income and vulnerable sections of the population. To improve nutritional status, a pilot programme on nutri-farms for introducing new crop varieties rich in micronutrients such as iron-rich pearl millet, protein-rich maize and zinc-rich wheat was implemented as a

Table 1.9. Minimum support prices (MSPs) – fixed by the Government of India (₹/quintal). (From Gok, 2016.)

		Year					
Commodity	Variety	2010–11	2011–12	2012–13	2013–14	2014–15	2015–16
***Kharif* crops**							
Paddy	Common	1000	1080	1250	1310	1360	1410
	Grade 'A'	1030	1110	1280	1345	1400	1450
Jowar	Hybrid	880	980	1500	1500	1530	1570
	Maldandi	900	1000	1520	1520	1550	1590
Bajra		880	980	1175	1250	1250	1275
Maize		880	980	1175	1310	1310	1325
Ragi		965	1050	1500	1500	1550	1650
Arhar (Tur)		3000[a]	3200[a]	3850	4300	4350	4425[b]
Moong		3170[a]	3500[a]	4400	4500	4600	4650[b]
Urad		2900[a]	3300[a]	4300	4300	4350	4425[b]
Cotton	Medium staple	2500[c]	2800[c]	3600	3700	3750	3800
	Long staple	3000[d]	3300[d]	3900	4000	4050	4100
Groundnut in shell		2300	2700	3700	4000	4000	4030
Sunflower seed		2350	2800	3700	3700	3750	3800
Soybean	Black	1400	1650	2200	2500	2500	–
	Yellow	1440	1690	2240	2560	2560	2600[e]
Sesame	–	2900	3400	4200	4500	4600	4700
Nigerseed	–	2450	2900	3500	3500	3600	3650
***Rabi* crops**							
Wheat		1120[f]	1285	1350	1400	1450	
Barley	–	780	980	980	1100	1150	
Gram	–	2100	2800	3000	3100	3175	

Continued

Table 1.9. Continued.

Commodity	Variety	Year 2010–11	2011–12	2012–13	2013–14	2014–15	2015–16
Masur (lentil)	–	2250	2800	2900	2950	3075	
Rapeseed/mustard	–	1850	2500	3000	3050	3100	
Safflower	–	1800	2500	2800	3000	3050	
Toria	–	1780	2425	2970	3020	3020	
Other crops							
Copra	Milling	4450	4525	5100	5250	5250	5550
(calendar year)	Ball	4700	4775	5350	5500	5500	5830
De-husked coconut (calendar year)		1200	1200	1400	1425	1425	1500
Jute		1575	1675	2200	2300	2400	2700
Sugarcane[g]		139.12	145	170	210	220	230

[a]Additional incentive at the rate of ₹500/quintal of tur, urad and moong (pulses) sold to procurement agencies was payable during the harvest/arrival period of 2 months.
[b]Bonus of ₹200/quintal is payable over and above the MSP.
[c]Staple length of 24.5–25.5 mm and micronaire value of 4.3–5.1. (Micronaire is a measurement of the thickness of the cell walls of a cotton fibre.)
[d]Staple length of 29.5–30.5 mm and micronaire value of 3.5–4.3.
[e]Single MSP has been fixed irrespective of the variety.
[f]An additional incentive bonus of ₹50/quintal was payable over the MSP.
[g]Fair and remunerative price.

Table 1.10. Productivity (kg/ha) of major crops in India vis-à-vis world average and country with highest yield. (From Agricultural Statistics at a Glance 2013; Kharif and Rabi Price Policy Reports, Commission for Agricultural Costs and Prices, as quoted in GoI, 2014.)

Crop/commodity	World average (TE 2011/12)[a]	India (TE 2012)[a]	Country with highest yield (TE 2012)[a]
Cereals			
Paddy	4,397	3,514	6,661 (China)
Wheat	3,094	3,000	7,360 (UK)
Maize	5,097	2,321	8,858 (USA)
Pulses			
Chickpea	917	912	1,663 (Ethiopia)
Pigeon pea	786	681	1,320 (Myanmar)
Oilseeds			
Groundnut	1,626	1,212	4,069 (USA)
Rapeseed/ mustard	1,855	1,163	3,588 (UK)
Cotton	769	517	1,920 (Australia)
Sugarcane	70,470	69,227	125,587 (Peru)

[a]TE, Triennium ending.

sub-scheme of the RKVY in 2013/14 in the 100 districts of nine states most affected by malnutrition, with an outlay of ₹200 crores (GoI, 2014).

While the continued robustness of Indian agriculture is significant in the context of food security and climate change, some major concerns remain. Growth rates of productivity are far below global standards; productivity levels of rice and wheat have declined after the Green Revolution of the 1980s. Another issue is soil degradation due to declining fertilizer-use efficiency.

Strengthening the agriculture sector is crucial for poverty alleviation, ensuring food security, increasing employment opportunities and enhancing rural incomes. Further, with 10.4% of total households in rural areas being headed by a woman (Census 2011), it is essential to formulate policies, and package technologies and services keeping in view the productive role played by women in all facets of the agriculture sector. Experience from BRICS (Brazil, Russia, India, China and South Africa) countries indicates that a 1% growth in agriculture is at least two to three times more effective in reducing poverty than the same growth emanating from non-agriculture sectors.

Currently, India is in an anomalous situation of being largely self-sufficient, with large stocks of food grains on the one hand and registering high food inflation on the other, which is largely due to the government becoming the single largest buyer. In this scenario of bumper production and stocks, a paradigm shift in the role of the government in all aspects of food grain production and distribution is necessary.

Major challenges include: (i) low productivity levels; (ii) soil degradation due to declining fertilizer-use efficiency; (iii) market distortions that prevent the creation of a national common market; (iv) the changing role of government in production and distribution in the current scenario of bumper production and stocks; and (v) phased shifting to direct transfer of fertilizer and food subsidies.

1.10 Towards an Innovative Strategy

Agriculture will continue to be the backbone of economies in Africa and South Asia in the foreseeable future. As most of the poor in the SAT are farmers and landless labourers, strategies for reducing poverty, hunger and malnutrition should be driven primarily by the needs of the rural poor; and should aim to build and diversify their livelihood sources. Substantial gains in land, water and labour productivity as well as better management of natural resources are essential to reverse the downward spiral of poverty and environmental degradation. Apart from the problems of equity, poverty and sustainability (and hence, the need for greater investment in SAT areas), studies have shown that research and development (R&D) investments in less-favoured semi-arid environments could provide high marginal profits in terms of generating new sources of economic growth. Renewed effort and innovative R&D strategies are needed to address these challenges, such as INRM that has been evolving within the 15 international agricultural research centres (IARCs) of the CGIAR. The basic role of the 15 IARCs is to develop innovations for improving agricultural productivity and natural resource management for addressing the problems of poverty, food insecurity and environmental degradation in developing countries. This effort has generated multiple and sizeable benefits (welfare, equity, environmental) (Kassam *et al.*, 2004). But much remains to be done in sub-Saharan Africa and less-favoured areas of South Asia.

Factors such as low soil fertility, inappropriate soil and water management practices causing land degradation, lack of improved varieties, pest and disease attack, resource-poor farmers, declining land:man ratio and poor rural communities, who are unable to meet even minimum standards of health and nutrition, add to the burgeoning problem of rural poverty (Wani *et al.*, 2001). The adoption of the new paradigm in rainfed agriculture has shown that with proper management of natural resources the systems productivity can be enhanced and poverty can be reduced without causing further degradation of the natural resource base.

Thereby, ICRISAT has focused on designing an innovative scaling-up mission mode approach, called Bhoochetana, for harnessing the untapped potential of rainfed agriculture on a large scale. The concept, its evolution, key features and refinements are explained in Chapter 2 and its successful execution in the state of Karnataka in India is discussed in subsequent chapters of this book.

Notes

[1] Drylands are defined by their scarcity of water. They are zones where precipitation is counterbalanced by evaporation from surfaces and transpiration by plants (evapotranspiration). The United Nations Environment Programme (UNEP) defines drylands as tropical and temperate areas with an aridity index of less than 0.65. The drylands can be further classified into four subtypes: (i) dry subhumid lands; (ii) semi-arid lands; (iii) arid lands; and (iv) hyper-arid lands. Some authorities such as the United Nations Convention to Combat Desertification (UNCCD) consider hyper-arid lands as deserts, although a number of the world's deserts include both hyper-arid and arid climate zones. The UNCCD excludes hyper-arid zones from its definition of drylands.

[2] Drylands cover 41.3% of the Earth's land surface, including 15% of Latin America, 66% of Africa, 40% of Asia and 24% of Europe. Worldwide there is a significantly greater proportion of drylands in developing countries (72%), and the proportion increases with aridity: almost 100% of all hyper-arid lands are in the developing world. Nevertheless, the USA, Australia and several countries in southern Europe also contain significant dryland areas.

[3] Reduction in the producing capacity of land due to wind and water erosion of soil, loss of soil humus, depletion of soil nutrients, secondary salinization, diminution and deterioration of vegetation cover as well as loss of biodiversity is referred to as land degradation.

References

Asian Development Bank (ADB) (2014) *Key Indicators for Asia and the Pacific 2014*, 45th edn. ADB, Mandaluyong City, Philippines.

Dimes, J., Twomlow, S., Rusike, J., Gerard, B., Tabo, R., Freeman, A. and Keatinge, J.D.H. (2005) Increasing research impacts through low-cost soil fertility management options for Africa's drought-prone areas. In: Omanya, G.O. and Pasternak, D. (eds) *Sustainable Agriculture Systems for the Drylands. Proceedings of the International Symposium for Sustainable Dryland Agriculture Systems, 2–5 December 2003, Niamey, Niger*. International Crops Research Institute for the Semi-Arid Tropics, Niamey, Niger, pp. 82–92.

Falkenmark, M. and Rockström, J. (2004) *Balancing Water for Humans and Nature: The New Approach in Ecohydrology*. Earthscan, London.

FAOStat (2005) Food and Agriculture Organization of the United Nations (FAO) Statistical Year Book 2005–6. Available at: http://faostat.fao.org/ (accessed 1 November 2005).

Food and Agriculture Organization of the United Nations (FAO) (2002) World Agriculture: Towards 2015/30. *Technical Interim Report*. Available at: http://www.fao.org/docrep/007/y5738e/y5738e06.htm (accessed 16 March 2016).

Food and Agriculture Organization of the United Nations (FAO) (2010a) AQUASTAT Database. Available at: http://www.fao.org/nr/aquastat (accessed 12 January 2011).

Food and Agriculture Organization of the United Nations (FAO) (2010b) FAOSTAT Database. Available at: http://faostat3.fao.org/home/E (accessed 1 August 2011).

Government of India (GoI) (2014) Economic Survey 2013-14. Ministry of Finance, GoI, New Delhi. Available at: http://indiabudget.nic.in/index.asp (accessed 9 August 2014).

Government of India (GoI) (2016) Farmer Portal: Minimum Support Prices. Available at: http://farmer.gov.in/mspstatements.aspx (accessed 16 March 2016).

International Crops Research Institute for the Semi-Arid Tropics (ICRISAT) (2005) Identifying systems for carbon sequestration and increased productivity in semi-arid tropical environments. A Project Completion Report funded by the National Agricultural Technology Project, Indian Council of Agricultural Research. ICRISAT, Patancheru, Andhra Pradesh, India.

International Crops Research Institute for the Semi-Arid Tropics (ICRISAT) (2010) *ICRISAT Strategic Plan to 2020: Inclusive Market-Oriented Development for Smallholder Farmers in the Tropical Drylands.* ICRISAT, Patancheru, Andhra Pradesh, India.

International Fertilizer and Development Center (IFDC) (2006) Agricultural production and soil nutrient mining in Africa: implications for resource conservation and policy development. *IFDC Technical Bulletin*. IFDC, Alabama, USA.

Irz, X. and Roe, T. (2000) Can the world feed itself? Some insights from growth theory. *Agrekon* 39(3), 513–528.

Joshi, P.K., Jha, A.K., Wani, S.P., Laxmi Joshi and Shiyani, R.L. (2005) Meta-analysis to assess impact of watershed program and people's participation. *Comprehensive Assessment Research Report 8*. Comprehensive Assessment Secretariat, International Water Management Institute, Colombo, Sri Lanka.

Kassam, A.H., Gregersen, H.M., Ferres, E., Javier, Q., Harwood, R.R., de Jancrym, A. and Cernea, M.M. (2004) A framework for enhancing and guarding the relevance and quality of science: the case of the CGIAR. *Experimental Agriculture* 40, 1–21.

Marothia, D.K. (2013) *Impact Assessment of Lucheba Watershed, China*. International Crops Research Institute for the Semi-Arid Tropics, Patancheru, Telangana, India.

Molden, D., Frenken, K., Barker, R., de Fraiture, C., Bancy Mati, Svendsen, M., Sadoff, C. and Finlayson, M.C. (2007) Trends in water and agricultural development. In: Molden, D. (ed.) *Water for Food: Water for Life – A Comprehensive Assessment of Water Management in Agriculture*. Earthscan, London and International Water Management Institute, Colombo, Sri Lanka, pp. 57–89.

Ncube, B., Dimes, J.P., Twomlow, S.J., Mupangwa, W. and Giller, K.E. (2006) Raising the productivity of smallholder farms under semi-arid conditions by use of small doses of manure and nitrogen: a case study of participatory research. *Nutrient Cycling in Agroecosystems* 77, 53–67.

Pathak, P., Sahrawat, K.L., Wani, S.P., Sachan, R.C. and Sudi, R. (2009) Opportunities for water harvesting and supplemental irrigation for improving rain-fed agriculture in semi-arid areas. In: Wani, S.P., Rockström, J. and Oweis, T. (eds) *Rainfed Agriculture: Unlocking the Potential*. Comprehensive Assessment of Water Management in Agriculture Series. CAB International, Wallingford, UK, pp. 197–221.

Ramankutty, N., Foley, J.A. and Olejniczak, N.J. (2002) People and land: changes in global population and croplands during the 20th century. *Ambio* 31(3), 251–257.

Rego, T.J., Wani, S.P., Sahrawat, K.L. and Pardhasaradhi, G. (2005) *Macro-benefits from Boron, Zinc, and Sulphur Application in Indian SAT: A Step for Grey to Green Revolution in Agriculture*. Global Theme on Agroecosystems Report no. 16. International Crops Research Institute for the Semi-Arid Tropics, Patancheru, Andhra Pradesh, India.

Rego, T.J., Sahrawat, K.L., Wani, S.P. and Pardhasaradhi, G. (2007) Widespread deficiencies of sulfur, boron, and zinc in Indian semi-arid tropical soils: on-farm crop responses. *Journal of Plant Nutrition* 30(10), 1569–1583.

Rockström, J. and Falkenmark, M. (2000) Semiarid crop production from a hydrological perspective: gap between potential and actual yields. *Critical Reviews in Plant Science* 19(4), 319–346.

Rockström, J., Hatibu, N., Oweis, T. and Wani, S.P. (2007) Managing water in rain-fed agriculture. In: Molden, D. (ed.) *Water for Food, Water for Life: A Comprehensive Assessment of Water Management in Agriculture*. Earthscan, London and International Water Management Institute, Colombo, Sri Lanka, pp. 315–348.

Rosegrant, M., Ximing, C., Cline, S. and Nakagawa, N. (2002) The role of rain-fed agriculture in the future of global food production. Environment and Production Technology Division (EPTD) Discussion Paper No. 90. EPTD, International Food Policy Research Institute, Washington, DC. Available at: http://ageconsearch.umn.edu/bitstream/16053/1/ep020090.pdf (accessed 16 March 2016).

Rusike, J., Twomlow, S.J., Freeman, H.A. and Heinrich, G.M. (2006) Does farmer participatory research matter for improved soil fertility technology development and dissemination in southern Africa? *International Journal of Agricultural Sustainability* 4(3), 176–192.

Ryan, J.G. and Spencer, D.C. (2001) *Challenges and Opportunities Shaping the Future of the Semi-Arid Tropics and their Implications*. International Crops Research Institute for the Semi-Arid Tropics, Patancheru, Andhra Pradesh, India.

Sahrawat, K.L., Wani, S.P., Rego, T.J., Pardhasaradhi, G. and Murthy, K.V.S. (2007) Widespread deficiencies of sulphur, boron and zinc in dryland soils of the Indian semiarid tropics. *Current Science* 93(10), 1–6.

Singh, P., Agarwal, P.K., Bhatia, V.S., Murthy, M.V.R., Pala, M., Oweis, T., Benli, B., Rao, K.P.C. and Wani, S.P. (2009) Yield gap analysis: modelling of achievable yields at farm level. In: Wani, S.P., Rockström, J. and Oweis, T. (eds) *Rainfed Agriculture: Unlocking the Potential*. Comprehensive Assessment of Water Management in Agriculture Series. CAB International, Wallingford, UK, pp. 81–124.

Snapp, S. and Heong, K.L. (2003) Scaling up and out. In: Pound, B., Snapp, S., McDougall, C. and Braun, A. (eds) *Managing Natural Resources for Sustainable Livelihoods: Uniting Science and Participation*. International Development Research Centre (IDRC) and Earthscan, London, pp. 67–83.

Stoorvogel, J.J. and Smaling, E.M.A. (1990) *Assessment of Soil Nutrient Depletion in Sub-Saharan Africa: 1983–2000, 1: Main Report. Report No. 28*. The Winand Staring Centre for Integrated Land, Soil and Water Research, Wageningen, The Netherlands.

Thirtle, C., Beyers, L., Lin, L., McKenzie-Hill, V., Irz, X., Wiggins, S. and Piesse, J. (2002) The impacts of changes in agricultural productivity on the incidence of poverty in developing countries. *Department for International Development (DFID) Report No. 7946*. DFID, London.

United Nations (2014) *World Urbanization Prospects: The 2014 Revision, Highlights* (ST/ESA/SER.A/352). Department of Economic and Social Affairs, Population Division, United Nations, New York.

United Nations Environment Programme (UNEP) (1997) *World Atlas of Desertification*, 2nd edn. UNEP, Rome.

Uttam, D., Bantilan, C. and Kumar, R. (2014) Dynamics of rural livelihoods in Bangladesh and India: insights from village dynamics studies villages. Presented at the pre-conference mini-symposium on Rapid Transformation of Rural Economies in South Asia: Insights from Village Dynamics Studies, organized as part of the 8th Conference of the Asian Society of Agricultural Economists (ASAE), held on 14 October 2014 at the BRAC Centre for Development Management (BRAC-CDM), Savar, Dhaka, Bangladesh.

Walker T. (2010) *Updating and Reviewing Future Challenges and Opportunities for Agricultural R&D in the Semi-Arid Tropics for ICRISAT Strategic Planning to 2020.* International Crops Research Institute for the Semi-Arid Tropics, Patancheru, Andhra Pradesh, India.

Wangkahart, T., Pathak, P., Seehaban, P., Bhotinam, A., Chueychoom, P., Wani, S.P., Rao, A.V.R.K and Sudi, R. (2012) Integrated land and water management for controlling land degradation and improving agricultural productivity in Northeast Thailand. In: *Community Watershed Management for Sustainable Intensification in Northeast Thailand.* International Crops Research Institute for the Semi-Arid Tropics, Patancheru, Andhra Pradesh, India, pp. 30–66.

Wani, S.P., Pathak, P., Tam, H.M., Ramakrishna, A., Singh, P. and Sreedevi, T.K. (2001) Integrated watershed management for minimizing land degradation and sustaining productivity in Asia. In: Zafar Adeel (ed.) *Integrated Land Management in Dry Areas. Proceedings of a Joint United Nations University (UNU)–Chinese Academy of Sciences (CAS) International Workshop, 8–13 September 2001, Beijing, China.* UNU, Tokyo, pp. 207–230.

Wani, S.P., Pathak, P., Jangawad, L.S., Eswaran, H. and Singh, P. (2003a) Improved management of Vertisols in the semi-arid tropics for increased productivity and soil carbon sequestration. *Soil Use and Management* 19, 217–222.

Wani, S.P., Pathak, P, Sreedevi, T.K., Singh, H.P. and Singh, P. (2003b) Efficient management of rainwater for increased crop productivity and groundwater recharge in Asia. In: Kijne, J.W., Barker, R. and Molden, D. (eds) *Water Productivity in Agriculture: Limits and Opportunities for Improvement.* CAB International, Wallingford, UK and the International Water Management Institute, Colombo, Sri Lanka, pp. 199–215.

Wani, S.P., Singh, H.P., Sreedevi, T.K., Pathak, P., Rego, T.J., Shiferaw, B. and Iyer, S.R. (2003c) Farmer-participatory integrated watershed management: Adarsha Watershed, Kothapally, India. An innovative and up-scalable approach. A case study. In: Harwood, R.R. and Kassam, A.H. (eds) *Research Towards Integrated Natural Resources Management: Examples of Research Problems, Approaches and Partnerships in Action in the CGIAR.* Interim Science Council, Consultative Group on International Agricultural Research, Washington, DC, pp. 123–147.

Wani, S.P., Meera Reddy, Sreedevi, T.K. and Damle Raju (eds) (2006a) *Corporate Science and Technology Institutions – Partnerships for Inclusive and Sustainable Development and Economic Growth: Proceedings of the CII-ICRISAT Workshop, 27 February 2006.* International Crops Research Institute for the Semi-Arid Tropics, Patancheru, Andhra Pradesh, India.

Wani, S.P., Ramakrishna, Y.S., Sreedevi, T.K., Long, T.D., Thawilkal Wangkahart, Shiferaw, B., Pathak, P. and Kesava Rao, A.V.R. (2006b) Issues, concepts, approaches and practices in integrated watershed management: experience and lessons from Asia. In: *Integrated Management of Watersheds for Agricultural Diversification and Sustainable Livelihoods in Eastern and Central Africa: Lessons and Experiences from Semi-Arid South Asia. Proceedings of the International Workshop held at International Crops Research Institute for the Semi-Arid Tropics (ICRISAT), Nairobi, 6–7 December 2004.* ICRISAT, Patancheru, Andhra Pradesh, India, pp. 17–36.

Wani, S.P., Rockström, J., Venkateswarlu, B. and Singh, A.K. (2011) New paradigm to unlock the potential of rainfed agriculture in the semi-arid tropics. In: Lal, R. and Stewart, B.A. (eds) *World Soil Resources and Food Security. Advances in Soil Science*. CRC Press, Boca Raton, Florida, pp. 419–469.

World Bank (2000) Spurring agricultural and rural development. In: *Can Africa Claim the 21st Century?* World Bank, Washington, DC, pp. 170–207.

World Bank (2005) *Agricultural Growth for the Poor: An Agenda for Development.* International Bank for Reconstruction and Development/World Bank, Washington, DC.

World Bank (2008) *World Development Report. Agriculture for Development.* World Bank, Washington, DC.

World Bank (2009) *World Development Report. Reshaping Economic Geography.* World Bank, Washington, DC.

World Health Organization (WHO) (2005) *Integrating Poverty and Gender into Health Programmes*. Available at: http://apps.who.int/iris/bitstream (accessed on 28 July 2016).

Evolution of Bhoochetana

2

Suhas P. Wani*

International Crops Research Institute for the Semi-Arid Tropics (ICRISAT), Patancheru, India

2.1 Introduction

Long-term experiments at the International Crops Research Institute for the Semi-Arid Tropics (ICRISAT) campus since 1976 as well as a number of studies in different countries (Rockström *et al.*, 2007; Wani *et al.*, 2008) clearly showed that current farmers' field yields were lower than their potential yields by two- to fivefold. These studies also demonstrated that there exists a large potential to increase farmers' crop yields by adopting available technologies. However, large yield gaps are largely due to lack of knowledge about the improved management practices for increasing productivity for the farmers and not due to lack of technologies (Wani *et al.*, 2008). If we can bridge the knowledge gap and make the necessary inputs needed for implementing improved management practices (seeds, fertilizers, credit, etc.) on farmers' fields, productivity can be substantially increased by bridging the yield gaps. With this knowledge and pilot studies in Adarsha Watershed, Kothapally, India, as well as other watersheds in different parts of the country, it was demonstrated that yields from farmers' fields can be substantially increased by up to 240%, providing farmers have the right information and inputs at the right time at the right price. By adopting a holistic approach, yield gaps even on small farmers' fields were successfully bridged and farmers benefited with increased productivity and profitability with the help

*s.wani@cgiar.org

of scientific knowledge provided to the farmers (Wani *et al.*, 2009). In order to ensure benefits for a large number of farmers and make a major impact, there is a need to scale up the pilot studies, for which a number of actors need to be involved in the process before we can achieve the impact. The scaling up of soil-test-based nutrient management practices as an entry point, along with other improved management practices such as improved crop cultivars, land and water management practices and pest management practices, could benefit the farmers if we could develop a scaling-up model using the available technologies to benefit the farmers through empowerment.

2.2 How Bhoochetana Evolved

Agriculture is an important contributor to the growth of Karnataka, India, and is still the mainstay for a large proportion (60%) of the population in the state. Agriculture and allied sectors' contribution to Karnataka's gross domestic product was around 43% in 1980/81; however, that came down to 26% in 2001/02, and 16.8% during 2007–2010. The agricultural annual growth rate in the state during the period 2000–2008 oscillated between negative to 2% and therefore was a big concern for the state government as 60% of the population depended on this sector for their livelihoods.

The strategic partnership between the Government of Karnataka and ICRISAT, Patancheru, started way back in 2003 with the implementation of the World Bank-aided Sujala watershed project for demonstrating a science-based integrated genetic and natural resource management (IGNRM) approach for enhancing the productivity of dryland crops. The consortium of the Watershed Development Department (WDD), Department of Agriculture (DoA), state agricultural universities (SAUs) and ICRISAT was formed, and widespread deficiencies of zinc, boron and sulfur were identified in six districts which served as an entry point. Farmer participatory action research showed increased crop yields up to 345% in sunflower (*Helianthus annuus*), 230% in finger millet (*Eleusine coracana*), 240% in groundnut (*Arachis hypogaea*), 150% in maize (*Zea mays*), 116% in soybean (*Glycine max*) and 27% in sorghum (*Sorghum bicolor*). Farmers revealed that crop yields increased up to 58% even during the unfavourable year of 2008.

The consortium approach emerged from the lessons learnt from long-term watershed-based research led by ICRISAT and national partners (Wani *et al.*, 2003). This interdisciplinary research, over the years, has shaped up into an IGNRM approach at ICRISAT (Twomlow *et al.*, 2006; Wani *et al.*, 2012). This approach was first adopted during the year 1999, when the Asian Development Bank came forward to support ICRISAT's idea of testing the consortium model in the Adarsha watershed in Kothapally village, Ranga Reddy district, Telangana. First, the aim was to minimize the gap between the research findings and on-farm developments. Secondly, the purpose was also to adopt the learning loop in the planning of strategic research based on the participatory research and development system. The integrated watershed management model has demonstrated that with proper management of natural resources the system's productivity can be enhanced and poverty can be reduced without causing any further degradation of the natural resource base. Based on these experiences, it was known that the potential of rainfed agriculture in the state is withheld because of the widespread deficiencies of micro- and secondary nutrients as well as imbalanced use of fertilizers applied by the farmers in the state along with unimproved crop management practices.

After a series of discussions, and a brainstorming session among the policy makers, bureaucrats and researchers, the Government of Karnataka decided to develop a mission project to address the issue of improving the productivity of dryland agriculture across the state, which was in fact a right and timely decision. The innovative integrated watershed management model has demonstrated that with proper management of natural resources the system's productivity can be enhanced and poverty can be reduced without causing further degradation of the natural resource base (Wani *et al.*, 2003; Sreedevi *et al.*, 2004; Garg and Wani, 2012). The scaling up of these innovations has been attempted in countries such as India, Vietnam, Thailand and China. At benchmark watershed locations in Madhya Pradesh at Lalatora, widespread micronutrient deficiencies were observed and subsequently corrections of deficiencies through application of appropriate nutrients resulted in increased crop yields as well as increased incomes for the farmers. From 2002, the Government of undivided Andhra Pradesh requested ICRISAT to undertake productivity enhancement initiatives in APRLP (Andhra Pradesh Rural Livelihood Programme) in five districts of the state (Anantapur, Kurnool, Mahabubnagar, Nalgonda and Prakasam). In this initiative, ICRISAT further refined the consortium approach and undertook soil health mapping in the selected five districts and reconfirmed the occurrence of widespread deficiencies of zinc, boron and sulfur in farmers' fields. Based on this information, soil-test-based nutrient recommendations resulted in increased crop productivity up to 120% under a rainfed situation and this provided the base for the Sujala initiative later in the state of Karnataka.

ICRISAT was asked to propose a plan to increase the productivity of rainfed crops in the state. The project proposed was to be implemented in a mission mode and it was christened 'Bhoochetana' meaning rejuvenation of land. The Bhoochetana project was launched to improve the livelihoods of the dryland farmers by increasing the agricultural productivity by 20% during a period of 4 years (2009/10 to 2012/13). The sole objective of the mission was to improve the livelihoods of the dryland farmers as well as the state's economy by increasing the agricultural productivity through new scientific technologies, innovative approaches and methods. Bhoochetana project outputs in the dryland agriculture zones of the state were expanded later to cover all the 30 districts including irrigated agriculture.

The Honourable Chief Minister of the Government of Karnataka launched the mission Bhoochetana in an inaugural ceremony held at Haveri on 23 May 2009. At this ceremony the Honourable Minister for Agriculture, Government of Karnataka, and the Deputy Director General (Research) of ICRISAT and other policy makers were also present. Launching of the project by the Chief Minister was aimed at enhancing awareness of the Bhoochetana mission initiative among farmers in the selected districts and also among the department officials in order to increase the chance of success. Another objective of such a launch by the Honourable Chief Minister was to deliver a message to all concerned that this is a mission supported by the government and all top policy makers will be involved in monitoring the progress to ensure the success of the programme.

The goal of the Bhoochetana project was to make a difference in the lives of dryland farmers in 24 districts of Karnataka, including the six districts covered under the Sujala–ICRISAT initiative, by increasing average crop productivity by 20% in 4 years. Later the project area was expanded to cover all 30 districts and during the fourth year, irrigated crops were also included.

The project was implemented in a mission mode across the state by adopting a consortium approach. This approach created a favourable environment at all levels for implementers. The DoA, Karnataka, WDD, the University of Agricultural Sciences (Bengaluru, Raichur and Dharwad) and ICRISAT, as a consortium leader, played a crucial role in the successful implementation of the project at the field level.

The Bhoochetana project was a demand-driven project as needed by the Government of Karnataka, with clear and measureable outputs in terms of increased productivity by 20% in 4 years in all the districts. The deliverables were worked out based on knowledge about the benefits from the improved technologies, adoption rates as well as varying socio-economic conditions and operating institutional mechanisms in the DoA. At the same time, an agreement was also made with the government that Bhoochetana mission project can deliver 5% increased productivity per annum in the agriculture sector. However, government also had to provide the needed timely support through enabled policies, institutional mechanisms as well as financial resources required for implementation of the programme. To ensure speedy decision making in the mission project, the State Level Coordination Committee (SLCC) (chaired by the Additional Chief Secretary and Development Commissioner) comprising senior members from SAUs and different line departments[1] of the government was constituted, which met frequently in order to ensure success of the programme.

The project was launched in six districts of the state where the World Bank supported the Sujala watershed project, which was implemented during 2005–2008. Bhoochetana was initiated in 2009 in these six districts – Kolar, Chikkaballapur, Tumkur, Chitradurga, Haveri and Dharwad. In these districts a total area of 2.25 lakh ha (0.225 million ha) was covered. The main crops were groundnut, finger millet, maize and soybean. In the first year (2009/10) of the project, 1440 villages were covered benefiting 2 lakh (0.2 million) farmers. In the second year (2010/11) of the project, 16 districts (including the six first-phase districts) were covered. In these districts a dryland area of 12 lakh ha (1.2 million ha) and 5030 villages was covered. In the second year, up to 8 lakh (0.8 million) farmers were covered. Subsequently during the third year of the project, all 30 districts of the state were covered and irrigated crops were also included during the fourth year.

The project focused on activities such as: (i) soil testing; (ii) farmers' registration; (iii) training; (iv) awareness generation; (v) wall writings in selected villages depicting soil fertility status and crop-specific best management practices; (vi) farmer field school; (vii) farm facilitators (FFs) and lead farmers; (viii) hiring of godowns (warehouses) at the cluster village for stocking of inputs and transportation of inputs from *Raithu Samparka Kendras* (RSKs)[2] to the cluster village; (ix) distribution of inputs at 50% subsidy; (x) seed treatment; and (xi) selection of major crops. These activities were carried out in all the villages where the project was implemented.

For effective implementation of the programme, guidelines were developed based on learning in the first 2 years and all communications were compiled in the form of guidelines in the local language Kannada. This helped the field team implement the project activities in a much more organized way through a science-led participatory approach.

The project focused on conducting soil analysis and taking action on its recommendations at field level. Around 11,000 samples were collected from farmers' fields in several taluks of each district covering six districts during 2008 and later on, all districts; and 92,904 soil samples were collected by adopting a stratified soil sampling method standardized earlier (Sahrawat *et al.*, 2008). Soil-testing reports along with soil maps using geographical information system (GIS) were interpolated with soil nutrient status data for the benefit of policy makers for the district and a comprehensive soil atlas was released (Wani *et al.*, 2011). This was a major contribution to the state.

During the formulation of the project and based on the earlier experiences in the state, it was observed that farmers were not getting the right information at the right time and as a result, they continued to adopt traditional practices for cultivating their crops. To overcome this shortfall, while preparing the project, this issue was addressed by suggesting identification of para-extension workers, FFs and lead farmers in the state to bridge the gap for the extension system. For every 500 ha, one FF who was a practising farmer from the villages themselves was identified.

In all the project villages, suitable extension methods were used effectively involving a large number of farmers, FFs and lead farmers who were trained suitably by the master trainers from the SAUs and the DoA. The training of master trainers by ICRISAT's team of scientists to deliver uniform messages using developed training modules, and organizing field days and field visits was the most effective and logical approach to explain the benefits of new technologies and their contribution to enhancing agricultural production to the farming community.

For promotion of new technologies and practices, it is vital to have a proper farmer selection process crafted and adopted at field level. The Bhoochetana programme also had a set of farmer selection criteria (more details in Chapter 4 of this book).

The project has been implemented in a mission mode. To deliver timely outputs a three-tier coordination mechanism was developed and put into operation. The taluk, district and SLCC successfully contributed professionally to the execution of the project interventions at ground level (more details in Chapter 6 of this book).

The outcome of organized and integrated efforts was positive and there was progressive growth in the state agriculture sector. In 2009/10, agricultural growth increased by 5.9% from 0.5% in 2008/09. Adoption of technology rate increased progressively with the increasing number of districts covered (six in 2009, 16 in 2010 and 30 in 2011 and 2012). During the period 2009–2011, 8.5 lakh (0.85 million) farmers, who cultivated 12 lakh ha (1.2 million ha) of dryland area in 16 districts, benefited from this scheme through an increase in yield between 21% and 66% in demonstration plots. During the first 3 years (2009–2011), 2.17 lakh t (0.217 million t) of quality seeds were made available at subsidized rates to 81 lakh (8.1 million) farmers for improving crop productivity.

The average increase in yield in the year 2009/10 for major crops was good at 50% in sample plots. The state experienced enhancement in yield in the treated areas for maize (44%), finger millet (35–65%), groundnut (32–41%) and soybean (39%). Similarly, about 23–57% increase in yield was observed in treated plots as compared with non-treated plots during 2010/11 and 21–43% during 2011/12.

Continuous learning helped in initiating timely corrective measures in the programme and facilitated the process of taking better policy decisions. In Bhoochetana, also, a few good policy decisions were taken based on continuous learning from the field. For example, conducting video conferences to review the programme was a big policy decision and required frequent coordination of committee meetings at state, district and taluk levels. Other policy decisions were the appointment of a nodal officer at district level and the hiring of fertilizer storage godowns at village cluster level. Identification and training of FFs and lead farmers for Bhoochetana, and development of taluk-wise fertilizer recommendations based on soil analysis are some of many such enabling policies and decisions. These decisions taken by the government made the programme effective and systematic.

2.3 Concept

The concept of '4 Cs' was adopted in the project. The *first C* is *Consortium* of research, education and field-based agencies to implement this programme effectively at ground level. The *second C* is *Convergence* within the department schemes and other programmes. The *third C* is *Capacity building* of the consortium partners, FFs, lead farmers and other stakeholders. The *fourth C* is *Collective action* at all levels during programme implementation in a mission mode.

Apart from adopting the concept of '4 Cs', efforts were also made to ensure that the '4 Es' were also achieved at ground level, namely *Efficiency*, *Economic gain*, *Equity* and *Environmental protection*. These are the important pillars of sustainable and inclusive development.

2.4 Identification of Constraints for Harnessing the Potential and Addressing Demand and Supply Needs

In Bhoochetana, identification of constraints was undertaken in a scientific way. As a first step, soil sampling and diagnosing nutrient status of farmers' fields was completed. To characterize the fertility status of soils, 92,904 soil samples were collected from farmers' fields in watersheds, spread over 30 districts in Karnataka. A detailed soil fertility atlas was prepared at block level (comprising a few villages), which showed nutrient deficiencies scattered differently over the region (Fig. 2.1). Maps served the purpose to develop site-specific nutrient recommendations and guide nutrient input mobilization and use. Block-level fertilizer recommendations, including deficient secondary and micronutrients, were developed in contrast to current blanket recommendations for macronutrients at state level. The DoA was empowered to adopt and disseminate soil test results and site-specific fertilizer recommendations through traditional and innovative extension means.

Other appropriate technologies in addition to integrated nutrient management application of deficient micronutrients along with organic manures, biofertilizers and biocontrol were identified to be implemented under Bhoochetana. The prominent technologies were: (i) *in situ* soil and water conservation techniques comprising a conservation furrow system, cultivation across a slope, and a broad-bed and furrow system; (ii) improved varieties of finger millet (GPU 28, L 5, MR 1),

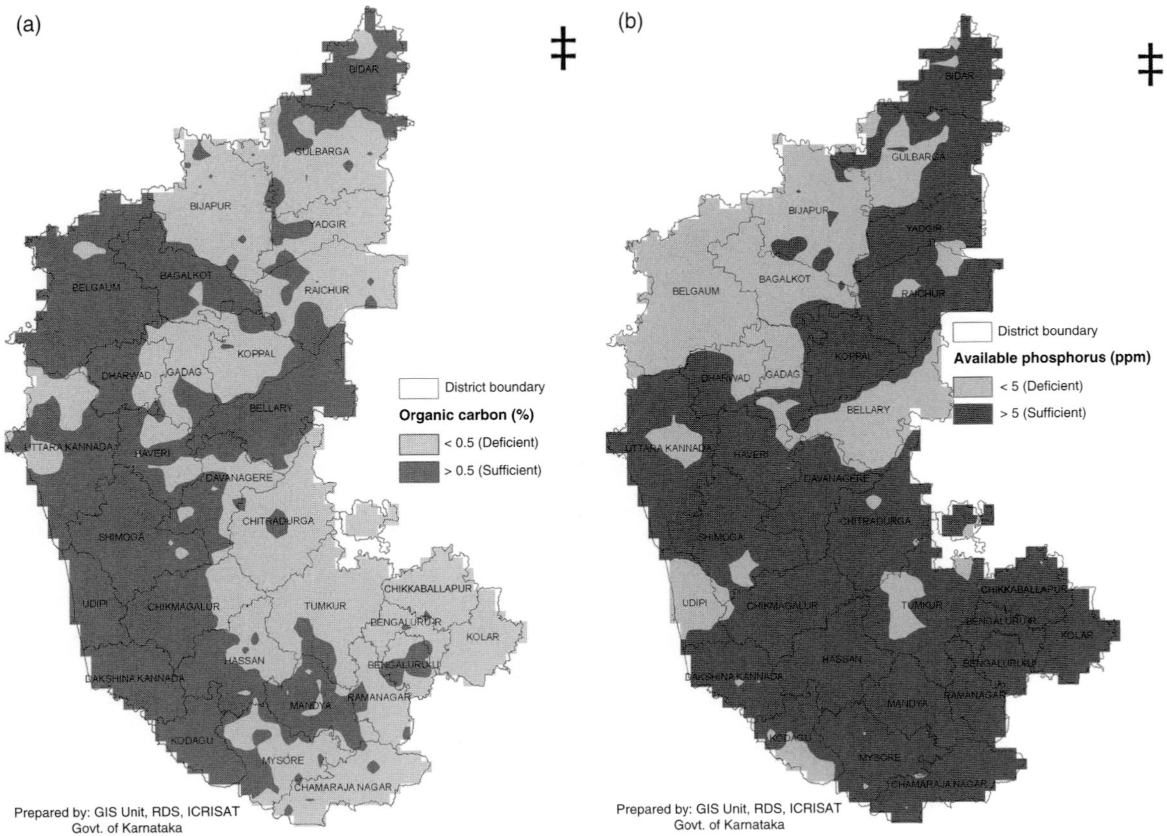

Fig. 2.1. Soil maps indicating nutrient deficiency/sufficiency in Karnataka: (a) organic C; (b) available P; (c) available K; (d) available S; (e) available B; and (f) available Zn. (From Wani *et al.*, 2011.)

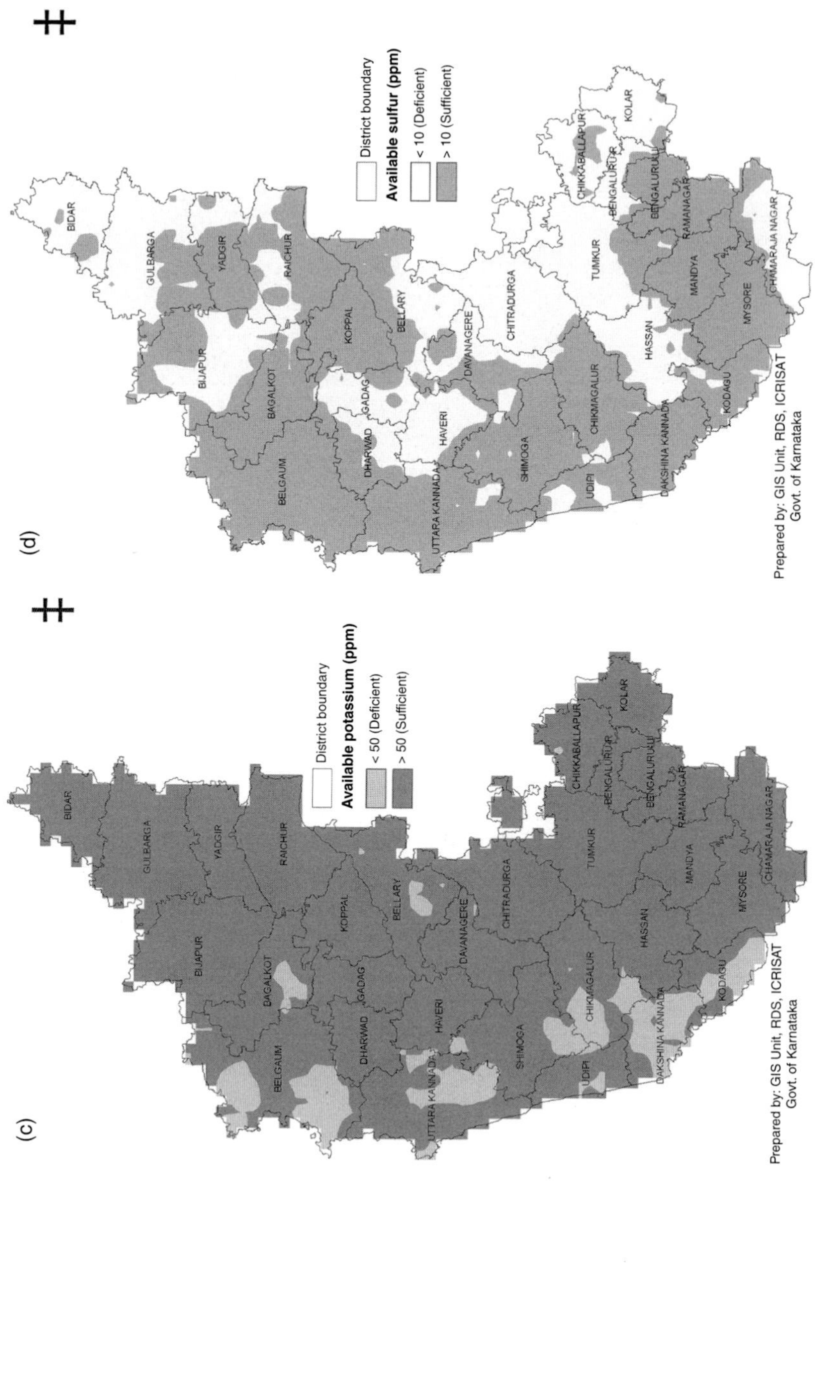

Fig. 2.1. Continued.

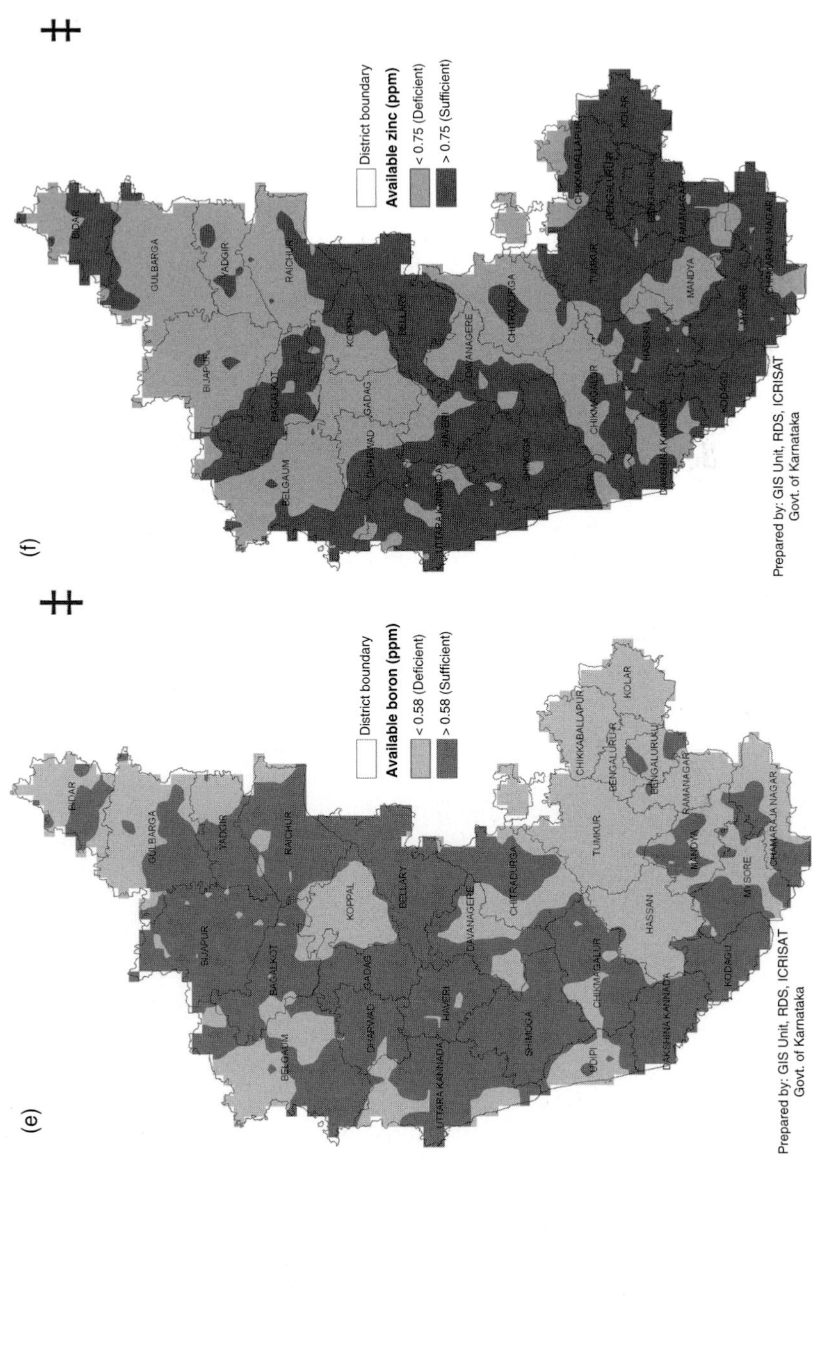

Fig. 2.1. Continued.

groundnut (GPBD 4, ICGV 91114, Kadiri 1375, Kadiri 6), soybean (JS 335, JS 9305), maize (various hybrids) and sunflower (various hybrids); (iii) biomass generation for soil fertility management; and (iv) low-cost and eco-friendly integrated pest management technologies. To ensure availability of inputs for all the farmers, RSKs in the state were linked with specific inputs' suppliers approved by the DoA to ensure good quality and rates.

2.5 Process Steps

- The ICRISAT consortium model of community watershed management provides technological options for the management of: (i) runoff, water harvesting and *in situ* conservation of rainwater for groundwater recharging and supplemental irrigation; (ii) appropriate nutrient and soil management practices; (iii) a waterway system; (iv) crop production technology; and (v) appropriate farming systems with income-generating microenterprises for improving livelihoods, while protecting the environment. The current model of watershed management as adopted by the ICRISAT watershed consortium team involves environment-friendly options and the use of new science tools (Sreedevi *et al.*, 2004; Wani *et al.*, 2008).
- This model was scaled up to enhance agricultural productivity of millions of hectares of rainfed areas in the state of Karnataka in partnership with the DoA and SAUs. The programme was called 'Bhoochetana' (soil rejuvenation). Bhoochetana seeks to unlock the potential of rainfed agriculture towards improving rural livelihoods and achieving food and nutrition security while protecting the environment. In Bhoochetana, soil health assessment is used as an entry point to plan science-based interventions which can result in tangible benefits for the farmers by converging sustainable technologies to increase productivity of farm households with an effective integrated watershed management approach.
- Through convergence of various government programmes and schemes implemented by the consortium, the project adopted a mission approach. For better planning, execution and monitoring, the Government of Karnataka constituted a high-powered committee chaired by the Additional Chief Secretary and Development Commissioner. The committee reviewed the performance of the project every fortnight. It also played a crucial role in making the project successful in the state.
- The project has been implemented on a mission mode and different levels of coordination have been established, starting with clusters of villages in each taluk linking up with Taluk Level Coordination Committees (TLCCs), District Level Coordination Committees (DLCCs) and SLCC (see

Fig. 2.2). Communication was very regular and shared through video conferences and e-mails to speed up the process at field level.

- Bhoochetana was implemented strategically over 4 years in the first phase to improve agricultural productivity, rural incomes and nutrition. It is part of a broader process of innovation and learning. With effective monitoring and evaluation processes, the knowledge acquired from the initial year was used to scale up the model. This was done to create larger impacts in the entire state. The process occurred in an iterative and interactive cycle, as the experience from scaling up feeds back into new ideas and learning (more details in Chapter 12 of this book).
- The unique mechanism of scaling up with comprehensive planning, review and monitoring along with new institutions such as FFs, lead farmers and RSKs, and supporting policies enabled the consortium to cover large areas in the state (Fig. 2.3). The project started with six districts and in 4 years covered a significant area in all 30 districts of the state.

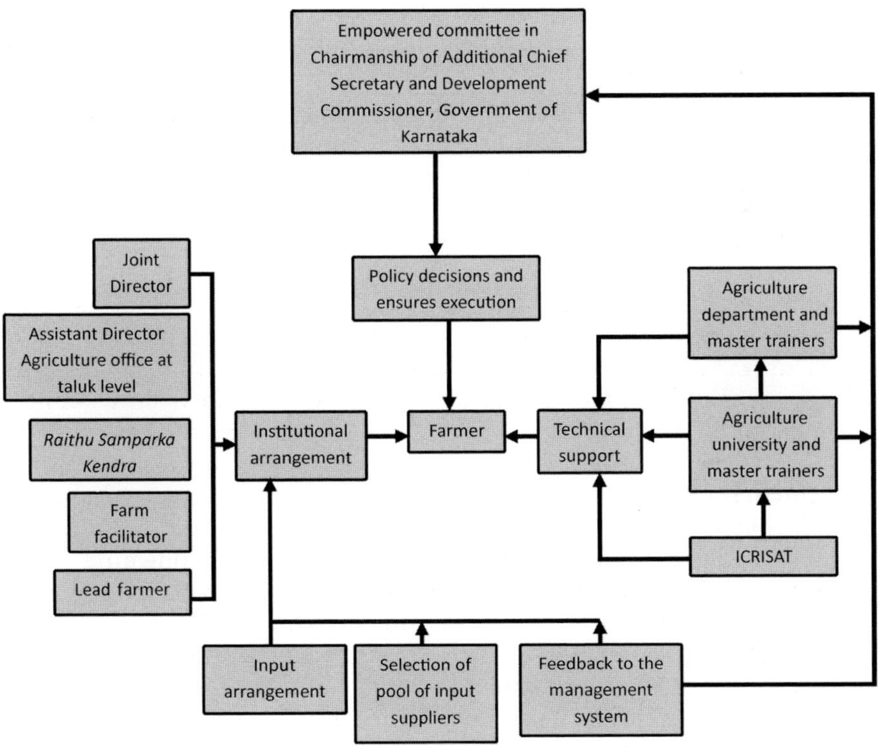

Fig. 2.2. Flow diagram based on discussion with different stakeholders about the Bhoochetana programme and its functioning.

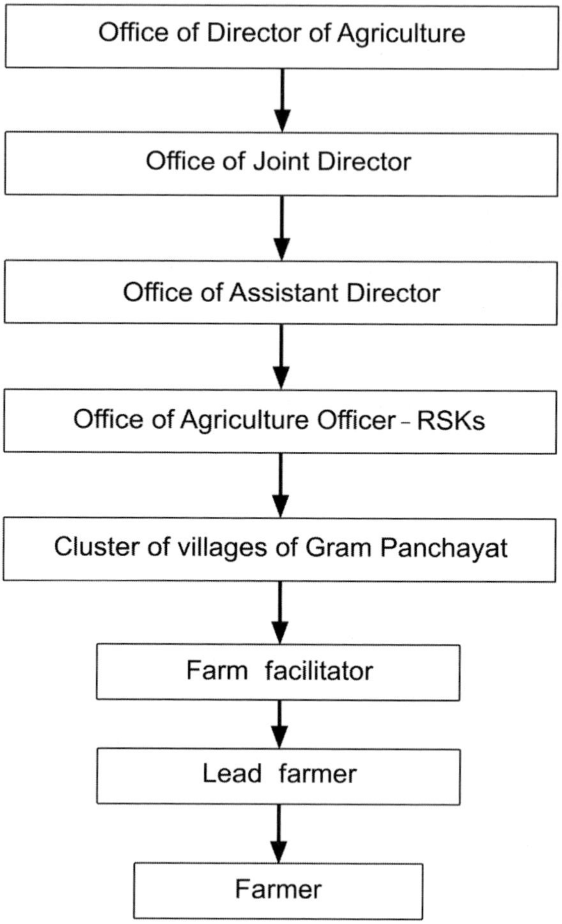

Fig. 2.3. Flow of information in Bhoochetana. Gram Panchayat means village council.

2.6 Key Tasks Performed

Soil testing: Soil testing is the most important and crucial feature of the programme. It was carried out in all the project districts in a phased manner (more details in Chapter 3 of this book).

Farmers' registration: Registration of farmers was a new concept in such a large-scale programme for better monitoring and coverage. After review, it emerged as a need of the programme (more details in Chapter 4 of this book).

Training: There was great demand for training and the field team along with the ICRISAT team conducted timely and good quality training sessions at field, district and state levels. ICRISAT contributed to capacity building of master trainers at state and district levels. University staff, district officials of the DoA and *Krishi Vigyan Kendras* (KVKs) form the field level team to conduct training programmes (more details in Chapter 4 of this book).

Awareness: Announcement by beating drums and *jatras* (carnivals) were organized at village level to generate mass awareness and communicate to the farmers about the programme concept and its benefits (more details in Chapter 4 of this book).

Wall writings: The concept of wall writing was picked up from the watershed department and was used as an effective communication mechanism in Bhoochetana. Wall writings were displayed in selected villages depicting the soil fertility status and crop-specific best management practices. The results were very encouraging. Two wall writings are written in every village at a cost of ₹2000 per wall writing (more details in Chapter 4 of this book).

Farmer field school: One farmer field school per FF was planned to demonstrate production technologies and for effective transfer of technology through technology demonstration in the field at a cost of ₹10,000 per school. This concept helped the programme to facilitate adaptation at field level. Farmers who run farmer field schools were known as facilitators and thus the FF concept evolved (more details in Chapter 4 of this book).

Farm facilitators and lead farmers: The FFs and lead farmers are the backbone of the programme at ground level. These service providers were hired mainly on a task basis at village level only. Their work was defined in each season. The support services of FFs were utilized for a period of 120 days for transfer of technologies at village level at a cost of ₹150/day initially. Lead farmer services were utilized at village level mainly for demonstration of new technology on their fields and sharing with the farmers at village level (more details in Chapter 4 of this book).

Hiring godowns for stocking of inputs: The concept of cluster-level storage of inputs emerged after interaction with the farmers. Apart from the RSK, the inputs were stocked at cluster-village level. Incentives for farmers to adopt micronutrients at the field level were supported and provided (more details in Chapter 9 of this book).

Distribution of inputs: Gypsum, micronutrients, biopesticides, biofertilizers and plant protection chemicals were made available at village and RSK level at a subsidized rate. This policy encouraged greater adoption of micronutrients in the project villages (more details in Chapter 9 of this book).

Seed treatment: A massive seed treatment campaign using biopesticides was organized at village level. Seed treatment chemicals were also provided to farmers. During discussion in village meetings the importance of biopesticides for disease management and monitoring of pests was explained to farmers. They were also told about alternative solutions if they do not have pesticides or chemicals for disease control and pest attacks (more details in Chapter 9 of this book).

Selection of major crops: Major crops were identified in the selected 30 target districts of Karnataka considering the annual crop statistics published by the Government of Karnataka for enhancing productivity of major dryland crops in each selected district. Cropping targets for *kharif* (rainy season) and *rabi* (post-rainy season) were scaled in a staggered manner annually as planned from 2009 to 2012 seasons. High-yielding, short-duration varieties of major rainfed crops were evaluated by a large number of farmers in six districts through the Sujala–ICRISAT project during crop seasons from 2005 to 2008. Farmers preferred some of these crop varieties based on economic yield and acceptable quality of grain or pod and fodder in their region (more details in Chapter 8 of this book).

2.7 Key Drivers

The mission adopted the principle of '4 Cs' (i.e. Consortium, Convergence, Capacity building and Collective action). This involved:

- Formation of a consortium of development agencies such as line departments of the state government and FFs, along with academic and research institutions that generate new technologies and knowledge for improving the livelihoods of the rural poor and in dryland areas.
- Convergence of all schemes of DoA (state and central levels) into Bhoochetana.
- Creation of a dedicated Bhoochetana cell at DoA headquarters to deal with implementation, planning and monitoring activities.
- A demand-driven approach – farmers were to register and pay 50% of the cost of inputs.
- Building capacity of DoA staff and a strong cadre of local FFs to adopt science-based development in the state, with the help of master trainers from SAUs.
- Addressing the mission goal through the '4 Es' (i.e. Efficiency, Economic gain, Equity and Environmental protection) – these are the important pillars of sustainable and inclusive development in the country.
- Ensuring timely supply, availability and access to necessary vital inputs such as knowledge-based soil nutrient management options, micronutrients, good quality seeds and other best soil practices, and also providing the necessary financial incentive to

undertake best-bet options for increasing agricultural productivity through RSKs.

- Adopting improved best-bet management practices on a large scale and sharing knowledge through trained FFs and lead farmers.
- Mapping soil nutrient deficiencies in the remaining 24 districts, which was the starting point for scaling up soil-analysis-based integrated nutrient management practices for sustainable growth, in dryland areas of Karnataka.
- Demonstrating and popularizing other best-bet management practices such as rainwater management, integrated pest management options and organic matter improvement practices to support long-term sustainability and enhanced productivity.
- Establishing village seed banks for crop cultivars by training farmers to ensure that they get a timely supply of seeds at reasonable prices.
- Well-planned, time-bound targets for covering productivity enhancement in 30 districts (as during the third year the programme was expanded to cover all 30 districts), soil sampling and nutrient analysis mapping, and capacity building of stakeholders during the project period.
- Conducting crop cutting experiments for estimating crop yields by a joint team of officials from DoA, the Department of Economics and Statistics and scientists from the University of Agricultural Sciences, along with ICRISAT technicians. As a result, uniform crop sampling procedures were adopted across all the districts.
- Identification of all farmers registered/taking inputs from RSKs and applying them in their designated fields and sowing selected major crops. This was ascertained through RSK bills and FFs, who facilitated farmers in the village to register/get the inputs. At taluk level, the Assistant Director of Agriculture (ADA) or Agricultural Officer (AO) ensured that the total list of those identified farmers was prepared, along with ICRISAT Research Technician and FFs or lead farmers in the villages.
- Preparing a comprehensive list of farmers at district level to facilitate further monitoring and evaluations.
- At taluk level, ICRISAT staff/AO/ADA made at least two field visits during the cropping season to randomly select farmers' fields having crops at the end of the vegetative phase and flowering or maturity phase.
- In these phases, field photos showing crop growth differences in individual farmer's fields were obtained, as a record for verification.
- At the time of crop harvest, the office of Joint Director of Agriculture prepared a farmers' list for crop sampling, randomly selecting fields which also had farmers' management treatment in the same farmer's field.[3]

2.8 Holistic Participatory Approach

The first integrated approach in agriculture was the introduction of Farming System Research, which was started in the 1970s by field practitioners who saw that technologies recommended as a result of agricultural research investment were, in general, inappropriate to the priorities and circumstances of small farmers (Collinson, 2000a, b). However, many believe that this approach was not successful due to the fact that researchers failed to properly understand the realities and practices of the intended beneficiaries. Therefore, science-led participatory approaches evolved over time and attempts were made to enlist farmers as active codevelopers of new technology rather than mere passive providers of information and evaluation as earlier approaches had done.

The new approach attempts to blend science along with farmers' existing wisdom. It recognizes that among other things, dissemination efforts should focus on replicating the social and organizational processes involved in bringing about technological change rather than concentrating on the transfer of technology. The concept of scaling is crucial to the participatory integrated approach. It has profound implications for evaluation of an integrated approach and participatory research in general. In the past the main focus has been on technologies rather than the processes. However, the process of technology dissemination is most crucial to reach the scale and farmers should be the partners.

The new technologies, including natural resource management, pest management as well as improved germplasm introduced into poorer and more complex rainfed systems, require farmers to learn much more, and are not so easily replicated and promulgated (Douthwaite *et al.*, 2001; Wani *et al.*, 2003). Thus, it has led to the realization that solutions to complex problems cannot be solved on-station but need to be developed *in situ* in farmers' fields, taking full advantage of farmer's knowledge and innovative abilities. These location-specific solutions/interventions need a strong network of stakeholders to achieve impact-oriented development. The best way of achieving this is through a consortium of organizations with a multidisciplinary team of scientists to address inclusivity and equity aspects of food production along with knowledge intertwined with inputs and market linkages. These location-specific solutions are cost-effective and are generally applicable processes that can be introduced into new areas.

2.9 Outcomes of Bhoochetana

Bhoochetana in Karnataka caught the attention of officials from the Ministry of Agriculture during the planning meetings in Delhi; other states recognized the progress as well. The example of Karnataka was mentioned during central planning meetings. Based on the success, the DoA, Andhra Pradesh sent a delegation to Karnataka. Members of the delegation interacted with officials and policy makers for the development of an agricultural budget as well as for an understanding of the nuances of Bhoochetana. After deliberations with the DoA, the Andhra Pradesh Bhoochetana programme was planned for all districts of the undivided state of Andhra Pradesh, with technical backstopping from ICRISAT. This itself is a good example of scaling-out strategy by covering an area of 700 ha in 2011 covering seven districts, 7000 ha in 2012 covering 14 districts, and 84,330 ha in 2013 covering all 22 districts of undivided Andhra Pradesh. Other governments of states such as Tamil Nadu and Maharashtra have also interacted with ICRISAT for details and project proposals on Bhoochetana.

The results of the Bhoochetana programme were presented in many national and international forums (e.g. the World Water Forum held in France in 2012 and CGIAR's Fund Council Meeting in New Delhi in 2013) by scientists as well as senior officials of the Government of Karnataka. The success was global and as a result, a senior policy maker from the Philippines, Vice Governor D. Salvanio visited Karnataka and had discussion with policy makers, technocrats and farmers to understand the process and impact of the Bhoochetana programme. Looking at the progress on the ground and his interactions with the DoA officials and farmers, he recommended that Ilocus Sur, a province in the Philippines, should undertake and implement Bhoochetana, with technical support from ICRISAT. As a result, since 2013, two projects on 'Bhoochetana Principles and Practices for Productivity Enhancement' (locally called *Yamang Lupa*) are being implemented by the Bureau of Agriculture in three pilot sites covering an area of 10,000 ha. The provincial Government of Ilocos Sur with ICRISAT's technical support is implementing this as well.

The success of the Bhoochetana programme changed the mindset of different stakeholders, including policy makers, in approaching the problem through science-led solutions. The policy makers understood the need for developing such programmes to benefit the state of Karnataka in general and smallholders in particular. In 2012, the state had initiated the discussion on bringing the international expertise to provide solutions to agriculture and allied sectors with the aim of addressing the problem through a systems approach with the help of SAUs, KVKs and line departments. As a result, in 2013, a programme called '*Bhoosamrudhi*' (land prosperity) was launched to address problems holistically covering agriculture, horticulture, animal husbandry and other allied sectors, together with the technical support from the eight international research institutions along with SAUs led by ICRISAT. Initially, this initiative was implemented in 2013 in four districts representing four revenue divisions covering an area of 320,000 ha and extended to another four districts in 2015 covering an area of 320,000 ha.

Following Bhoochetana and *Bhoosamrudhi* in Karnataka, a unique model was developed in Andhra Pradesh to increase the state's gross domestic product to the level of double-digit growth with overall development of the primary sector. The Government of Andhra Pradesh has launched the Primary Sector Mission (later named as *Rythu Kosam*) in all the 13 districts by converging agriculture, horticulture, livestock, fisheries, marketing and rural development with the technical support from a ICRISAT-led consortium since January 2015 (Fig. 2.4). In addition to convergence at the state level, another innovative mechanism was formulated for convergence by entrusting the Primary Sector Strategy implementation responsibility to the Joint Collector at district level along with allocation of resources with accountability to deliver double-digit growth by implementing the identified growth engines in different sectors. It is an innovative approach to break the existing silos and achieve convergence for attaining efficiency and impacts at ground level. As in Karnataka, Andhra Pradesh too extended higher level policy support to implement the programme in all 13 districts. The major objective was to establish sites of learning with an area of 10,000 ha in each district to demonstrate innovative technologies to improve the productivity and income.

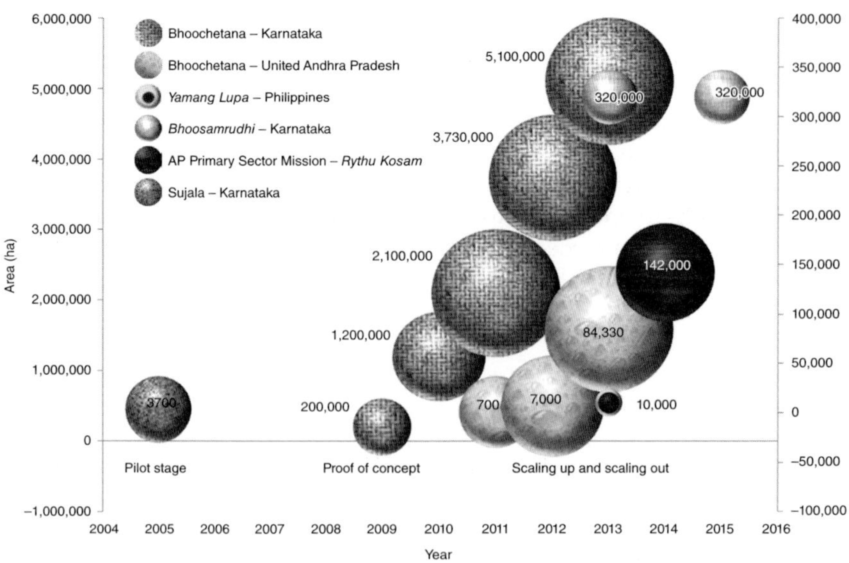

Fig. 2.4. An example of scaling up and scaling out of the Bhoochetana approach.

2.10 Champions at Higher Policy Level

Bhoochetana was championed by higher level policy makers – right from the Development Commissioner, the Chief Secretary, the Minister for Agriculture up to the Chief Minister of the state. Championing by the higher policy makers ensured timely actions by the department as well as effective implementation in the state, which helped to achieve the impact on a larger scale.

2.11 Rejuvenated Extension System

Bhoochetana has developed an innovative institutional arrangement to rejuvenate the extension system as well as to empower farmers through RSKs. This initiative has also helped to create a new institutional arrangement such as creation of the 'Bhoochetana cell' in the state to deal with agricultural extension services and input delivery. Since the inception of the initiative, FFs and lead farmers are the new extension agents who are effectively disseminating the knowledge to the community, which has made huge impacts on the state's agricultural scenario. After realizing the importance of FFs in the extension system, this concept was adopted by other departments of the Government of Karnataka such as the Departments of Horticulture and Sericulture to implement other schemes such as *Suvarna Bhoomi Yojane* in the state (more details in Chapter 6 of this book).

2.12 Regular Monitoring and Evaluations

Detailed meetings to devise work plans at the district level were conducted annually at ICRISAT. District-wise planning at district and taluk levels was undertaken by the DoA in the state, wherein district and taluk areas to be covered, target crops, input requirements as well as financial requirements were discussed and planned every season. Such planning enabled the mobilization of human resources as well as other necessary inputs (more details in Chapters 4 and 6 of this book).

2.13 Lessons Learnt

The Inclusive Market Oriented Development (IMOD) approach has been used in all the stages of project activities, including the planning phases of this project. The biggest strength of this project is the partnership approach, which primarily focuses on knowledge development as well as economic development, with soil, water and crop as strong intermediaries. The key factor of IMOD that has significantly influenced the project so far is the inclusiveness approach, which primarily focuses on the economic development of smallholders of the community. It is learnt that a better approach can help in organizing the communities for effective utilization of new innovative technologies to increase productivity of dryland agriculture. Inclusiveness is the key to achieving success. Thus, clear and timely information flow of written communication to lower levels of the implementing authorities has to be ensured. For this purpose, smallholders and marginal farmers need to be enabled through training, field exposure visits and demonstrations, ensuring availability of necessary inputs and handholding support.

Involving partners and farmers from the planning and implementation of activities would help immensely as that provides comfort levels as well as builds trust for their active participation in implementing project activities. This also helped them to gain an understanding of the project objectives, which led to stronger ties and opportunities to test and improve the system productivity.

Continued communication and interaction between DoA staff, consortium partners, FFs and farmers was a key factor for creating successful outcomes in Bhoochetana. During the project, the annual planning and review meeting seemed sufficient for implementation of project activities. However, intensive contact by the DoA staff, consortium partners and farmers contributed to the strong ownership of the project. A holistic integrated system approach with multi-stakeholder partnership is needed to unlock the potential of rainfed agriculture. The Bhoochetana initiative has got attention from different state and national governments. In 2011, the Krishi Karman Award for the highest productivity of coarse cereals through Bhoochetana was presented to the Government of Karnataka. This award was instituted to recognize the state's contribution to increase the food production in the country. The Agriculture Leadership Award – 2011 for best performing state by *Agriculture Today* (the national agriculture magazine) was presented to the Government of Karnataka.

Among the consortium members, some partners try to protect their turfs intensively by raising questions on several aspects. In such cases, scientific evidence-based clarifications are essential to remove doubts from the minds of other members/policy makers. Initially, resistance is very much evident as such initiatives try to shake up the people to move out of their comfort zones and people resist change through questioning the strategies, interventions, etc. However, once success is achieved and the credit is shared with such partners, they become willing and helpful. The leader should have patience, expertise as well as the knack to handle such situations politely but firmly without compromising on the strategy and needed interventions agreed upon to meet the project goals.

2.14 Scaling up and Scaling out the Bhoochetana Initiative

The concept of scaling is very crucial for the impact of technology. Scaling-up and scaling-out concepts are defined below:

- Scaling up is institutional expansion from grass-root organizations to policy makers; development institutions and other stakeholders are key to building an enabling environment for impact.
- Scaling out is innovation spread from farmer to farmer and community to community, within the same stakeholder groups.

Both scaling up and scaling out are linked because as a change spreads further geographically, the greater are the chances of influencing those at higher levels, and likewise, as one goes to higher institutional levels then the greater are the chances for horizontal spread (Douthwaite *et al.*, 2003).

The well-established institutional arrangements along with strong policy guidance helped to scale up this approach not only within the state but also outside the state to benefit millions of farmers. The productivity enhancement approach started during 2005 in the form of the Sujala watershed initiative covering communities in six districts of Karnataka on 3700 ha by adopting the consortium approach. The initiative which started with 13 watersheds in 2005 was scaled out to 47 watersheds in 2008 for demonstrating productivity enhancement measures.

Bhoochetana started with an area of 0.2 million ha in six districts covering 0.2 million farmers in 1440 villages. The impact of this approach in the first year itself was evident in increased crop yield and that led to the expansion of Bhoochetana to 16 districts to demonstrate new innovations to increase crop yields. As a result, in 2010, Bhoochetana was implemented in 16 districts covering an area of 1.2 million ha in 5030 villages and 0.8 million farmers benefited from the programme. Similarly, after realizing the impact, the focus of Bhoochetana was shifted from purely rainfed to irrigated crops as well. This led to expansion of Bhoochetana further to the entire state of Karnataka covering 30 districts on 2.1 million ha in 2011 and 3.73 million ha in 2012. In 2013, Bhoochetana was implemented in 5.1 million ha covering both rainfed and irrigated crops with the help of 9700 FFs in transferring the technology at the farmers' doorstep.

2.15 Epilogue

'Bhoochetana', a mission project, was taken up by the Government of Karnataka based on the brainstorming session to kick-start the agricultural growth in the state, which was stagnant and hovered around negative to

2% during 2001–2008. The mission project was conceived by ICRISAT, based on the demand of the Government of Karnataka and formulated with a clear strategy and measurable outputs with appropriate requirement of the financial/human/institutional resources for the success of the project. For ensuring internalization of the initiative, high-level policy makers as well as decision makers were involved in various activities of the programme, which increased ownership among the different stakeholders and finally it became a farmer-driven programme rather than the government programme. Different stages of the mission programme were: (i) preparation of the detailed project activities; (ii) identifying measurable indicators and outputs expected along with the detailed strategic plan for implementation at different levels with the appropriate resources needed; (iii) clear roles for the partners; (iv) understanding possible bottlenecks in the implementation at ground level; and (v) formulating a quick decision-making process through the SLCC. The SLCC members adopted regular monitoring and evaluation using the specified format with key performance indicators, and subsequently regular video conferencing with all the 30 districts from the headquarters ensured detailed monitoring and evaluation as well as identification of the issues and the difficulties faced on ground implementation by the district officials. Transparent monitoring and evaluation as well as eligibility for all the farmers in the state to participate in the programme along with appropriate knowledge and timely availability of inputs ensured the success of the programme.

Increased crop productivity by 20–66% with tangible benefits during the first year in six districts of Karnataka, covering 0.2 million ha area, served as a very good booster for enhancing ownership by the officials of the DoA for further scaling up. With these results on the ground, the implementing agency as well as the policy makers and ICRISAT team also realized the power of science-led development for unlocking the potential of rainfed agriculture in the developing countries. The programme expanded by leaps and bounds and by 2012 Bhoochetana covered 4.4 million farmers with area coverage of 3.75 million ha. For individual farmers, the benefit–cost ratio of 2:1 to 14:1 with gross value of increased productivity accrued to be ₹1260 crores (US$230 million) in different districts for different crops during the 4 years. The Government of Karnataka was recognized through the success of Bhoochetana and received the Krishi Karman Award and Agriculture Leadership Award in 2011 from the Government of India.

Notes

[1] A line department is a department in the government which provides technical support in the nature of estimates, measurements and the supervision of works carried out.
[2] *Raithu Samparka Kendra* (Farmer Contact Centre) is a local level administrative set up to facilitate farmers with inputs and information sharing.

[3] The Joint Director of Agriculture is responsible for overall supervision of agricultural activities in the district. He/she is assisted by ADAs and AOs at taluk level and Assistant Agricultural Officers at RSK level.

References

Collinson, M.P. (2000a) FSR – understanding farmers and their farming. In: Collinson, M.P. (ed.) *A History of Farming Systems Research*. Food and Agriculture Organization of the United Nations, Rome and CABI Publishing, Wallingford, UK, pp. 5–12.

Collinson, M.P. (2000b) Evolving typologies for agricultural R&D. In: Collinson, M.P. (ed.) *A History of Farming Systems Research*. Food and Agriculture Organization of the United Nations, Rome and CABI Publishing, Wallingford, UK, pp. 51–58.

Douthwaite, B., Keatinge, J.D.H. and Park, J. (2001) Why promising technologies fail: the neglected role of user innovation during adoption. *Research Policy* 30(5), 819–836.

Douthwaite, B., Delve, R., Ekboir, J. and Twomlow, S.J. (2003) Contending with complexity: the role of evaluation in implementing sustainable natural resource management. *International Journal of Agricultural Sustainability* 1, 51–68.

Garg, K.K. and Wani, S.P. (2012) Opportunities to build groundwater resilience in the semi-arid tropics. *Groundwater* 51(5), 679–691.

Rockström, J., Hatibu, N., Oweis, T. and Wani, S.P. (2007) Managing water in rain-fed agriculture. In: Molden, D. (ed.) *Water for Food, Water for Life: A Comprehensive Assessment of Water Management in Agriculture*. Earthscan, London and International Water Management Institute, Colombo, Sri Lanka, pp. 315–348.

Sahrawat, K.L., Rego, T.J., Wani, S.P. and Pardhasaradhi, G. (2008) Stretching soil sampling to watershed: evaluation of soil-test parameters in a semi-arid tropical watershed. *Communications in Soil Science and Plant Analysis* 39, 2950–2960.

Sreedevi, T.K., Shiferaw, B. and Wani, S.P. (2004) *Adarsha Watershed in Kothapally: Understanding the Drivers of Higher Impact*. Global Theme on Agroecosystems Report No. 10, International Crops Research Institute for the Semi-Arid Tropics, Patancheru, Andhra Pradesh, India.

Twomlow, S., Shiferaw, B., Cooper, P. and Keatinge, J.D.H. (2006) *Integrating Genetics and Natural Resource Management for Technology Targeting and Greater Impact of Agricultural Research in the Semi-Arid Tropics*. International Crops Research Institute for the Semi-Arid Tropics, Patancheru, Andhra Pradesh, India.

Wani, S.P., Pathak, P., Sreedevi, T.K, Singh, H. and Singh, P. (2003) Efficient management of rainwater for increased crop productivity and groundwater recharge in Asia. In: Kijne, J.W., Barker, R. and Molden, D. (eds) *Water Productivity in Agriculture: Limits and Opportunities for Improvement*. CAB International, Wallingford, UK, pp. 199–215.

Wani, S.P., Joshi, P.K., Raju, K.V., Sreedevi, T.K., Wilson, J.M., Shah Amita, Diwakar, P.G., Palanisami, K., Marimuthu, S., Jha, A.K., Ramakrishna, Y.S., Meenakshi Sundaram, S.S. and D'Souza, M. (2008) *Community Watershed as a Growth Engine for Development of Dryland Areas. A Comprehensive Assessment of Watershed Programs in India*. Global Theme on Agroecosystems Report No. 47. International Crops Research Institute for the Semi-Arid Tropics, Patancheru, Andhra Pradesh, India.

Wani, S.P., Sreedevi, T.K., Rockström, J. and Ramakrishna, Y.S. (2009) Rainfed agriculture – past trends and future prospects. In: Wani, S.P., Rockström, J. and Oweis, T. (eds) *Rainfed Agriculture: Unlocking the Potential.* CAB International, Wallingford, UK, pp. 1–35.
Wani, S.P., Sahrawat, K.L., Sarvesh, K.V., Baburao, M. and Krishnappa, K. (2011) *Soil Fertility Atlas for Karnataka, India*. International Crops Research Institute for the Semi-Arid Tropics, Patancheru, Andhra Pradesh, India.
Wani, S.P., Sarvesh, K.V., Krishnappa, K., Dharmarajan, B.K. and Deepaja, S.M. (eds) (2012) *Bhoochetana: Mission to Boost Productivity of Rainfed Agriculture through Science-led Interventions in Karnataka*. International Crops Research Institute for the Semi-Arid Tropics, Patancheru, Andhra Pradesh, India.

Soil Nutrient Mapping for On-farm Fertility Management

3

K.L. Sahrawat,[†] Suhas P. Wani, Girish Chander,* G. Pardhasaradhi and K. Krishnappa

International Crops Research Institute for the Semi-Arid Tropics (ICRISAT), Patancheru, India

3.1 Introduction

Feeding the projected population of 9.1 billion globally and 1.6 billion in India by 2050 is one of the greatest challenges of the century, and in this endeavour to ensure future food security, efficient soil nutrient management is crucial (Wani *et al.*, 2003; Sahrawat *et al.*, 2010; Chander *et al.*, 2013). Since the era of the Green Revolution in India in the late 1960s, the focus has been on only three macronutrients, namely nitrogen (N), phosphorus (P) and potassium (K), and this has brought nutrient imbalances and widespread deficiencies of micro and secondary nutrients such as sulfur (S), boron (B) and zinc (Zn) in addition to macronutrients (Wani *et al.*, 2009; Sahrawat and Wani, 2013; Chander *et al.*, 2014). Most farmers and stakeholders are not aware of soil fertility issues and management alongside water and crop management, which is the main reason for large yield gaps in the semi-arid tropics (SAT). In order to ensure future food security and the future of smallholder farmers, science-led interventions are needed to bridge the yield gaps in the SAT. Some pilot initiatives such as the International Crops Research Institute for the Semi-Arid Tropics (ICRISAT)–Andhra Pradesh Rural Livelihood Programme (APRLP) initiative in Andhra Pradesh and the Bhoochetana initiative in Karnataka have shown that soil nutrient mapping is the best entry point

*g.chander@cgiar.org

activity to enhance productivity and livelihoods through soil-need-based fertility management (Wani *et al.*, 2011; Chander *et al.*, 2013; Sahrawat and Wani, 2013). This chapter therefore focuses on soil fertility management issues and the need of soil nutrient mapping for informed decisions.

3.2 Soil Infertility – A Major Constraint in Addition to Water Shortage

In rainfed production systems, the importance of water shortage and associated stress effects on crops can hardly be overemphasized, especially in the SAT regions (Bationo *et al.*, 2008; Pathak *et al.*, 2009; Passioura and Angus, 2010; Rockström *et al.*, 2010). However, apart from water shortage, soil infertility is the issue for crop production and productivity enhancement in much of the SAT regions of the world, and SAT regions of India are no exception (El-Swaify *et al.*, 1985; Twomlow *et al.*, 2008; Wani *et al.*, 2009; Sahrawat and Wani, 2013; Chander *et al.*, 2014).

Equally importantly, apart from the deficiencies of major nutrients, especially N and P, deficiencies of secondary nutrients especially of S and micronutrients have been reported with increasing frequencies from the intensified irrigated production systems (Kanwar, 1972; Pasricha and Fox, 1993; Takkar, 1996; Scherer, 2001, 2009; Fageria *et al.*, 2002; Singh, 2008). While in the irrigated production systems, the deficiencies of various plant nutrients have been diagnosed through soil and plant testing and managed through the fertilization of crops, but little attention has been paid to diagnosing the deficiencies of secondary nutrients such as S and micronutrients in dryland rainfed production systems especially in SAT regions of India (Sahrawat *et al.*, 2007, 2010; Sahrawat and Wani, 2013).

In the past, little attention has been devoted to survey and determine the fertility status of farmers' fields with an overall objective to diagnose the nutrient problems in the rainfed production systems, which is a prerequisite for developing effective nutrient management strategies for enhancing agricultural productivity in these areas.

Moreover, we have observed that lack of adequate analytical laboratory infrastructure to provide high-throughput analytical research support, coupled with lack of awareness of the mining of secondary and micronutrients in production systems, is constraining the cause of upgrading rainfed agriculture (Wani *et al.*, 2009; Sahrawat, 2013; Sahrawat and Wani, 2013). Information on the soil fertility status of farmers' fields is needed not only for enhancing crop productivity through balanced nutrient management, but also to promote judicious use of costly external inputs of nutrients and to enhance the efficiency of scarce water resources in developing countries like India (Sahrawat, 2006; Wani, 2008; Sahrawat and Wani, 2013; Chander *et al.*, 2014).

This apparent paradox of lack of application of adequate amount of nutrients from external inputs is rather inexplicable (Katyal, 2003; Bationo *et al.*, 2008) despite the common knowledge that the soil resource base in the rainfed systems of the SAT regions is relatively fragile and marginal compared with that under irrigated production systems (El-Swaify *et al.*, 1985; Rego *et al.*, 2003; Sahrawat *et al.*, 2007, 2010; Sahrawat and Wani, 2013).

In Indian rainfed systems, water management for reducing water shortage has been the primary focus of research and developmental activities in these areas, and soil infertility has been largely rather ignored (El-Swaify *et al.*, 1985; Wani *et al.*, 2003; Sahrawat *et al.*, 2010) or has not been addressed in a comprehensive and integrated manner along with soil and water conservation practices (Wani *et al.*, 2009; Rockström *et al.*, 2010). However, it has been observed that even in water-limiting environments there is indeed potential to enhance agricultural productivity through efficient management of soil, water and nutrients in an integrated manner (Twomlow *et al.*, 2008; Wani *et al.*, 2009; Sahrawat *et al.*, 2010; Chander *et al.*, 2013).

For achieving the potential of productivity in water-limited environments, a concept of water-limited potential yield seems quite appropriate as this forms the basis to reach the attainable yield in these environments through management of various constraints other than only water shortage (Passioura, 2006; Singh *et al.*, 2009).

For example in Australia, farmers have adopted the notion of water-limited potential yield as a benchmark for crop yield and if farmers find that their crops are performing below the benchmark, they look for the reasons and attempt to improve their management accordingly (Passioura and Angus, 2010). It must be emphasized that in the concept of water-limited potential yield in the rainfed systems, natural resource management in general and soil fertility management in particular need to be paid due attention alongside water stress management in view of the fragile nature of the soil resource base (Wani *et al.*, 2009; Sahrawat *et al.*, 2010; Sahrawat and Wani, 2013).

In addition there is a commonly held belief among researchers and agriculturists that at relatively low yields of crops in the rainfed systems of India, only the deficiencies of major nutrients (especially those of N and P) are important for the SAT Indian soils (El-Swaify *et al.*, 1985; Rego *et al.*, 2003). As a result of this belief, very little attention has been devoted to diagnosing the extent of deficiencies of the secondary nutrients such as S and micronutrients in various crop production systems on millions of small and marginal farmers' fields (Rego *et al.*, 2005, 2007; Sahrawat *et al.*, 2007, 2010).

However, there is no denying the fact that the productivity of the SAT soils is low due to water shortages. Although low fertility is also an issue, in practice the deficiencies of major nutrients (N and P) are considered important and the role of secondary and micronutrients in enhancing water use efficiency in various rainfed production systems is neglected. Also, even the input of major nutrients to dryland production systems is rather meagre compared with that in the irrigated systems (Rego *et al.*, 2005; Wani *et al.*, 2009). Because of low productivity of the rainfed crops, it is assumed that the uptake and mining of secondary and micronutrient reserves in soils is much less than in irrigated production systems (Rego *et al.*, 2003).

Nevertheless for sustained increase in dryland productivity, soil and water conservation measures need to be integrated with balanced plant nutrition, and the choice of crops and/or cultivars, and their management (Wani *et al.*, 2003; Passioura, 2006; Passioura and Angus, 2010; Sahrawat *et al.*, 2010). The ongoing farmer participatory integrated watershed management programme of the ICRISAT, Patancheru, India, provided an appropriate opportunity to implement a balanced nutrient management strategy alongside soil and water conservation practices in farmers' fields in the Indian SAT. For achieving efficient and judicious use of nutrients through fertilizer inputs, assessing the soil's inherent nutrient status is considered a prerequisite (Sahrawat, 2006; Sahrawat and Wani, 2013).

3.3 Identifying a Suitable Entry Point Activity (EPA)

The choice of an appropriate knowledge-based entry point activity (EPA) for building rapport with the community cannot be overemphasized as an EPA is capable of providing a head start to a community-based programme such as watershed management for overall rural development. During our watershed work over a decade, we learnt that it is useful to consider the following points while selecting an appropriate EPA for integrated community watershed management:

- The EPA should be knowledge-based and should not involve direct cash payment through the project in the village.
- The EPA should have a high success probability (> 80–90%), and be based on proven research results.
- The EPA should involve a participatory research and development approach, and community members should preferably be involved in undertaking the activity in watersheds.
- An EPA should result in the measurable tangible economic benefits to the farming community with a relatively high benefit–cost ratio.
- The EPA preferably should be simple and easy for the participating farmers to undertake its participatory evaluation.

- Most importantly, the EPA should benefit the majority of farmers in the watershed.
- The EPA should have a reliable and cost-effective approach to assess the constraints.

Based on our experience with watershed work and considering the above-stated requirement for an EPA, we felt that for building rapport with the community, good participatory rural appraisal (PRA) and knowledge about local natural resources can be used to identify a knowledge-based EPA. The knowledge-based EPAs were found to be superior to the subsidy- or cash-based EPA for enabling community participation of higher order (cooperative and collegiate) rather than in a contractual mode (Dixit *et al.*, 2007).

Indeed, there is much need to innovate new methods to share knowledge with primary stakeholders as the traditional methods of extension have not been successful (Olson, 1971; Sreedevi *et al.*, 2004; Wani, 2008; Wani *et al.*, 2009).

3.4 Use of Soil Testing as an EPA

With the purpose of soil fertility augmentation, and for judicious use of nutrient inputs from external sources, it was indeed considered most appropriate to introduce and evaluate the concept of soil-test-based use of plant nutrients in our ongoing on-farm research in watersheds. The objective was to diagnose the deficiencies of all nutrients, including major, secondary and micronutrients in a comprehensive manner through high-volume soil analysis in the ICRISAT Central Analytical Services Laboratory, and based on soil-test-based nutrient management develop a balanced nutrient management strategy to sustainably enhance systems' productivity by increasing rainwater use efficiency.

Soil testing indeed was thought as a most appropriate tool for on-farm soil fertility management; and further, the integration of soil-test-based balanced nutrient management with the implementation of soil and water conservation practices was considered a prerequisite for sustainably increasing the productivity in rainfed areas of the SAT (Sahrawat and Wani, 2013).

The use of soil testing was introduced in a joint ICRISAT–APRLP watershed project as a science-based activity to diagnose the nutrient deficiencies and associated soil fertility problems prior to conducting on-farm productivity enhancement trials. A farmer-participatory stratified random sampling methodology was employed for collecting soil samples from farmers' fields (Sahrawat *et al.*, 2008). During 2002–2004 seasons, soil testing was employed to diagnose the nutrient deficiencies in the farmers' fields in three districts (Mahabubnagar, Nalgonda and Kurnool) of Andhra Pradesh, India. The results of soil analysis were shared with the participating farmers; and based on the results, recommendations were formulated for balanced nutrient management and the nutrients found deficient were added. The results were presented in the local language along with the necessary interpretative details and shared with the farmers in group meetings at the block level in a district.

The results of soil analyses showed that 81–99% of soil samples were deficient in Zn, B and S (Rego *et al.*, 2007). Past research experience at ICRISAT also emphasized that carefully conducted PRA along with the knowledge of local practices followed by farmers could help diagnose the constraints for identifying knowledge- and constraint-based EPA (Sreedevi *et al.*, 2004; Wani, 2008).

Follow-up on-farm trials on the comparative evaluation of farmers' input treatment with that based on soil-test-based balanced nutrient management, conducted during three seasons (2002–2004) under the ICRISAT–APRLP joint project showed significant responses of crops over the farmers' input treatment (Rego *et al.*, 2007). The results further demonstrated that an appropriate EPA such as soil testing could indeed ensure tangible economic benefit to individual farmers. As indicated earlier, the identification of major constraints limiting crop production and their alleviation ensured tangible economic benefits to individual farmers, and thereby triggered farmers' interest to participate in the project activities. Other researchers have also reported on the importance of using natural resource management as an EPA in community-based projects such as watershed management (Olson, 1971; Sreedevi *et al.*, 2004; Wani, 2008; Wani *et al.*, 2009).

However, it must be stated that since 1997 the natural resources management group at the ICRISAT centre in India, along with its partners, has been conducting systematic and detailed studies on the diagnosis and management of nutrient deficiencies in the semi-arid regions of Asia with emphasis on the semi-arid regions of India. It started with detailed analysis of farmers' fields in the Milli watershed at Lalatora in Madhya Pradesh where analysis of soil samples for micronutrients was deliberately included as a part of the baseline characterization of the site (Sahrawat *et al.*, 2010).

ICRISAT and its partners have been working with the Government of Karnataka to sustainably enhance the productivity of rainfed areas of the state covering all the 30 districts. The strategy used to enhance agricultural production and productivity is based on the principle of 'Bhoochetana' or rejuvenating the soil, by enhancing its fertility. In this mission mode programme, soil testing was used as an EPA to diagnose the nutrient disorders and manage them via a balanced nutrient management approach in farmer participatory manner.

3.5 Soil Sampling Methodology and Soil Analysis

First and foremost, a soil sampling methodology was developed and standardized to collect representative soil samples in a watershed. Following several seasons' work experience in watersheds in Andhra Pradesh, the methodology based on the principle of stratified random sampling was found most appropriate for collecting soil samples to represent an entire watershed. The standardized methodology takes into consideration several factors in the watershed, including soil types, topography of the land, major crops grown, and farmers' land-holding size (for details see Sahrawat *et al.*, 2008). For example, for an effective soil sampling of land in an undulating landscape, farmers' fields were divided into three groups based on the position on the toposequence: top, middle and bottom, depending on the elevation and drainage pattern. We separated different soil types in each category. For soil sampling of the entire watershed, we randomly selected 20% of farmers in each position on the toposequence, taking into consideration the farm size, types of soils and crops grown (Sahrawat *et al.*, 2008).

The soil sampling programme of watersheds in various states was initiated in 2002, and has been continuing since. The main feature of this programme is farmer participation in the soil sampling. The soil sampling methodology was demonstrated to farmers in groups; and following this, the participating farmers themselves collected the soil samples from their respective fields. Using stratified random sampling methodology (Sahrawat *et al.*, 2008) eight to ten cores of surface (0–15 cm depth) soils were collected to make one composite sample.

The soil samples were air dried and using a wooden hammer were turned into a powder that could pass through a 2-mm sieve. For organic carbon (C) analysis, the soil samples were ground to pass through a 0.25-mm sieve. Prepared samples were analysed for various fertility characteristics in the ICRISAT Central Analytical Services Laboratory (Sahrawat and Wani, 2013).

For soil analysis, pH was measured by a glass electrode using a soil-to-water ratio of 1:2. Organic C was determined using the Walkley–Black method (Nelson and Sommers, 1996). Exchangeable K was determined using the ammonium acetate method (Helmke and Sparks, 1996). Available S was measured using 0.15% calcium chloride ($CaCl_2$) as an extractant (Tabatabai, 1996; Sahrawat *et al.*, 2009), and available P (Olsen-P) was measured using sodium bicarbonate ($NaHCO_3$) as an extractant (Olsen and Sommers, 1982). Available Zn was extracted by DTPA reagent (Lindsay and Norvell, 1978) and available B was extracted by hot water (Keren, 1996). Details of the methods used for testing soil samples for various fertility parameters are given in Sahrawat *et al.* (2007, 2010).

3.6 Knowledge Sharing with Farmers

The soil test results of soil samples collected from farmers' fields were shared with farmers in their own language via various modes of communication, including wall writing and soil health cards (Wani, 2008). The need to apply nutrients found deficient, as part of a balanced nutrient management strategy for enhancing productivity, was also discussed with the participating farmers.

Based on the results of soil samples collected from farmers' fields, recommendations were developed at block level for balanced nutrient management of crops in farmers' fields. For this, critical limits in the soil for various plant nutrients were used (Table 3.1) to separate deficient soil samples from the non-deficient ones (Sahrawat, 2006; Rego *et al.*, 2007; Sahrawat *et al.*, 2007) for follow-up on-farm crop response studies.

Table 3.1. Critical limits in the soil of plant nutrient elements to separate deficient samples from non-deficient samples. (Data gleaned from various literature sources, for details see Rego *et al.*, 2007; Sahrawat *et al.*, 2007.)

Plant nutrient	Critical limit (mg/kg)
Sodium bicarbonate-extractable P	5
Ammonium acetate-extractable K	50
Calcium chloride-extractable S	10
Hot water-extractable B	0.58
DTPA-extractable Zn	0.75

For practical utilization of the soil-test-based nutrient management, we have already mapped, using the geographical information system (GIS)-based extrapolation methodology, the deficiencies of all nutrients including especially those of S, B and Zn along with soil fertility parameters pH, electrical conductivity (EC) (indicator of soluble salts) and organic C in all the 30 districts of Karnataka state, India (Wani *et al.*, 2011). Finally, the soil-test-based fertilizer application has been made web-based so that the recommendations can be downloaded and made available nutrient-wise to farmers using colour codes depicting the deficiency or sufficiency of a nutrient. Such information can be easily used by smallholders, and the farmers can be kept updated regularly with the latest results on the website.

For soil parameter mapping covering the entire state of Karnataka, a total of 92,904 soil samples were collected from farmers' fields in watersheds in all the 30 districts of Karnataka. Soil test results of the soil samples analysed for pH, EC, organic C and extractable (available) major, secondary and micronutrients, from all the 30 districts of Karnataka at the block (or *mandal*) level, were used for mapping individual soil fertility parameters. Detailed mapping of soil fertility parameters, and discussion and interpretation of the results for use by researchers, agriculturists and farmers is provided in a separate treatise (Wani *et al.*, 2011). Such maps can be extended and used by farmers in a cluster of villages to plan the application of deficient plant nutrients to production systems. However, a summary of the results on various soil fertility parameters is provided in Table 3.2. The values of fertility parameters pH, EC, organic C, extractable or available P, K, S, B and Zn in terms of range, mean and the percentage of samples deficient (for organic C and available P, K, S, B and Zn) are summarized district-wise for all the 30 districts of Karnataka (Table 3.2). As mentioned earlier, the maps for soil fertility parameter for the 30 districts at the block or *mandal* level are available in Wani *et al.* (2011).

3.7 Correlations Among Soil Parameters

Soil characteristics, especially those related to soil fertility, are interrelated among themselves. We studied the relationships among various soil fertility parameters (pH, EC, organic C and available (extractable) P, K, S, Zn and B) using 92,904 soil samples collected from the 30 districts of Karnataka.

Soil pH is an important property to influence different soil parameters through adsorption and precipitation reactions of nutrients, modifying uptake by influencing activities of microorganisms and influencing the abilities of plants to absorb ions. Correlation studies showed a positive relationship of pH with EC, K and B, while the pH was negatively correlated with organic C, P, S and Zn (Table 3.3). Soil organic C and EC were found to have a positive relationship between them and with rest of the soil fertility parameters. Therefore, there is a need to manage optimum amounts of soil organic C to regulate adequate supplies of essential plant nutrients. There were positive significant correlations between available P and K, P and S, P and Zn, K and Zn, K and B, S and Zn, S and B, and Zn and B, while there were negative significant correlations between P and B, and K and S (Table 3.3).

The positive correlations among various extractable or available nutrients are due to the fact that all these nutrients are in the deficient range, and hence there is hardly any scope for antagonistic relationships, which have usually been reported in the literature, for example between Zn and P (Sahrawat, 2006). Clearly, the positive role of soil organic C status on available nutrient elements is obvious in these soils, which are low in organic matter without very low inputs of plant nutrients from external sources as mineral or organic fertilizers (Bationo *et al.*, 2008; Sahrawat and Wani, 2013).

Table 3.2. Summary of results on chemical characteristics[a] of soil samples collected from farmers' fields covering all the 30 districts of Karnataka.

				Available				
District[b]	pH	EC (dS/m)	Organic C (%)	P (ppm)	K (ppm)	S (ppm)	Zn (ppm)	B (ppm)
Bagalkot (2,440) – range	6.3–8.9	0.11–1.99	0.18–1.23	0.6–6.2	17–74	4.1–39.9	0.50–10.69	0.12–12.78
– Mean	7.8	0.35	0.62	2.3	60	11.7	0.92	0.70
– Percentage of samples deficient	–	–	36	97	28	59	55	69
Belgaum (4,560) – range	4.7–8.9	0.04–5.11	0.02–2.62	0.0–15.3	0–169	0.2–460.0	0.02–3.48	0.01–3.29
– Mean	7.3	0.44	0.64	2.1	52	152.2	0.66	0.59
– Percentage of samples deficient	–	–	29	95	52	2	68	74
Bellary (2,100) – range	6.2–9.0	0.10–2.25	0.20–1.24	0.6–6.2	16–74	4.1–41.4	0.52–13.81	0.12–18.02
– Mean	7.4	0.40	0.63	2.9	55	11.1	1.27	1.20
– Percentage of samples deficient	–	–	32	90	33	67	19	36
Bengaluru Rural (4,448) – range	4.2–9.5	0.01–9.96	0.01–1.50	0.0–543.8	9–1,414	0.5–2,299.1	0.05–235.00	0.02–5.12
– Mean	6.3	0.28	0.41	18.0	100	6.8	1.50	0.37
– Percentage of samples deficient	–	–	73	21	23	90	29	79
Bengaluru Urban (2,680) – range	4.4–8.7	0.02–2.20	0.03–3.00	0.7–351.5	2–580	0.8–335.0	0.03–5.79	0.02–6.86
– Mean	6.7	0.19	0.49	43.0	125	29.3	1.30	0.60
– Percentage of samples deficient	–	–	58	10	14	6	37	60
Bidar (2,375) – range	5.5–9.5	0.03–4.04	0.12–1.98	0.6–118.6	18–2,297	1.0–181.3	0.16–18.00	0.10–6.18
– Mean	7.6	0.24	0.59	8.4	208	7.3	0.85	0.55
– Percentage of samples deficient	–	–	40	48	1	83	62	66
Bijapur (2,791) – range	6.1–9.4	0.05–78.00	0.02–1.50	0.1–91.9	24–2,613	0.9–4,647.4	0.12–10.40	0.02–18.22
– Mean	8.3	0.40	0.42	3.8	209	24.4	0.50	0.93
– Percentage of samples deficient	–	–	70	81	3	77	89	43
Chamarajanagar (1,640) – range	5.1–9.7	0.02–8.00	0.04–1.85	0.2–121.6	20–766	0.4–119.4	0.14–6.40	0.02–3.80
– Mean	7.7	0.29	0.41	10.0	188	6.3	0.73	0.58

– Percentage of samples deficient	–	–	76	37	4	87	67	62
Chikkaballapur (2,257) – range	4.5–9.9	0.01–16.62	0.07–1.42	0.2–430.8	4–1,650	0.5–470.0	0.06–21.50	0.06–1.98
– Mean	6.9	0.19	0.39	18.0	95	9.1	1.15	0.38
– Percentage of samples deficient	–	–	78	37	34	80	52	80
Chikmagalur (4,140) – range	2.9–9.8	0.01–1.89	0.01–2.45	0.5–129.2	1–304	1.0–2,425.0	0.01–6.75	0.02–55.44
– Mean	6.5	0.13	0.62	17.6	82	31.7	0.59	1.46
– Percentage of samples deficient	–	–	48	15	44	34	77	43
Chitradurga (1,489) – range	4.7–10.1	0.01–4.11	0.03–1.36	0.2–480.0	12–1,953	0.8–291.8	0.08–40.50	0.04–6.94
– Mean	7.8	0.23	0.40	7.0	137	7.3	0.64	0.63
– Percentage of samples deficient	–	–	76	54	15	86	80	64
Dakshina Kannada (1,418) – range	4.8–8.3	0.01–1.38	0.04–3.63	0.1–364.2	1–336	0.2–613.6	0.01–8.94	0.01–22.08
– Mean	5.5	0.09	1.26	12.6	46	38.5	0.84	1.66
– Percentage of samples deficient	–	–	2	29	71	21	65	44
Davangere (2,968) – range	4.2–9.9	0.01–6.74	0.04–2.70	0.2–95.4	11–480	0.9–99.7	0.04–4.80	0.02–3.00
– Mean	7.0	0.22	0.49	14.0	108	10.4	0.69	0.54
– Percentage of samples deficient	–	–	59	30	12	76	74	64
Dharwad (1,129) – range	5.1–9.3	0.03–1.91	0.17–1.99	0.2–207.0	36–2,344	1.4–715.0	0.24–24.30	0.10–12.48
– Mean	7.4	0.24	0.65	9.3	220	9.7	0.98	0.82
– Percentage of samples deficient	–	–	31	53	1	79	44	39
Gadag (1,270) – range	5.1–9.6	0.04–5.53	0.04–1.41	0.0–82.8	27–1,145	0.4–223.3	0.06–7.98	0.10–9.62
– Mean	8.2	0.27	0.41	5.3	185	7.1	0.42	0.88
– Percentage of samples deficient	–	–	75	65	2	85	92	34
Gulbarga (now Kalaburagi) (3,640) – range	4.9–9.8	0.05–34.50	0.04–2.50	0.2–88.7	19–1,722	0.4–12,647.9	0.10–5.18	0.02–24.90
– Mean	8.0	0.34	0.49	5.7	266	28.1	0.53	0.63
– Percentage of samples deficient	–	–	60	64	1	83	86	71

Continued

Table 3.2. Continued.

District[b]	pH	EC (dS/m)	Organic C (%)	Available				
				P (ppm)	K (ppm)	S (ppm)	Zn (ppm)	B (ppm)
Hassan (10,274) – range	3.9–9.7	0.03–3.60	0.04–5.71	0.2–363.0	9–1,394	0.2–515.1	0.06–41.90	0.02–4.08
– Mean	6.3	0.24	0.58	19.4	116	8.4	1.12	0.32
– Percentage of samples deficient	–	–	48	23	18	82	50	91
Haveri District (1,532) – range	5.1–10.5	0.03–2.34	0.08–3.60	0.1–143.0	25–3,750	0.3–120.3	0.20–34.10	0.08–8.44
– Mean	7.7	0.18	0.51	12.4	133	7.0	0.81	0.71
– Percentage of samples deficient	–	–	55	42	5	85	60	46
Kodugu (1,160) – range	4.0–7.8	0.01–2.06	0.28–1.26	1.2–15.5	0–223	1.1–206.5	0.03–37.30	0.03–11.75
– Mean	5.6	0.07	1.15	7.0	53	12.7	4.13	1.21
– Percentage of samples deficient	–	–	0	59	68	74	24	28
Kolar (2,161) – range	4.6–10.2	0.02–13.00	0.04–1.50	0.0–182.0	9–1,144	0.7–141.2	0.14–14.40	0.04–1.82
– Mean	7.0	0.16	0.38	20.3	87	7.0	1.31	0.34
– Percentage of samples deficient	–	–	81	31	34	85	32	87
Koppal (2,499) – range	5.2–9.8	0.01–5.70	0.03–2.90	0.0–214.6	24–708	0.3–1,482.5	0.01–20.09	0.01–2.98
– Mean	7.7	0.26	0.45	19.6	147	82.5	0.84	0.30
– Percentage of samples deficient	–	–	65	7	2	22	59	87
Mandya (5,479) – range	4.5–8.9	0.01–3.10	0.01–1.26	1.5–27.2	7–164	1.0–278.3	0.01–4.86	0.01–3.98
– Mean	6.8	0.39	0.59	15.1	103	43.3	0.62	0.60
– Percentage of samples deficient	–	–	43	14	6	27	71	65
Mysore (4,860) – range	3.2–9.3	0.01–3.20	0.03–1.26	0.4–15.7	3–168	0.9–1,459.8	0.01–19.80	0.03–14.73
– Mean	6.8	0.18	0.43	10.1	129	59.7	2.13	0.68
– Percentage of samples deficient	–	–	69	25	3	13	26	60
Raichur (3,343) – range	4.8–9.8	0.02–56.90	0.03–1.60	0.0–169.6	13–1,797	0.8–49,083.7	0.12–15.24	0.01–34.34
– Mean	8.2	0.60	0.42	11.1	202	177.2	0.66	1.17
– Percentage of samples deficient	–	–	71	48	4	64	79	39

Ramanagara (3,068) – range	3.2–8.4	0.03–1.71	0.03–3.00	0.5–378.2	3–631	0.3–2,675.0	0.01–9.52	0.01–20.68
– Mean	6.4	0.16	0.41	25.4	104	175.0	1.05	0.32
– Percentage of samples deficient	–	–	70	5	15	13	48	88
Shimoga (6,140) – range	3.8–8.2	0.01–2.32	0.07–3.15	0.7–90.5	2–175	0.5–99.5	0.07–20.00	0.01–31.76
– Mean	5.6	0.13	0.71	8.8	80	15.8	1.03	0.80
– Percentage of samples deficient	–	–	23	41	46	34	36	36
Tumkur (3,041) – range	2.8–10.0	0.01–14.00	0.04–2.08	0.1–204.0	11–1,470	0.1–128.4	0.14–17.26	0.03–3.60
– Mean	6.6	0.13	0.39	5.9	92	5.5	0.89	0.33
– Percentage of samples deficient	–	–	77	65	34	92	50	91
Udupi (1,000) – range	5.4–7.0	0.10–0.59	0.36–0.99	1.5–14.2	20–169	3.1–25.5	0.12–4.18	0.11–3.55
– Mean	6.0	0.26	0.81	3.6	71	10.3	0.94	0.52
– Percentage of samples deficient	–	–	4	85	34	54	51	69
Uttara Kannada (4,980) – range	3.5–8.4	0.01–5.00	0.08–9.58	0.1–47.1	0–199	0.1–470.0	0.02–26.40	0.02–290.00
– Mean	5.5	0.12	0.56	6.4	64	81.6	0.95	4.05
– Percentage of samples deficient	–	–	46	41	45	28	53	48
Yadgir (1,982) – range	5.0–10.0	0.03–8.78	0.01–1.19	0.0–97.3	14–1,558	0.9–237.4	0.12–14.80	0.02–4.60
– Mean	7.9	0.35	0.40	9.6	204	26.8	0.49	0.66
– Percentage of samples deficient	–	–	74	48	5	72	90	58
Karnataka State (92,904) – range	3.5–10.0	0.03–8.78	0.01–9.58	0.0–543.8	0–3,750	0.9–237.4	0.00–235.00	0.02–4.60
– Mean	6.8	0.25	0.54	12.5	115	44.4	1.01	0.87
– Percentage of samples deficient	–	–	52	41	23	52	55	62

[a]EC, Electrical conductivity; C, carbon; P, phosphorus; K, potassium; S, sulfur; Zn, zinc; B, boron.
[b]The values in parentheses are the number of farmers' fields sampled.

Table 3.3. Correlations among different soil fertility parameters in soil samples collected from farmers' fields in Karnataka (n = 92,904).[a]

Parameter	EC	Organic C	Available P	Available K	Available S	Available Zn	Available B
pH	0.19**	–0.20**	–0.06**	0.40**	–0.03**	–0.10**	0.06**
EC		0.03**	0.05**	0.17**	0.15**	0.02**	0.20**
Organic C			0.05**	0.01	0.00	0.17**	0.15**
Available P				0.15 **	0.05**	0.19**	–0.01**
Available K					–0.05**	0.03**	0.10**
Available S						0.04**	0.04**
Available Zn							0.05**

[a]**, Significant at 1% level.

3.8 General Discussion and Conclusions

It is well recognized that water-shortage-related plant stress is the primary constraint to crop production and productivity in the rainfed systems in the SATs, and consequently the importance of water shortage has globally been rightly emphasized (Wani *et al.*, 2002, 2003; Pathak *et al.*, 2009). However, apart from water shortage, severe soil infertility is another problem in the rainfed systems (Sanchez *et al.*, 1997; Zougmore *et al.*, 2003; Rego *et al.*, 2007; Lal, 2008; Bekunda *et al.*, 2010; Sahrawat *et al.*, 2010) and managing water stress alone cannot sustainably enhance the productivity of rainfed systems; hence for achieving sustainable gains in rainfed productivity both water shortage and soil fertility problems need to be simultaneously addressed through effective natural resource management practices (Wani *et al.*, 2009; Sahrawat *et al.*, 2010; Chander *et al.*, 2014).

Most probably for the first time, a large number of farmers' fields in the SAT regions of India were sampled and analysed for organic C and extractable or available nutrients in an effort to diagnose the prevalence of major and micronutrient deficiencies. The results of the analyses of soil samples from the farmers' fields demonstrated that the soils in rainfed areas are indeed infertile and they are not only deficient in major nutrients, especially N (soil organic C status used as an index for available N) and P, but are low in organic matter reserve. The most revealing results, however, were the widespread nature of the deficiencies of S, B and Zn (Rego *et al.*, 2007; Sahrawat *et al.*, 2007, 2010; Sahrawat and Wani, 2013).

A summary of results of on-farm responses of several field crops to applications of deficient nutrients together with N and P demonstrated that balanced nutrient management has indeed the potential to significantly enhance the productivity of a range of crops, and improve grain and straw quality in the SAT regions under rainfed conditions (Sahrawat and Wani, 2013).

Our results from on-farm trials during the past decade suggest that a soil-test-based nutrient management approach can be an important EPA and also a mechanism to diagnose and manage soil fertility in practical agriculture (Wani, 2008). Soil and plant tests have long been used as tools to diagnose and manage soil fertility problems in the intensified irrigated systems and commercial crops, including fruit and vegetable crops, to maximize productivity (Dahnke and Olson, 1990; Black, 1993; Mills and Jones, 1996; Reuter and Robinson, 1997). However, soil testing has not been used to diagnose and manage nutrient problems in farmers' fields in the SAT regions at a large scale (Sahrawat and Wani, 2013).

The critical limits for P, K, S, B and Zn in the soil (see Table 3.1) seem to provide a fair basis for separating deficient soils from those that are not deficient. Soils below the critical limits of the nutrients evaluated responded to the applications of nutrients, although the overall crop response was regulated by the rainfall received during the cropping season (Rego *et al.*, 2007; Sahrawat *et al.*, 2007, 2010). Soil-test-based nutrient application also allows judicious and efficient use of nutrient inputs at the local and regional levels (Black, 1993; Sahrawat *et al.*, 2010; Sahrawat and Wani, 2013).

For widespread adoption and use of soil testing for the diagnosis and management of plant nutrient deficiencies in the rainfed systems of the SAT regions, there is a need to strengthen the soil testing facilities at the local and regional levels for science-based management and maintenance of soil fertility, a prerequisite for sustainable increase in productivity of the rainfed systems (Sahrawat *et al.*, 2007, 2010; Sahrawat, 2013; Sahrawat and Wani, 2013).

For enhancing the overall agricultural productivity and crop quality of the rainfed systems, the choice of crops and adapted cultivars along with soil, water and nutrient management practices need to be integrated at the farm level (Wani *et al.*, 2009; Sahrawat *et al.*, 2010). To achieve this, research and extension support and backstopping along with capacity building of all the stakeholders need to converge (Wani, 2008; Sahrawat *et al.*, 2010). It is in this context that ICRISAT and its research partners most appropriately advocate the integration of genetics and natural resource management for technology targeting and greater impact of agricultural research in the SATs (Twomlow *et al.*, 2008; Chander *et al.*, 2013).

Acknowledgements

We thank the APRLP of Andhra Pradesh, the Department of International Development, UK, the World Bank Special Fund for Watershed Development, the Government of Karnataka and the Sir Dorabji Tata Trust for financial support. Thanks are also due to all farmers who participated in the studies reported in this chapter.

References

Bationo, A., Kihara, J., Vanlauwe, B., Kimetu, J., Waswa, B.S. and Sahrawat, K.L. (2008) Integrated nutrient management: concepts and experience from sub-Saharan Africa. In: Aulakh, M.S. and Grant, C.A. (eds) *Integrated Nutrient Management for Sustainable Crop Production*. The Haworth Press, New York, pp. 467–521.

Bekunda, M., Sanginga, N. and Woomer, P.L. (2010) Restoring soil fertility in sub-Saharan Africa. *Advances in Agronomy* 108, 183–286.

Black, C.A. (1993) *Soil Fertility Evaluation and Control*. Lewis Publishers, Boca Raton, Florida.

Chander, G., Wani, S.P., Sahrawat, K.L., Pal, C.K. and Mathur, T.P. (2013) Integrated plant genetic and balanced nutrient management enhances crop and water productivity of rainfed production systems in Rajasthan, India. *Communications in Soil Science and Plant Analysis* 44, 3456–3464.

Chander, G., Wani, S.P., Sahrawat, K.L., Dixit, S., Venkateswarlu, B., Rajesh, C., Rao, P.N. and Pardhasaradhi, G. (2014) Soil test based nutrient balancing improved crop productivity and rural livelihoods: case study from rainfed semi-arid tropics in Andhra Pradesh, India. *Archives of Agronomy and Soil Science* 60, 1051–1066.

Dahnke, W.C. and Olson, R.A. (1990) Soil test calibration, and the recommendation. In: Westerman, R.L. (ed.) *Soil Testing and Plant Analysis*, 3rd edn. Soil Science Society of America Series 3. Soil Science Society of America, Madison, Wisconsin, pp. 45–71.

Dixit, S., Wani, S.P., Rego, T.J. and Pardhasaradhi, G. (2007) Knowledge-based entry point and innovative up-scaling strategy for watershed development projects. *Indian Journal of Dryland Agriculture and Development* 22(1), 22–31.

El-Swaify, S.A., Pathak, P., Rego, T.J. and Singh, S. (1985) Soil management for optimized productivity under rainfed conditions in the semi-arid tropics. *Advances in Soil Science* 1, 1–64.

Fageria, N.K., Baligar, V.C. and Clark, R.B. (2002) Micronutrients in crop production. *Advances in Agronomy* 77, 185–268.

Helmke, P.A. and Sparks, D.L. (1996) Lithium, sodium, potassium, rubidium, and cesium. In: Sparks, D.L. (ed.) *Methods of Soil Analysis, Part 3. Chemical Methods.* Soil Science Society of America Book Series no. 5. Soil Science Society of America and American Society of Agronomy, Madison, Wisconsin, pp. 551–574.

Kanwar, J.S. (1972) Twenty-five years of research in soil, fertilizer and water management in India. *Indian Farming* 22(5), 16–25.

Katyal, J.C. (2003) Soil fertility management – a key to prevent desertification. *Journal of the Indian Society of Soil Science* 51, 378–387.

Keren, R. (1996) Boron. In: Sparks, D.L. (ed.) *Methods of Soil Analysis, Part 3. Chemical Methods*. Soil Science Society of America Book Series no. 5. Soil Science Society of America and American Society of Agronomy, Madison, Wisconsin, pp. 603–626.

Lal, R. (2008) Soils and sustainable agriculture. A review. *Agronomy for Sustainable Development* 28, 57–64.

Lindsay, W.L. and Norvell, W.A. (1978) Development of a DTPA test for zinc, iron, manganese and copper. *Soil Science Society of America Journal* 42, 421–428.

Mills, H.A. and Jones, J.B., Jr (1996) *Plant Analysis Handbook II: A Practical Sampling, Preparation, and Interpretation Guide*. MicroMacro Publishing, Inc., Athens, Georgia.

Nelson, D.W. and Sommers, L.E. (1996) Total carbon, organic carbon, and organic matter. In: Sparks, D.L. (ed.) *Methods of Soil Analysis, Part 3. Chemical Methods*. Soil Science Society of America Book Series no. 5. Soil Science Society of America and American Society of Agronomy, Madison, Wisconsin, pp. 961–1010.

Olsen, S.R. and Sommers, L.E. (1982) Phosphorus. In: Page, A.L. (ed.) *Methods of Soil Analysis, Part 2*, 2nd edn. Agronomy Monograph 9. American Society of Agronomy and Soil Science Society of America, Madison, Wisconsin, pp. 403–430.

Olson, M. (1971) *The Logic of Collective Action: Public Goods and Theory of Groups*, revised edition. Schocken Books, New York.

Pasricha, N.S. and Fox, R.L. (1993) Plant nutrient sulfur in the tropics and subtropics. *Advances in Agronomy* 50, 209–269.

Passioura, J.B. (2006) Increasing crop productivity when water is scarce – from breeding to field management. *Agricultural Water Management* 80, 176–196.

Passioura, J.B. and Angus, J.F. (2010) Improving productivity of crops in water-limited environments. *Advances in Agronomy* 106, 37–74.

Pathak, P., Sahrawat, K.L., Wani, S.P., Sachan, R.C. and Sudi, R. (2009) Opportunities for water harvesting and supplemental irrigation for improving rainfed agriculture in semi-arid areas. In: Wani, S.P., Rockström, J. and Oweis, T. (eds) *Rainfed Agriculture: Unlocking the Potential.* CAB International, Wallingford, UK, pp. 197–221.

Rego, T.J., Rao, V.N., Seeling, B., Pardhasaradhi, G. and Kumar Rao, J.V.D.K. (2003) Nutrient balances – a guide to improving sorghum and groundnut-based dryland cropping systems in semi-arid tropical India. *Field Crops Research* 81, 53–68.

Rego, T.J., Wani, S.P., Sahrawat, K.L. and Pardhasaradhi, G. (2005) *Macro-benefits from Boron, Zinc and Sulfur Application in Indian SAT: A Step for Gray to Green Revolution in Agriculture*. Global Theme on Agroecosystems Report no. 16. International Crops Research Institute for the Semi-Arid Tropics, Patancheru, Andhra Pradesh, India.

Rego, T.J., Sahrawat, K.L., Wani, S.P. and Pardhasaradhi, G. (2007) Widespread deficiencies of sulfur, boron, and zinc in Indian semi-arid tropical soils: on-farm crop responses. *Journal of Plant Nutrition* 30, 1569–1583.

Reuter, D.J. and Robinson, J.B. (eds) (1997) *Plant Analysis: An Interpretation Manual,* 2nd edn. Commonwealth Scientific and Industrial Research Organisation (CSIRO), Melbourne, Australia.

Rockström, J., Karlberg, L., Wani, S.P., Barron, J., Hatibu, N., Oweis, T., Bruggeman, A., Farahani, J. and Qiang, Z. (2010) Managing water in rainfed agriculture – the need for a paradigm shift. *Agricultural Water Management* 97, 543–550.

Sahrawat, K.L. (2006) Plant nutrients: sufficiency and requirements. In: Lal, R. (ed.) *Encyclopedia of Soil Science,* 2nd edn. Taylor and Francis, Philadelphia, Pennsylvania, pp. 1306–1310.

Sahrawat, K.L. (2013) Need to strengthen analytical support in agricultural research. *Current Science* 105, 13–14.

Sahrawat, K.L. and Wani, S.P. (2013) Soil testing as a tool for on-farm soil fertility management: experience from the semi-arid zone of India. *Communications in Soil Science and Plant Analysis* 44, 1011–1032.

Sahrawat, K.L., Wani, S.P., Rego, T.J., Pardhasaradhi, G. and Murthy, K.V.S. (2007) Widespread deficiencies of sulphur, boron and zinc in dryland soils of the Indian semi-arid tropics. *Current Science* 93, 1428–1432.

Sahrawat, K.L., Rego, T.J., Wani, S.P. and Pardhasaradhi, G. (2008) Stretching soil sampling to watershed: evaluation of soil-test parameters in a semi-arid tropical watershed. *Communications in Soil Science and Plant Analysis* 39, 2950–2960.

Sahrawat, K.L., Murthy, K.V.S. and Wani, S.P. (2009) Comparative evaluation of Ca chloride and Ca phosphate for extractable sulfur in soils with a wide range in pH. *Journal of Plant Nutrition and Soil Science* 172, 404–407.

Sahrawat, K.L., Wani, S.P., Pardhasaradhi, G. and Murthy, K.V.S. (2010) Diagnosis of secondary and micronutrient deficiencies and their management in rainfed agroecosystems: case study from Indian semi-arid tropics. *Communications in Soil Science and Plant Analysis* 41, 346–360.

Sanchez, P.A., Shepherd, K.D., Soul, M.J., Place, F.M., Buresh, R.J., Izac, A.M.N., Mokwunye, A.U., Kwesiga, F.R., Nderitu, F.G. and Woomer, P.L. (1997) Soil fertility replenishment in Africa: an investment in natural resource capital. In: Buresh, R.J., Sanchez, P.A. and Calhoun, F. (eds) *Replenishing Soil Fertility in Africa*. Soil Science Society of America Special Publication no. 51. Soil Science Society of America, Madison, Wisconsin, pp. 1–46.

Scherer, H.W. (2001) Sulphur in crop production. *European Journal of Agronomy* 14, 81–111.

Scherer, H.W. (2009) Sulfur in soils. *Journal of Plant Nutrition and Soil Science* 172, 326–335.

Singh, A.K. (2008) Soil resource management – key to food and health security. *Journal of the Indian Society of Soil Science* 56, 348–357.

Singh, P., Pathak, P., Wani, S.P. and Sahrawat, K.L. (2009) Integrated watershed management for increasing productivity and water-use efficiency in semi-arid tropical India. *Journal of Crop Improvement* 23, 402–429.

Sreedevi, T.K., Shiferaw, B. and Wani, S.P. (2004) *Adarsha Watershed in Kothapally: Understanding the Drivers of Higher Impact*. Global Theme on Agroecosystems Report no. 10. International Crops Research Institute for the Semi-Arid Tropics, Patancheru, Andhra Pradesh, India.

Tabatabai, M.A. (1996) Sulfur. In: Sparks, D.L. (ed.) *Methods of Soil Analysis, Part 3. Chemical Methods*. Soil Science Society of America Book Series no. 5. Soil Science Society of America and American Society of Agronomy, Madison, Wisconsin, pp. 921–960.

Takkar, P.N. (1996) Micronutrient research and sustainable agricultural productivity. *Journal of the Indian Society of Soil Science* 44, 563–581.

Twomlow, S., Shiferaw, B., Cooper, P. and Keatinge, J.D.H. (2008) Integrating genetics and natural resource management for technology targeting and greater impact of agricultural research in the semi-arid tropics. *Experimental Agriculture* 44, 235–256.

Wani, S.P. (2008) Taking soil science to farmers' doorsteps through community watershed management. *Journal of the Indian Society of Soil Science* 56, 367–377.

Wani, S.P., Pathak, P., Tam, H.M., Ramakrishna, A., Singh, P. and Sreedevi, T.K. (2002) Integrated watershed management for minimizing land degradation and sustaining productivity in Asia. In: Adeel, Z. (ed.) *Integrated Land Management in Dry Areas. Proceedings of a Joint UNU-CAS International Workshop, 8–13 September 2001, Beijing, China*. United Nations University (UNU), Tokyo, pp. 207–230.

Wani, S.P., Pathak, P., Jangawad, L.S., Eswaran, H. and Singh, P. (2003) Improved management of Vertisols in the semiarid tropics for increased productivity and soil carbon sequestration. *Soil Use and Management* 19, 217–222.

Wani, S.P., Sreedevi, T.K., Rockström, J. and Ramakrishna, Y.S. (2009) Rainfed agriculture – past trends and future prospects. In: Wani, S.P., Rockström, J. and Oweis, T. (eds) *Rainfed Agriculture: Unlocking the Potential*. CAB International, Wallingford, UK, pp. 1–35.

Wani, S.P., Sahrawat, K.L., Sarvesh, K.V., Baburao, M. and Krishnappa, K. (2011) *Soil Fertility Atlas for Karnataka, India*. International Crops Research Institute for the Semi-Arid Tropics, Patancheru, Andhra Pradesh, India.

Zougmore, R., Zida, Z. and Kambou, N.F. (2003) Role of nutrient amendments in the success of half-moon soil and water conservation practice in semiarid Burkina Faso. *Soil and Tillage Research* 71, 143–149.

4

Human Capacity Development to Adopt Best Practices

K.H. Anantha,* Suhas P. Wani, Girish Chander and Gajanan Sawargaonkar

International Crops Research Institute for the Semi-Arid Tropics (ICRISAT), Patancheru, India

4.1 Introduction

The concept of capacity development, which emerged during the 1980s, gained prominence in the 1990s and currently has wide usage in community development (Eade, 1997; UNDP, 1998; Bolger, 2000). The term capacity development is usually discussed as an approach to development and cooperation. Capacity development encompasses human resource development as an essential part of development (FAO, 1998). It is a process by which individuals, groups, organizations and societies enhance their abilities to identify and meet development challenges in a sustainable manner (UNDP, 1998). It is human resource development, which is a process of equipping individuals with the understanding of access to information, knowledge, training and skills that enables them to perform effectively. There is a direct relationship between capacity building and agricultural education.

Human capacity development is an essential component in the agriculture sector as it involves human resources to spread scientific knowledge to the farming community. A systematic way of capacity development in the agriculture sector requires an understanding of how different socio-economic components or processes are interlinked and how these influence each other. Capacity development at the individual or organizational level may not be of much use if there

*k.anantha@cgiar.org

are systematic impediments to performance, such as poor incentives or lack of access to resources. It is more than acquisition of skills by a particular individual, community organization or line department and includes a systematic dimension (UNDP, 1998).

This chapter highlights the processes and approaches of capacity development activities developed in a mission mode project implemented by the Government of Karnataka, India, from 2009 to boost productivity of rainfed agriculture in the state of Karnataka. Unique mechanisms have been developed to enhance the capacity of farmers, community and organizations.

4.2 Framework for Human Capacity Development

Human capacity development plays a key role in economic growth and development because human beings occupy the centre of the production, distribution and consumption chain. From a macroeconomics perspective, the improvement of human capital productivity facilitates technological innovations, increases returns to capital and makes economic growth more sustainable (Chimboza, 2012). Because of agriculture being a multidisciplinary subject, regular human capacity development assumes much more importance. In the current context of demographic pressure, climate change, economic and other considerations, capacity building in agriculture is the need of the hour. Indian agriculture needs to be rejuvenated with new scientific research/knowledge and a new extension system. With increasing risks and uncertainties, agricultural ventures must be attractive, profitable and sustainable to induce economic growth. These major attributes can be achieved through scientific research to develop quality inputs, improve agronomic practices and develop good management skills. Likewise, agricultural extension services ought to be upgraded to provide the education needed to modernize production practices and change past ways and perceptions of agriculture as a means of subsistence to a feasible business opportunity.

It is imperative to provide scientific research-based knowledge in order to enhance the capacity of not only programme implementers but also extension agents as well as farmers. The declining rate of extension personnel in the agriculture sector has contributed to low level adoption of science-led innovations thereby adversely affecting agricultural growth in the state (Government of Karnataka, 2006). Low technology in the agriculture sector has hindered the production of high-value products that generates employment and income (Government of Karnataka, 2011).

During 2009, the Government of Karnataka initiated implementation of an innovative agricultural development programme called 'Bhoochetana' with the aim to enhance agricultural productivity for improving rural livelihoods in the rainfed system through a science-led integrated approach. This initiative provided an opportunity to train youths particularly in the agriculture sector to perform extension activities. The Government of Karnataka is collaborating with many international agricultural organizations/agencies and institutions to organize and conduct training programmes and awareness campaigns to develop young innovative/progressive farmers as extension experts for disseminating scientific knowledge.

This framework largely focuses on developing capacity at the bottom two levels – the organization and the individual/farmer – although consideration has been given to targeted initiatives (in a range of areas) at the other two levels (i.e. environmental and sectoral levels) (Fig. 4.1). For practical purposes, the framework focuses on building capacity of agricultural officers of the state, district, taluk/block and *Raithu Samparka Kendra* (RSK) as they are directly responsible for agricultural management in the state.

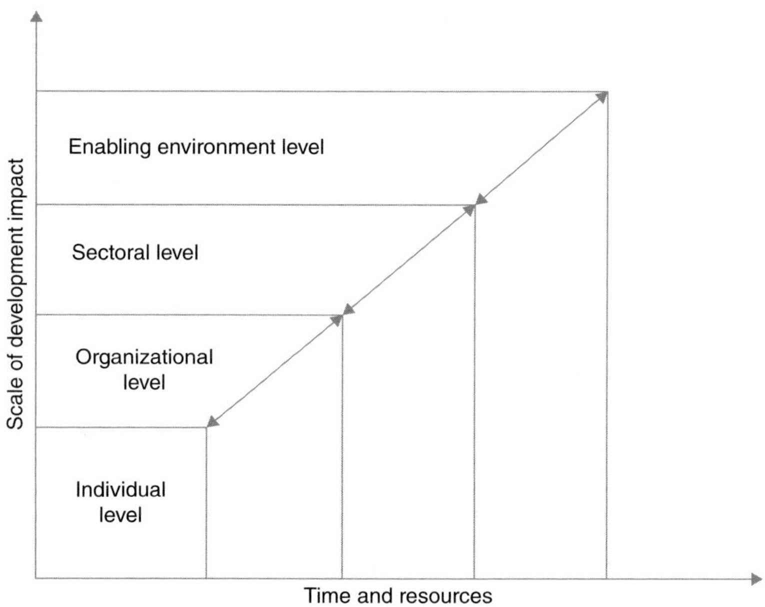

Fig. 4.1. Levels of capacity development.

4.3 Linking Knowledge to Action through Capacity Development

There is now emerging evidence that regenerative and resource-conserving technologies and practices can bring both environmental and economic benefits for farmers and communities (Wani *et al.*, 2003; Rockström *et al.*,

2007). The best evidence comes from countries of Asia and Africa where the concern is to increase food production in the areas where farming has been marginalized due to fragmentation and limited use of technologies (Rockström *et al.*, 2007). In these complex and remote lands, some farming communities adopting regenerative technologies have substantially improved agricultural yields (Wani *et al.*, 2003; Rockström *et al.*, 2007, 2010). The common element of these successes was that farmers have made use of resource-conserving technologies such as: (i) integrated pest management; (ii) soil and water conservation; (iii) integrated nutrient management; (iv) crop diversification; and (v) water harvesting (Wani *et al.*, 2003). Farmer groups/communities at the local level have become experts at managing farms as ecosystems and benefit from them. Moreover, there have been supportive and enabling external government and/or non-governmental organizations (NGOs), often working in new partnerships with new participatory approaches, which have reoriented their activities to focus on local needs (Wani *et al.*, 2012).

Bhoochetana is a science-led participatory approach, which emerged from the lessons learnt from long-term watershed-based research led by the International Crops Research Institute for the Semi-Arid Tropics (ICRISAT), Patancheru, India, and national partners (Wani *et al.*, 2003). This interdisciplinary research, over the years, has shaped up into an integrated genetic natural resource management approach within ICRISAT (Twomlow *et al.*, 2006). After realizing the importance and potential of combining disciplinary expertise in a complementary way, the idea of the consortium approach based on the multidisciplinary approach at the research station was adopted. The consortium is a convergence of agencies/actors/stakeholders that have a significant role to play in a watershed development project (Wani *et al.*, 2003). Facilitated by a leader/leading organization, member organizations prepare common plans and work towards achieving the agreed common objectives. This approach was first adopted during 1999. The Asian Development Bank came forward to support ICRISAT's idea of testing the consortium model in a watershed in Kothapally village of the Rangareddy district in Andhra Pradesh (now Telangana state) called the Adarsha watershed, to minimize the gap between research findings and on-farm development. The purpose was also to adopt the learning loop in planning of strategic research based on participatory research and development. The innovative integrated watershed management model has demonstrated that with proper management of natural resources the system's productivity can be enhanced and poverty can be reduced without causing further degradation of the natural resource base. The scaling up of these innovations has been attempted in countries such as India, Vietnam, Thailand and China.

The Bhoochetana mission project is a cost-share project with principles of convergence, consortium, capacity building and collective action, inspired by the Sujala–ICRISAT initiative (Wani *et al.*, 2012). Farmers participate in the project on an individual basis and carry out land rejuvenation on their own land, while the Department of Agriculture (DoA) provides extension services, seeds and fertilizers (micronutrients) with 50% incentives. The idea behind Bhoochetana was to support the interest in soil conservation already present among farmers. The project idea was developed during the brainstorming session organized by the government in 2008. In 2009, ICRISAT formed a consortium with state agricultural universities (SAUs), line departments such as the DoA, Watershed Development Department (WDD), the Directorate of Economics and Statistics (DES), community-based organizations and farmers for the project. The consortium collectively took part in planning and promoting the project activities. The project activities were designed for six districts during the first year and during the second year upscaled to 16 districts (including the six districts). After realizing the success, during the third year the project activities were scaled up to the entire state. The project 'Bhoochetana' then became a mission project for the government and the consortium provided technical support.

The innovativeness of this project includes identification of soil nutrient deficiency status and taluk-wise nutrient recommendations based on nutrition status supported by making available the necessary inputs at 50% subsidy. The major focus of the project is on both dry-land and irrigated crops. For timely and easy availability of inputs, advance positioning of inputs was done at the cluster-village-level and RSKs at every *hobli* (group of villages). These inputs (seeds, seed treatment chemicals, gypsum, micronutrients and biofertilizers) have been supplied as a package. To operationalize the project, the services of farm facilitators (FFs) to take technologies to the farmer's doorstep were utilized. This was followed by 1 week of institutional training to all the FFs to update their knowledge about the technologies. Village meetings were held and there was wide publicity through mass media and wall writings in each of the selected villages. Above all, the convergence of all the existing schemes of the DoA channelled through Bhoochetana was the innovative strategy. The DoA is formally responsible for the project, while the consortium supervises daily activities at different levels. Timelines are defined clearly for covering productivity enhancements in 30 districts; soil sampling and nutrient analysis mapping and capacity building of stakeholders during the project period are shown in Table 4.1.

4.4 Consortium and Collective Action for Human Capacity Development

Capacity development, particularly in agriculture, is a challenging task which requires multidisciplinary experts to impart knowledge. This is,

Table 4.1. Timeline for execution of activities in Bhoochetana districts.

		Activity coverage (%)		
Activity	Year	Districts 1–6	Districts 7–16	Districts 17–30
Productivity enhancement	2009	25		
	2010	50	33	
	2011	75	66	50
	2012	100	100	100
Nutrient status mapping	2009	100		
	2010		100	
	2011			100
	2012			
Capacity building	2009	100		
	2010		100	
	2011			100

therefore, possible through a consortium approach. A consortium is a combination of organizations/institutions who organize the development, dissemination and delivery of new technologies, strategies and methods for achieving holistic development. In Bhoochetana, one of the objectives was to strengthen the partnership by bringing SAUs, *Krishi Vigyan Kendras* (KVKs), line departments along with the international agricultural research institute (ICRISAT) for bridging the knowledge gap existing among farmers, researchers and line departments so as to achieve productivity enhancement (Raju *et al.*, 2013). The Bhoochetana programme is a culmination of efforts by the policy makers, bureaucrats, scientists, extension agents, farmers and other stakeholders. This initiative has shown the way of partnering for strengthening the agricultural production system.

In Bhoochetana, a consortium of knowledge-generating institutions like the SAUs (Bengaluru, Dharwad and Raichur) and ICRISAT along with knowledge-disseminating line departments, such as the DoA, WDD and KVKs, was formed with the aim to impart knowledge, skills and technologies to the farming community (Fig. 4.2). The role of the consortium partners was to enhance the awareness of stakeholders to adopt improved technologies to achieve sustainable crop production. Each consortium partner has a specific responsibility for enhancing the capacity of different stakeholders in the project. The Bhoochetana consortium has assigned specific roles and responsibilities to respective partners so as to avoid duplication and delay in implementation as well as delivering the outputs.

4.4.1 Department of Agriculture (DoA)

As a nodal agency, the DoA is responsible for overall coordination of the programme for smooth implementation. The DoA issues timely guidelines

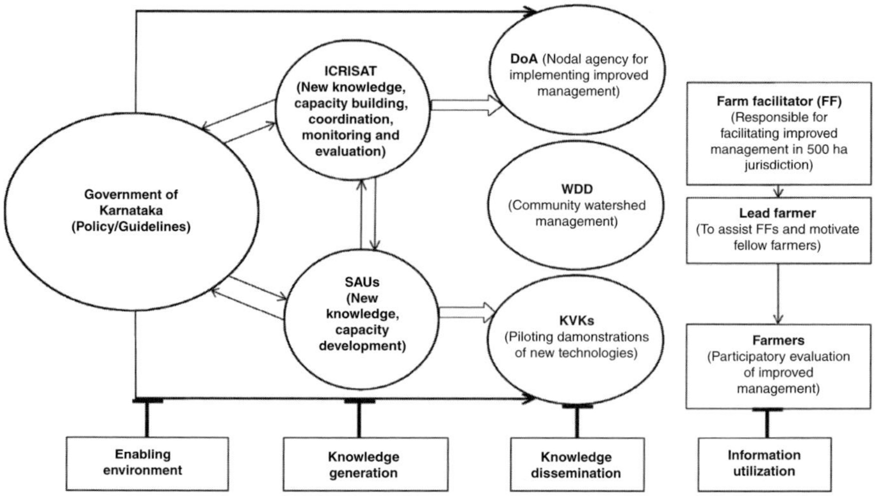

Fig. 4.2. Consortium of partners in the Bhoochetana programme in Karnataka. DoA, Department of Agriculture; FF, farm facilitators; KVKs, *Krishi Vigyan Kendras*; SAUs, state agricultural universities; WDD, Watershed Development Department.

and circulars with new information to enable field personnel to effectively implement the programme. In addition, it organizes the training of trainers with experts from ICRISAT, SAUs, lead NGOs and line department staff for internalizing nuances of various interventions to be undertaken in the respective districts.

4.4.2 State agricultural universities (SAUs)

In an effort to bridge the gap between research organizations and the farming community, the SAUs (Bengaluru, Dharwad and Raichur) were included in the consortium and shared the responsibilities of capacity development of various actors involved in the implementation of the Bhoochetana programme in the form of master trainers. The concept of master trainers was introduced with the main objective of periodically disseminating new knowledge uniformly across actors at various levels. The master trainers were trained through training of trainers by expert scientists of ICRISAT and partner organizations, covering various issues of agricultural operations. The purpose of training of master trainers on the same wavelength was to convey the same messages down the line in simple language. However, this process was not as effective as expected. Alternatively, KVKs volunteered to impart training and conduct field demonstrations. This initiative is considered to be a strength of the consortium approach which generally complements the strengths and weaknesses of partners.

4.4.3 *Krishi Vigyan Kendras* (KVKs)

The KVKs were visualized as a technical support agency to strengthen agriculture and they have contributed to awareness building, training programmes and sharing of new technologies with the farming community. The KVKs have enhanced the capacity of FFs through organizing training programmes on integrated nutrient management, integrated pest management, improved varieties and good agricultural practices, etc. In terms of sharing scientific knowledge, they have conducted field demonstrations. However, KVKs have constraints in terms of human resources to effectively address the problems as well as organizing training programmes and demonstrations. It was observed during the discussion with KVK staff that capacity building is essential even for KVK scientists and subject matter specialists on new technologies, strategies and methods (Wani *et al.*, 2013).

4.4.4 ICRISAT

Along with the DoA as the nodal agency, ICRISAT has a big role in facilitating and handholding for smooth implementation of the initiative. As a technical support provider, ICRISAT ensures new knowledge generation, sharing of knowledge and capacity development of various stakeholders, including policy makers, researchers, line departments, farmers and others. As a facilitator, ICRISAT is responsible for regular monitoring and evaluation, generation of reports and feedback on improvements in the initiative.

4.5 Building Human Capacity at Different Levels

Effective communication among partners is a key aspect, and dialogues and deliberations provide opportunities for partners and stakeholders to share ideas, values and knowledge to have real-time capacity building. Failure of previous projects is often attributed to lack of ability of the leading organization to engage in meaningful participatory capacity building processes. In a science-led participatory approach as adopted in Bhoochetana, effective participation of all stakeholders and willingness to share responsibility is the first and foremost requirement. In this context, building human capacity at various levels received the highest priority.

4.5.1 Team-building workshops at the state level

Team-building measures help in developing stronger partnerships and internalization of operating guidelines (Shambu Prasad *et al.*, 2006).

These are strong measures that bring partners together and build a trust among the consortium partners as there is a mechanism for sharing inputs, common communication and conflict resolution.

The messages from the top have a strong bearing down the line. Therefore, the human capacity building initiatives in the Bhoochetana programme were started from the top at state level. Team-building workshops were organized for the team members of the Bhoochetana project with the objective of establishing good coordination among team members from the newly scaled-up districts and discussing the technology implementation strategies in a team spirit. The participants in the team-building workshop usually included state-level policy makers, Joint Directors of Agriculture (JDAs), Assistant Directors of Agriculture (ADAs), nodal officers of different districts from the DoA, scientists from the SAUs (Bengaluru, Dharwad and Raichur) and scientists and scientific officers from ICRISAT (Fig. 4.3). Capacity building of partners, participating farmers and sensitization of policy makers helped in building partnerships and reducing transaction costs as the programme reaped greater benefits during the first year than expected.

Fig. 4.3. Participants at the brainstorming workshop on soil-test-based nutrient management including boron and other micronutrients.

4.5.2 Regular review and planning workshops

Regular review and planning workshops were organized at the end of the crop season to obtain first-hand information on programme implementation, benefits accrued and other experiences. Based on the learnings, planning would be carried out for the following season well in advance. The review and planning workshops were held at ICRISAT for the obvious reasons that the partners (policy makers, line department staff, university scientists, FFs, farmers and private sector partners) would be exposed to the new working environment and work culture as well as to the team culture. These workshops acted as a catalyst to bring about periodic changes in the planning process, implementation and monitoring of different interventions. The intensive review and planning workshops were aimed at instilling new thinking into the minds of all actors as almost all stakeholders had the opportunity to put forward their ideas in the open forum. This also provided an opportunity for all the stakeholders to learn innovative ideas and approaches so that they could implement the programme more meaningfully in their respective work areas. There was also an opportunity for private sector partners to get to know the requirements of the farming community, bureaucrats and policy makers as their presence would be seen as a blessing in disguise. Thus, the review and planning workshops were good learning opportunities for different actors to prepare evolved work plans.

The workshops were platforms for the senior officers to take bold decisions in consensus with their subordinates and partners. Innovative decisions were taken during these workshops on: (i) expanding the area coverage; (ii) input procurement, storage and distribution; and (iii) regular monitoring of progress. The expert guidance and advice from senior-level officials and scientists on finding solutions to various problems were very well valued by the partners.

4.5.3 District-level training

Capacity development programmes at the district level were targeted for heads of departments mainly from the DoA, the WDD and their team leaders managing different schemes at district level and representative officers at taluk/block level. During these training sessions, the participants were oriented to be on the same wavelength to understand implementation of the programme, input procurement, and storage, soil nutrient status in each taluk, etc. The module of the training sessions generally included lectures on specific topics by master trainers and interactive sessions for thorough understanding about implementation of the initiative. A summary of district-level training programmes conducted under Bhoochetana during the period 2009/10 to 2012/13 is given in Table 4.2. There was not only a gradual increase in the adoption of Bhoochetana activities in various districts of Karnataka but also increased participation of key decision makers at the district level (see Table 4.1). This is an indication of the high quality and relevance of the training programmes in their work.

Table 4.2. District-level training programmes organized under Bhoochetana during 2009–2013 in Karnataka.

Year	No. of districts	No. of training programmes	No. of participants
2009/10	6	19	1128
2010/11	15	38	3707
2011/12	30	77	4791
2012/13	30	46	5751

4.5.4 Taluk/block-level training

Trained taluk/block-level senior officials of the DoA, namely ADAs, organized training programmes for their line staff and implementing NGOs on: (i) soil nutrient recommendations; (ii) suitable high-yielding varieties; (iii) integrated pest management and disease management; (iv) livelihood options for the rural landless poor; and (v) best-bet management options for enhancing the productivity of the agricultural system. Under the Bhoochetana initiative, in each taluk/block, an average of ten taluk/block-level training programmes were organized annually, based on the need and convenience of participants (Table 4.3). Along with the required lectures and interactive sessions, the main focus was on hands-on training for implementing the interventions in the field. In these training sessions, leaflets/reference materials with recommendations for implementation in the field were provided for the participants.

Table 4.3. Taluk/block- and cluster-village-level training programmes organized under Bhoochetana during 2009–2013 in Karnataka, India.

		Taluk-level training		Cluster-village-level training	
Year	No. of districts	No. of programmes	No. of participants	No. of programmes	No. of participants
2009/10	6	68	930	1,806	33,996
2010/11	15	–	–	–	–
2011/12	30	318	23,446	3,873	63,912
2012/13	30	271	14,647	8,800	275,561

4.5.5 Institutional training for farm facilitators (FFs)

Appointment of educated or progressive FFs to guide fellow farmers in improved management was an innovative decision to fill in the gap in the extension system. Being from among the community, the FFs proved very effective in disseminating new technology as and when required. However, this could become feasible because of extensive initial institutional training of FFs. The critical focus of institutional training was to enhance the skill and knowledge of FFs through regular orientation programmes. Institutional training programmes were organized by the Deputy Director of the District Agricultural Training Centre (DATC) in every district. The training programmes usually ran for 5 days soon after the finalization of the district-level action plan for every season. The newly recruited FFs were given opportunities to learn the programme's implementation strategies, farmers' registration and other scientific innovations. In terms of science-led innovations, FFs would be trained on methods of soil, water and nutrient management, crop-wise nutrient recommendations, new seed varieties and seed production, integrated pest and disease management, farmer field schools, information on various schemes/programmes of the DoA and WDD, scientific crop-cutting experiments, etc. During the course of the mission project, institutional training has become an effective way of building capacity of FFs.

4.5.6 Cluster-village-level training

Cluster-village-level training was targeted at farmers to improve their knowledge and skills for impact on the ground. The DoA line staff, FFs and lead farmers served as guides to other fellow farmers in implementation of the locally required interventions. In village-level training, the demonstrations on lead farmers' fields were used for hands-on training on targeted interventions. The technologies, which were mainly focused in Bhoochetana, and in which most of the farmers were trained, included: (i) stratified soil sampling; (ii) participatory varietal selection; (iii) seed treatment; (iv) balanced nutrition; (v) vermicomposting; (vi) integrated nutrient management; (vii) landform management; and (viii) nursery raising. These training sessions were conducted with enthusiastic participation of farmers (see Table 4.3), which led to the success of the initiative and indicated that farmers are willing to practise new technologies. However, the issue is: how do we reach them and provide holistic solutions?

4.5.7 Farmer field schools

Farmer field schools are defined as 'schools without walls', which stemmed from adult education principles and evolved to become a distinct approach that builds on the process of group learning and community action. This participatory extension method recognizes the need to involve farmers in technology development and transfer. Participating farmers are encouraged to share their knowledge with other farmers, and are sometimes trained to teach the courses themselves, thus reducing the need for external support.

In Bhoochetana, the FFs took the lead in organizing planning meetings and conducting farmer field school sessions in a selected village with 20–30 farmers. Crop-specific information was imparted through short studies and long-term experiments. Capacity building and demonstrations in all aspects of crop cultivation were imparted to the participating farmers. Almost 20 weekly sessions were held to analyse the abiotic and biotic stresses on the plant. The observations recorded in the field were pooled and farmers arrived at the subsequent conclusions themselves.

Through farmer field schools, farmers were able to take the right crop management decisions at every stage of the crop and farmers developed the habit of visiting their fields regularly and monitoring crop pests and diseases. This extension system was found effective in imparting the required knowledge and skill of crop cultivation as there is frequent contact between farmers and FFs. Group dynamics, participatory group presentation and discussion were part of farmer field schools and as a result, a sense of cooperation and team coordination developed among farmers which were helpful in spreading and sharing the technologies with each other. The method capitalizes on existing local social networks, based on the belief that experienced and skilled farmers are the people best suited to train other farmers. The trainees familiarize themselves with the technology since they get practical training in the field where they have the opportunity to see and carry out the activities, learn from mistakes and receive advice. This approach proved a common and important vehicle for diffusing technology under Bhoochetana.

4.5.8 Field days

Field days are effective ways of information dissemination where interested farmers are invited to a particular field or plot and specific information about the technology is demonstrated and discussed. The session takes about 4–6 h and ranges from a structured presentation to an informal event where participants walk through the field plot at their own pace to view the demonstrations (Braun and Duveskog, 2008). During these field days farmers interact with FFs as well as with other farmers and exchange ideas and experiences. In Bhoochetana, field days were organized with the full support and participation of the DoA district-level staff during all seasons and for all important crops of the district. The average participants in each field day ranged between 50 and 75 farmers with high participation of men (Table 4.4). However, nearly 30% of participants were women farmers (i.e. the average number of women participants per field day ranged from 14 to 22). This indicates the increasing gender participation (equity) in agricultural development.

Table 4.4. Field days organized under Bhoochetana during 2010–2013 in Karnataka, India.

		No. of participants	
Year	No. of field days	Men	Women
2010/11	203	10,939	4,554
2011/12	596	30,543	9,980
2012/13	4,353	159,239	61,800

4.5.9 Awareness campaigns in the villages

There are various types of awareness methods used for effective dissemination of knowledge to farmers and to create awareness about new technologies, practices and methods of cultivation. These methods include wall writings, soil health cards, brochures and handouts, cycle *jathas* (a long march, usually aimed at spreading a message), street plays and *Bhoochetana rath* (Bhoochetana awareness vehicle). Until 2012–2013, nearly 60,000 wall writings were written on the walls of public properties such as bus stops and panchayat buildings, covering about 26,000 villages across 30 districts.

4.6 Tools to Create Awareness and Skill Development

4.6.1 Soil health cards

A total of 92,904 soil health cards were provided to individual farmers whose fields were sampled. The card was printed on both sides in the local language (Kannada) giving basic information about the farmer's field with details of individual nutrient status and critical limits of the field (Fig. 4.4). Also, the recommended dose of nutrients for each crop was included as well as the quantity of nutrients available in commercially marketed fertilizers.

4.6.2 Wall writings and media

Under the Bhoochetana initiative wall writings were done in each and every village to create awareness among the farmers. Wall writings were quite conspicuous with details of, for example, the soil health status, input quantities per hectare supplied to the farmer and components of inputs (i.e. micronutrients). Wall writings are an effective communication channel in rural areas which gained momentum in Bhoochetana on a large scale, helping farmers to understand their soil and agricultural practices, the objective of the programme and areas to be covered by the programme (Fig. 4.5). Additionally, thousands of brochures and handouts were published and distributed in each district on improved management practices, information on nutrient status, and nutrients recommended taluk-wise; brochures were widely distributed in all 30 districts (Fig. 4.6).

Print media news coverage was extensive to introduce the Bhoochetana programme to farmers and also provide information on activities during the season in all districts and contacts of FFs and lead farmers with individual farmers in selected villages.

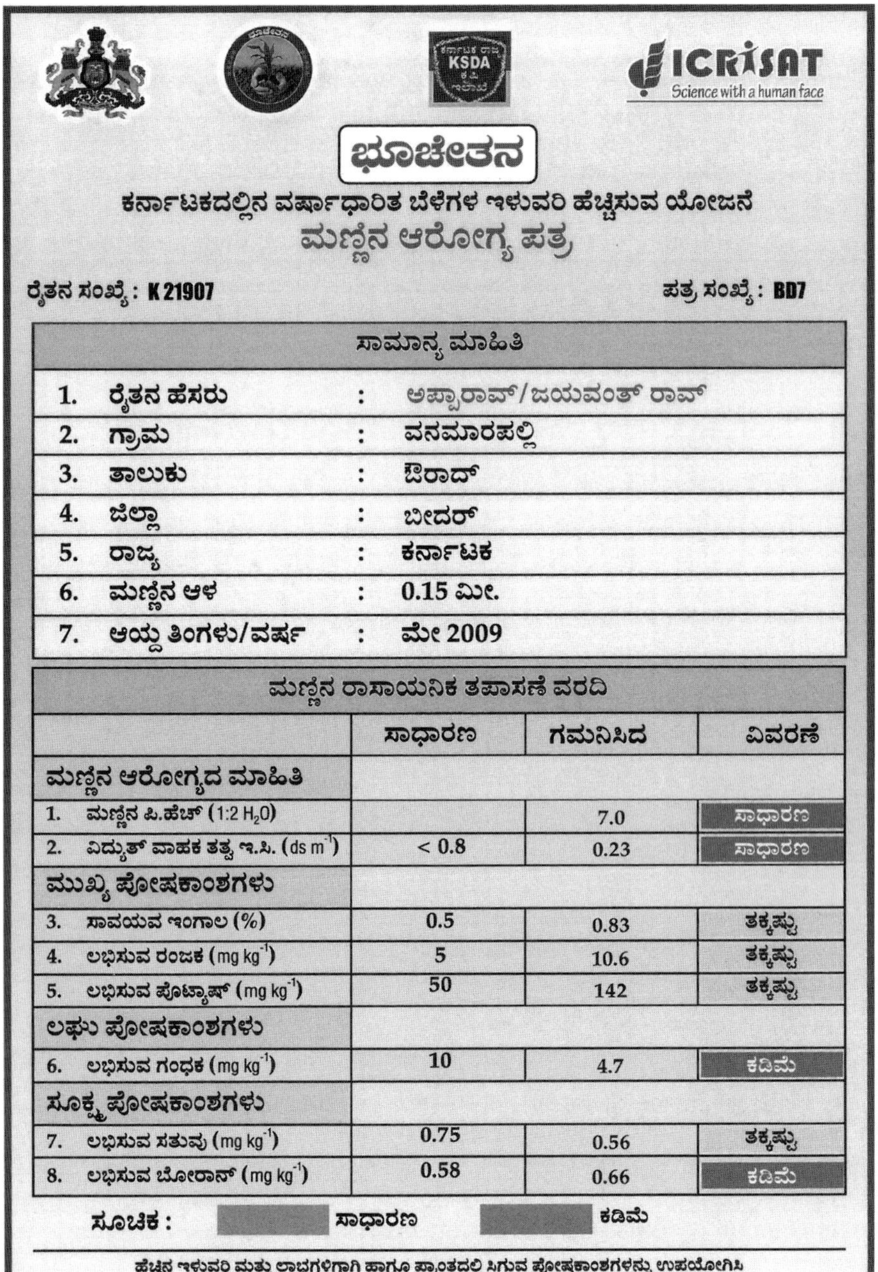

KSDA

ICRISAT Science with a human face

ಭೂಚೇತನ

ಕರ್ನಾಟಕದಲ್ಲಿನ ವರ್ಷಾಧಾರಿತ ಬೆಳೆಗಳ ಇಳುವರಿ ಹೆಚ್ಚಿಸುವ ಯೋಜನೆ

ಮಣ್ಣಿನ ಆರೋಗ್ಯ ಪತ್ರ

ರೈತನ ಸಂಖ್ಯೆ : K 21907 ಪತ್ರ ಸಂಖ್ಯೆ : BD7

ಸಾಮಾನ್ಯ ಮಾಹಿತಿ		
1. ರೈತನ ಹೆಸರು	:	ಅಪ್ಪಾರಾವ್/ಜಯವಂತ್ ರಾವ್
2. ಗ್ರಾಮ	:	ವನಮಾರಪಲ್ಲಿ
3. ತಾಲುಕು	:	ಔರಾದ್
4. ಜಿಲ್ಲಾ	:	ಬೀದರ್
5. ರಾಜ್ಯ	:	ಕರ್ನಾಟಕ
6. ಮಣ್ಣಿನ ಆಳ	:	0.15 ಮೀ.
7. ಆಯ್ದ ತಿಂಗಳು/ವರ್ಷ	:	ಮೇ 2009

ಮಣ್ಣಿನ ರಾಸಾಯನಿಕ ತಪಾಸಣೆ ವರದಿ

	ಸಾಧಾರಣ	ಗಮನಿಸಿದ	ವಿವರಣೆ
ಮಣ್ಣಿನ ಆರೋಗ್ಯದ ಮಾಹಿತಿ			
1. ಮಣ್ಣಿನ ಪಿ.ಹೆಚ್ (1:2 H_2O)		7.0	ಸಾಧಾರಣ
2. ವಿದ್ಯುತ್ ವಾಹಕ ತತ್ವ ಇ.ಸಿ. (ds m^{-1})	< 0.8	0.23	ಸಾಧಾರಣ
ಮುಖ್ಯ ಪೋಷಕಾಂಶಗಳು			
3. ಸಾವಯವ ಇಂಗಾಲ (%)	0.5	0.83	ತಕ್ಕಷ್ಟು
4. ಲಭಿಸುವ ರಂಜಕ (mg kg^{-1})	5	10.6	ತಕ್ಕಷ್ಟು
5. ಲಭಿಸುವ ಪೊಟ್ಯಾಷ್ (mg kg^{-1})	50	142	ತಕ್ಕಷ್ಟು
ಲಘು ಪೋಷಕಾಂಶಗಳು			
6. ಲಭಿಸುವ ಗಂಧಕ (mg kg^{-1})	10	4.7	ಕಡಿಮೆ
ಸೂಕ್ಷ್ಮ ಪೋಷಕಾಂಶಗಳು			
7. ಲಭಿಸುವ ಸತುವು (mg kg^{-1})	0.75	0.56	ತಕ್ಕಷ್ಟು
8. ಲಭಿಸುವ ಬೋರಾನ್ (mg kg^{-1})	0.58	0.66	ಕಡಿಮೆ

ಸೂಚಿಕ : ಸಾಧಾರಣ ಕಡಿಮೆ

ಹೆಚ್ಚಿನ ಇಳುವರಿ ಮತ್ತು ಲಾಭಗಳಿಗಾಗಿ ಹಾಗೂ ಪ್ರಾಂತದಲ್ಲಿ ಸಿಗುವ ಪೋಷಕಾಂಶಗಳನ್ನು ಉಪಯೋಗಿಸಿ ಬೆಳೆಸುವ ಬೆಳೆಗಳ ಉತ್ತಮ ತಳಿಗಳು ಉಪಯೋಗಿಸಲು ಮುಖ್ಯವಾಗಿ ಸೂಚಿಸಲಾಗಿದೆ.

Fig. 4.4. Soil health cards with details printed in Kannada on both sides of the sheet.

ಮಣ್ಣಿನ ತಪಾಸಣೆ ಆಧಾರಿತ ಗೊಬ್ಬರ ಶಿಫಾರಸ್ಸು (ಕೇ.ಜಿ/ಹೆಕ್ಟೆರಿಗೆ)							
ಬೆಳೆ	ಯೂರಿಯಾ	ಡಿ.ಎ.ಪಿ.	ಎಂ.ಓ.ಪಿ.	ಜಿಪ್ಸಮ್	ಝಿಂಕ್ ಸಲ್ಫೇಟ್	ಬೋರಾಕ್ಸ್ ಅಥವಾ ಆಗ್ರಿಬೋರ್	
ಏಕದಳ ಧಾನ್ಯಗಳು	50	38	0	200	25	2.5	1.25
ಜೋಳ	66	54	25	200	25	2.5	1.25
ಮುಸುಕಿನ ಜೋಳ	70	43	33	200	25	2.5	1.25
ಹತ್ತಿ	38	43	17	200	25	2.5	1.25
ರಾಗಿ	44	27	0	200	25	2.5	1.25
ಸಜ್ಜೆ	23	38	21	200	50	2.5	1.25
ಔಡಲ	44	27	21	200	25	2.5	1.25
ಸೂರ್ಯಕಾಂತಿ							
ದ್ವಿದಳ ಧಾನ್ಯಗಳು	7	65	21	200	25	2.5	1.25
ಸೋಯಾಬೀನ್ ಶೇಂಗಾ (ನೆಲಕಡಲೆ), ತೊಗರಿ, ಕಡಲೆ, ಉದ್ದು, ಹೆಸರು,	0	54	21	200	25	2.5	1.25
ಅಲಸಂದಿ							
ತರಕಾರಿ	138	65	50	200	25	2.5	1.25
ಟೊಮಾಟೊ	49	82	83	200	25	2.5	1.25
ಆಲೂಗಡ್ಡೆ	88	54	42	200	25	2.5	1.25
ಮೆಣಸಿನ ಕಾಯಿ	183	87	50	200	25	2.5	1.25

ಪೋಷಕಾಂಶದ ಕೊರತೆ ಇದ್ದಲ್ಲಿ ಸೂಚಿಸಿದ ಪ್ರಮಾಣವನ್ನು ಉಪಯೋಗಿಸಿ. ಸರಿಯಾದ ಪ್ರಮಾಣದಲ್ಲಿದ್ದರೆ ಸೂಚಿಸಿದ ಡೋಸಿನಲ್ಲಿ ಅರ್ಧ ಮಾತ್ರ ಉಪಯೋಗಿಸಿ.
ಏಕದಳ ಧಾನ್ಯಗಳು ಹಾಗೂ ತರಕಾರಿಗಳಿಗಾಗಿ ಸಾರಜನಕವನ್ನು 2 ಅಥವಾ 3 ಆವರ್ತಿ ಉಪಯೋಗಿಸಿ.
ಕೊಟ್ಟಿಗೆ ಗೊಬ್ಬರ 5 ಟನ್/ಹೆಕ್ಟೇರಿಗೆ ಉಪಯೋಗ ಸೂಕ್ತ.

ವರ್ಷಾಧಾರಿತ ಬೆಳೆಗಳಿಗೆ ಪೋಷಕಾಂಶ ಶಿಫಾರಸುಗಳು (ಕೇ.ಜಿ/ಹೆಕ್ಟೆರಿಗೆ)						
ಬೆಳೆ	ಸಾರಜನಕ	ರಂಜಕ	ಪೊಟ್ಯಾಶ್	ಗಂಧಕ	ಸತುವು	ಬೋರಾನ್
ಜೋಳ	60	35	0	30	10	0.5
ಮುಸಿಕಿನ ಜೋಳ	80	50	30	30	10	0.5
ಹತ್ತಿ	80	40	40	30	10	0.5
ರಾಗಿ	50	40	20	30	10	0.5
ಸಜ್ಜೆ	50	25	0	30	10	0.5
ಔಡಲ	35	35	25	30	10	0.5
ಸೂರ್ಯಕಾಂತಿ	50	25	25	30	10	0.5
ಸೋಯಾಬೀನ್	30	60	25	30	10	0.5
ಶೇಂಗಾ (ನೆಲಕಡಲೆ), ತೊಗರೆ, ಕಡಲೆ, ಉದ್ದು, ಹೆಸರು, ಅಲಸಂದಿ	20	50	25	30	10	0.5
ಟೊಮಾಟೊ	150	65	60	30	10	0.5
ಆಲೂಗಡ್ಡೆ	75	82	100	30	10	0.5
ಮೆಣಸಿನ ಕಾಯಿ	100	54	50	30	10	0.5
ಈರುಳ್ಳಿ	200	87	60	30	10	0.5

ವರ್ಮಿಕಂಪೊಸ್ಟ್ ಹಾಗೂ ಬದುಗಳ ಮೇಲೆ ಬೆಳೆಸಿದ ಗ್ಲೀರಿಸೀಡಿಯಾವನ್ನು ಉಪಯೋಗಿಸುವುದರಿಂದ ವರ್ಷಾಧಾರಿತ ಪ್ರಾಂತಗಳಲ್ಲಿ ಗೊಬ್ಬರದ ಮೇಲೆ ಆಗುವ ಖರ್ಚಿನಲ್ಲಿ ಕಡಿಮೆಯಾಗುವುದು ಹಾಗೂ ಹೆಚ್ಚಿನ ಇಳುವರಿ ಪಡೆಯಬಹುದು.

ಹೆಚ್ಚಿನ ಮಾಹಿತಿಗಾಗಿ ಸಂಪರ್ಕಿಸಿ : ಡಾ॥ ಸುಹಾಸ ಪಿ.ವಾಣಿ ಪ್ರಿನ್ಸಿಪಲ್ ಸೈಂಟಿಸ್ಟ್ (ಜಲಾನಯನ) ಮತ್ತು ರಿಜಿನಲ್ ಥೀಮ್ ಕೋಆರ್ಡಿನೇಟರ್ (ಏಸಿಯಾ)
ಆರ್.ಪಿ.1 ರೆಜಿಲಿಯಂಟ್ ಡ್ರೈಲ್ಯಾಂಡ್ ಸಿಸ್ಟಮ್ಸ್
ಇಕ್ರಿಸಾಟ್, ಪಟ್ಯಾನ್‌ಚೆರು - 502 324. ಆಂಧ್ರಪ್ರದೇಶ್, ಇಂಡಿಯಾ
ಫೋನ್ : +91 (40) 30713466, ಫ್ಯಾಕ್ಸ್ : +91 (40) 30713075,
ಈ-ಮೇಯಿಲ್ : s.wani@cgiar.org ~ www.icrisat.org

ICRISAT Science with a human face

Fig. 4.4. Continued.

(a) (b)

Fig. 4.5. Wall writings on Bhoochetana in Karnataka written in Kannada on: (a) Bhoochetana activities in Bidar district; and (b) integrated-pest-management technologies and their use in farmers' fields for enhancing productivity in Dharwad district.

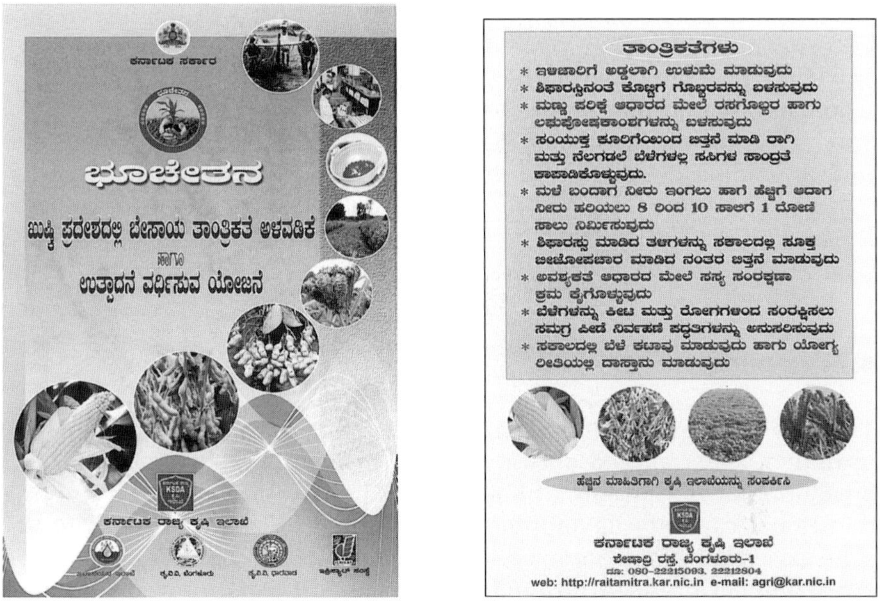

Fig. 4.6. Information brochures and leaflets on Bhoochetana in Kannada.

4.6.3 Internet

Considering the expanding reach of the Internet, information on Bhoochetana was extensively circulated on the Internet. Detailed objectives and the framework of the initiative, along with technical information on soil health mapping, recommendations and other best practices, were made available to various stakeholders.

4.6.4 Tablet-based knowledge sharing

Use of e-tablets is an innovative step towards improving the extension system under the Bhoochetana initiative; e-tablets have proved very effective to pass on real-time information on managing various issues as well as being useful for two-way interactions (see Chapter 5 of this book).

4.6.5 Farmer-to-farmer videos for sharing success stories

Considering the fact that farmers quickly follow the success of fellow farmers, video recording of success stories of various interventions were recorded and shared with other fellow farmers through Pico projectors. This technology proved to be very effective as it does not have any big requirements but rather few farmers can sit in small groups anywhere and watch the video and learn the way to success (more details are in Chapter 5 of this book). More than 50 short videos were recorded from four districts and are uploaded onto YouTube for wider dissemination. The topics covered include: (i) seed treatment; (ii) benefits of RSKs; (iii) soil testing; (iv) the importance of micronutrients; and (v) a seed germination test.

4.7 Summary

The new strategy for human capacity development in the Bhoochetana initiative, as explained above, has had a large impact on agricultural development in the state of Karnataka. The new strategies for reviving traditional capacity development methods paid rich dividends to the state as a whole. It is believed that if science-led research is to achieve a real impact on farm productivity and livelihoods, new methodologies for human capacity development have to be adapted along with a new extension system. The innovative approaches such as master trainers, institutional training for FFs and video conferences have a strong bearing on improving the knowledge flow, skill improvement as well as decision-making processes. Through regular capacity development programmes, a large number of stakeholders have benefited in improving their skills, knowledge and understanding about the programme in general and agriculture in particular.

For further improvement of these strategies some suggestions include:

- It is essential for the SAUs to be proactively involved in capacity development activities as well as in conducting demonstrations to disseminate field-based knowledge.
- The KVKs should be recognized and provided with responsibilities for demonstrations, capacity development and information/knowledge dissemination.
- Regular capacity development programmes at different levels should be organized for different stakeholders to enhance their understanding, knowledge and skills regarding agricultural development.

Acknowledgements

The authors sincerely thank the Government of Karnataka for funding support and the DoA for implementing the Bhoochetana programme in all 30 districts of the state. The support received from DoA staff at district, taluk and RSK levels, master trainers from the University of Agricultural Sciences (i.e. SAUs) (Bengaluru, Dharwad and Raichur), KVKs and ICRISAT is highly appreciated. Special thanks are due to FFs and farmers for their participation in the programme.

References

Bolger, J. (2000) *Capacity Development: Why, What and How*. Capacity Development/ Occasional Paper Series 1(1). Canadian International Development Agency, Policy Branch, Hull, Quebec, Canada.

Braun, R. and Duveskog, D. (2008) *The Farmer Field School Approach – History, Global Assessment and Success Stories*. Background Paper for the International Fund for Agricultural Development (IFAD) Rural Poverty Report 2011. IFAD, Rome.

Chimboza, A. (2012) *From Brain Drain to Brain Gain: Addressing Human Capital Needs for Post Crisis Zimbabwe's Capacity Building*. University of Pennsylvania, Philadelphia.

Eade, D. (1997) *Capacity Building: A People-Centred Approach to Development*. Oxfam, Oxford.

Food and Agriculture Organization of the United Nations (FAO) (1998) Knowledge and Information for Food Security in Africa: From Traditional Media to the Internet. Available at: http://www.fao.org/sd/CDdirect/CDan0017.htm (accessed 1 November 2014).

Government of Karnataka (2006) *Karnataka Agricultural Policy, 2006*. Department of Agriculture and Horticulture, Government of Karnataka, Bangalore, India.

Government of Karnataka (2011) *Karnataka Agriculture Budget – 2011–12*. Government of Karnataka, Bangalore, India.

Raju, K.V., Wani, S.P. and Anantha, K.H. (2013) *Bhoochetana: Innovative Institutional Partnerships to Boost Productivity of Rainfed Agriculture in Karnataka, India*. Resilient Dryland Systems Report No. 59. International Crops Research Institute for the Semi-Arid Tropics, Patancheru, Andhra Pradesh, India.

Rockström, J., Hatibu, N., Oweis, T. and Wani, S.P. (2007) Managing water in rain-fed agriculture. In: Molden, D. (ed.) *Water for Food, Water for Life: A Comprehensive Assessment of Water Management in Agriculture*. Earthscan, London and International Water Management Institute, Colombo, Sri Lanka, pp. 315–348.

Rockström, J., Karlberg, L,. Wani, S.P., Barron, J., Hatibu, N., Oweis, T., Bruggeman, A., Farahani, J. and Zhu Qiang (2010) Managing water in rainfed agriculture – the need for a paradigm shift. *Agricultural Water Management* 97, 543–550.

Shambu Prasad, C., Laxmi, T. and Wani, S.P. (2006) *Institutional Learning and Change (ILAC) at ICRISAT: A Case Study of the Tata–ICRISAT Project*. Global Theme on Agroecosystems Report No. 12. International Crops Research Institute for the Semi-arid Tropics (ICRISAT), Patancheru, Andhra Pradesh, India.

Twomlow, S.J., Steyn, J.T. and du Preez, C.C. (2006) Dryland farming in southern Africa. In: *Dryland Agriculture*, 2nd edn. Agronomy Monograph No. 23. American Society of Agronomy, Madison, Wisconsin, pp. 769–836.

United Nations Development Programme (UNDP) (1998) *Capacity Assessment and Development in a Systems and Strategic Management Context.* Technical Advisory Paper No. 3. Bureau for Development Policy, UNDP, New York.

Wani, S.P., Singh, H.P., Sreedevi, T.K., Pathak, P., Rego, T.J., Shiferaw, B. and Iyer, S.R. (2003) Farmer-participatory integrated watershed management: Adarsha Watershed, Kothapally, India. An innovative and up-scalable approach. A case study. In: Harwood, R.R. and Kassam, A.H. (eds) *Research Towards Integrated Natural Resources Management: Examples of Research Problems, Approaches and Partnerships in Action in the CGIAR.* Interim Science Council, Consultative Group on International Agricultural Research (CGIAR), Washington, DC, pp. 123–147.

Wani, S.P., Sarvesh, K.V., Krishnappa, K., Dharmarajan, B.K. and Deepaja, S.M. (eds) (2012) *Bhoochetana*: *Mission to Boost Productivity of Rainfed Agriculture through Science-led Interventions in Karnataka.* International Crops Research Institute for the Semi-Arid Tropics, Patancheru, Andhra Pradesh, India.

Wani, S.P., Sandeep Khanwalkar, Krishnappa, K., Raju, K.V. and Sarvesh, K.V. (2013) *Bhoochetana: Process Documentation – A Program that Helped Farmers in Coming out From Low Productivity and Poor Economic Status*. Resilient Dryland Systems Report No. 60. International Crops Research Institute for the Semi-Arid Tropics, Patancheru, Andhra Pradesh, India.

5

Digital Technologies for Agricultural Extension

Mukund D. Patil,* K.H. Anantha and Suhas P. Wani

International Crops Research Institute for the Semi-Arid Tropics (ICRISAT), Patancheru, India

5.1 Introduction

Ensuring global food security for the 9 billion people by 2050 and improving livelihoods sustainably with scarce and finite land and water is a challenging task (UNDP, 2009). The quantity of neither available water nor land has increased but water and land availability per capita has declined significantly due to the increase in human population. Eighty per cent of the world's cultivable area which contributes to feeding 60% of the total world population is rainfed. In developing countries in general, and India in particular, the growing population and shrinking landholding has increased pressure on natural resources to produce more. Although agriculture contributes a major share to the gross domestic product (GDP), the growth rate of the agricultural sector has reduced over recent years for various reasons. As per the latest estimates released by the Central Statistical Office of India the share of agriculture and allied sectors in GDP of India was 51.9% in 1950/51 and has declined to 13.7% in 2012/13 (GoI, 2013). At the same time, productivity has been stagnant or less than the potential (Singh *et al.*, 2009). One of the contributing factors is low investment along with low adoption of science-led interventions.

Lack of awareness among farmers about good agricultural management practices is also an important factor for stagnant productivity

*m.patil@cgiar.org

levels. Generally, farmers in developing countries follow traditional practices that are inadequate to sustain with changing climate and increasing population. The diverse nature of the semi-arid tropics of southern peninsular India calls for context- and situation-specific services to deal with emerging problems, including climate variation. In India, more than 81% of farmers cultivate an area of 2 ha or less (GoI, 2009). Thus, there is an increasing need for strong intermediaries that can facilitate information access for diverse smallholders. The agricultural extension system (AES) plays a very important role in enhancing the knowledge and skills of farmers for improving agricultural productivity. However, the present extension system appears to be inadequate to address the challenges faced by the farmers in India in the context of changing climate scenarios. In India, the extension system has limited penetration below the block level due to lack of trained humanpower at the grass-roots level to work in the villages serving the large number of smallholder farmers and ensure effective extension services.

It is important to rejuvenate the AES with innovative information communication technology (ICT) models for knowledge generation and dissemination. In this context, in this chapter we discuss new innovative digital technologies which are being used for agricultural knowledge dissemination in Karnataka under the Bhoochetana programme.

5.1.1 Strengthening the extension system through digital technologies

An innovative AES is necessary to ensure deep penetration of available agricultural technologies into the farming community. Introduction of farm facilitators (FFs) and lead farmers in the AES has provided the local point of information dissemination at village level. FFs are not agricultural graduates and information acquired by them during training sessions is adequate to address basic issues, such as use of soil-test-based nutrient management, promotion of improved crop cultivars, generic pests and diseases, and other field operations. However, this information may not be sufficient to address real-time issues, such as crop planning based on weather, correcting the nutrient deficiency during the crop growing period and identifying pests and diseases. Thus, there is a need to create a convenient channel for information exchange between FFs and development/research agencies.

Creative and effective ways of disseminating the information have been explored during the Bhoochetana programme in order to improve the adoption rate among the farming community. For example, information related to soil fertility status has been disseminated among the farmers through writing the information on the walls of schools or houses and through soil health cards. These ways are far more effective than the dissemination through group meetings. Information written on walls will be available for all the farmers in the villages. However, this information is not customized according to the crops grown or the landholding. In Bhoochetana, an e-tablet-based extension system was piloted as a tool of information dissemination and for monitoring the activities. This system is aimed at strengthening the local extension agent by providing a channel for information flow and to monitor the real-time agriculture status on the ground.

Digital technologies have been used in the AES, such as television/radio programmes, call centres, satellite programmes and a short message service (SMS)-based advisory system. The cellular telecommunication has good penetration in urban as well as rural areas of India. Thus cellular technology has become a very effective tool for marketing. Through mobile marketing, service providers can directly communicate with consumers; the main objective is to develop close and stronger interactions with the consumer and provide customized services. AESs are already adapting mobile communication technology to expand the outreach of improved technology.

5.1.2 Digital technology for agricultural extension: the Indian context

Dissemination of agricultural knowledge to the small and marginal farmers has been a weak link in Indian agriculture. Farmers require information about improved farming practices, weather, insect/pest control options and markets. This information should come at the time when it is actually needed. Information communication tools have provided the vital assistance to extension agents for supporting farmers' demands for information. However, large-scale implementation of digital technologies for agriculture is required to support millions of farmers in India. The Government of India and private companies are transforming the AES. The government has good human resources for agricultural knowledge generation, whereas private companies have good digital technologies. Thus, several technologies are being used for information dissemination.

The important media for mass communication are television and radio. These two tools are being used for dissemination of agricultural knowledge across the country. *Krishi Darshan* is one such example of a television programme initiated in 1967 and broadcasted by Doordarshan, India's public broadcasting service. Agriculture-related programmes are also being telecast by private broadcasters. The limitations for these programmes are short time slots, too much information, the given information is not applicable to all locations, and a mismatch between the period of information required and delivery.

A few of these limitations have been bridged by phone-in programmes, where farmers can ask questions specific to their situations. But still these programmes have short time slots and are not sufficient to cater for the need of millions of farmers. In fact, the Government of India has launched a dedicated channel for agriculture – 'Doordarshan Kisan' – in 2015.

Television and radio programmes are not location specific or crop specific, thus often they do not satisfy the need of farmers. These limitations were addressed by dedicated call centres for AES. The Government of India has initiated the Kisan Call Centre (KCC) facility to satisfy information request as per farmers' demand in 22 local languages. This service has been operating since 2004. There are 14 call centres across the country operated by 116 agriculture graduates. In addition to agriculture graduates, 123 experts located in different parts of the country at state agricultural universities, national agricultural research institutes, State Departments of Agriculture, Horticulture and others are also answering the calls at a second level. A common 11-digit toll-free number (1800-180-1551) has been allotted to KCC. This number is accessible through mobile phones and landlines of all telecom networks including private service providers. Replies to the farmers' queries are given in 22 local languages. The Government of India is bringing new technologies in to the AES. Currently, all the information communication tools for disseminating agricultural knowledge are brought under one umbrella 'mkisan' (www.mkisan.gov.in). The services provided through this website are Unstructured Supplementary Service Data (USSD) and SMS-based dissemination, pull-and-push SMS, interactive voice response system (IVRS), KCC and android-based applications.

In addition to the government, private agricultural companies are also providing innovative solutions for agricultural extension (see Box 5.1). For example, Indian Farmers Fertiliser Cooperative Limited – Kisan Sanchar Limited (IKSL) – has introduced voice messages for an agro-advisory system, Thomson Reuters introduced a mobile-based integrated agro-advisory system 'Reuters Market Light' and Digital Green has introduced the concept of video documentation of best farming practices carried out by a farmer and then showing it to farmers from the same region.

5.2 E-tablet-based Extension Tools

5.2.1 Devices and computing infrastructure

The extension system was piloted with Samsung Galaxy Tab 2. However, a tablet with a similar specification, including a 17.8 cm (7 inch) touch

Box 5.1. Key constituents of an ICT-assisted AES.

Key constituents are:

- awareness among farmers about available tools for getting the information;
- a one-stop application for all agriculture-related activities;
- regular updates of information;
- voice messages with a toll-free number;
- video documentation;
- a consortium of government agencies and agricultural research-and-development agencies for content generation and dissemination;
- weather-based agro-advisory information;
- fertilizer recommendation(s);
- mobile data collection;
- insect and pest monitoring and control measures;
- a two-way communication option; and
- market information.

screen, 3G and Wi-Fi connectivity, a voice-calling facility, a primary and secondary camera with good resolution, global positioning system (GPS), Bluetooth, an expandable memory and 1 GB RAM, is also suitable for a tablet-based extension system. Since this tablet will be used in farmers' fields, ruggedness of the tablet is also the most preferable feature. The proposed tablet runs on an android operating system. The price of an android tablet ranges from US$65 to US$450. However, low-cost tablet devices may not be suitable for outdoor conditions. In India, the Ministry of Human Resource Development had launched a low-cost tablet device, 'Akash'. As an effective teaching aid, the Indian government had procured these tablets and distributed them to students at a subsidized price. Now, a few manufacturers are competing to refine the first-generation low-cost tablet. Similarly, low-cost tablets developed by Bharat Electronics Limited for the Ministry of Rural Development were used for a door-to-door survey for the Socio-Economic and Caste Census 2011.

5.2.2 *Krishi Gyan Sagar* (KGS)

A tablet-based extension system *Krishi Gyan Sagar* (KGS) (literally meaning 'ocean of agricultural knowledge') was developed by the International Crops Research Institute for the Semi-Arid Tropics (ICRISAT), Patancheru, India, in collaboration with NUNC Systems. The concept of KGS is not restricted to Bhoochetana or India. It is a generic framework for a digital extension system that can be deployed in any part of the world. One of the key requirements while developing the application was that it should be in the local language as the main user group does not understand English. The KGS app in Bhoochetana is available in English as well as Kannada (the local language of Karnataka state, India).

KGS is designed to help in knowledge sharing from laboratory to farmers as well as information collection from farmers to laboratory (Fig. 5.1). The KGS has two platforms for two different user groups. The first part is an android app, which is designed as an information dissemination tool as well as data collection tool. FFs are the primary users for the android app in the tablet. Each FF has jurisdiction of about 500 ha area, which covers one or two villages. Once logged in, FFs can access information available in the KGS app and give advice to farmers. In addition, they can capture details of ongoing farming activity using various options available in the app. Availability of information in the app is restricted based on the jurisdiction of the logged-in user. The other platform of KGS is the web application. Both the android app and web app are backed with a common database server. The server receives data from remote users as well as a database administrator. This web app is more useful for policy makers and development agents for monitoring and report generation. Web app users can generate query-based reports from data captured by FFs at field level.

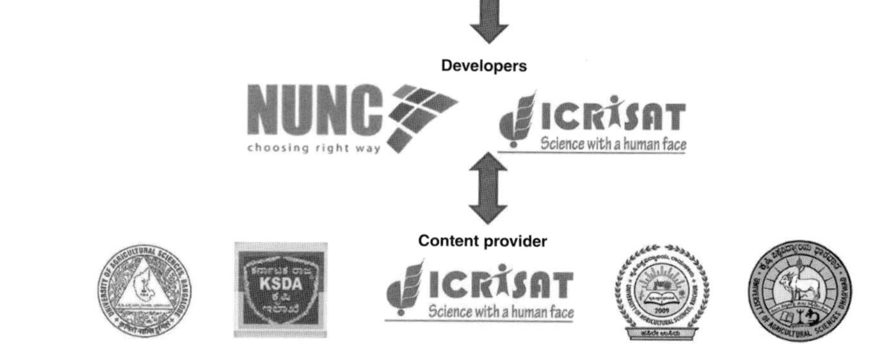

Fig. 5.1. The structure of the *Krishi Gyan Sagar* (KGS) system and roles of consortium partners.

Android app

FARMERS' REGISTRATION The farmer database is the base of AES. Farmers' details include their basic information and farm information (Fig. 5.2). The basic information is a one-time entry, though it can be edited at any time. All this information is captured by FFs while interacting with farmers. Earlier, this information was collected in paper form when farmers used to take inputs from local extension centres. The location information of each farmer is linked with the georeferenced soil fertility data, which is used for providing site-specific fertilizer recommendations.

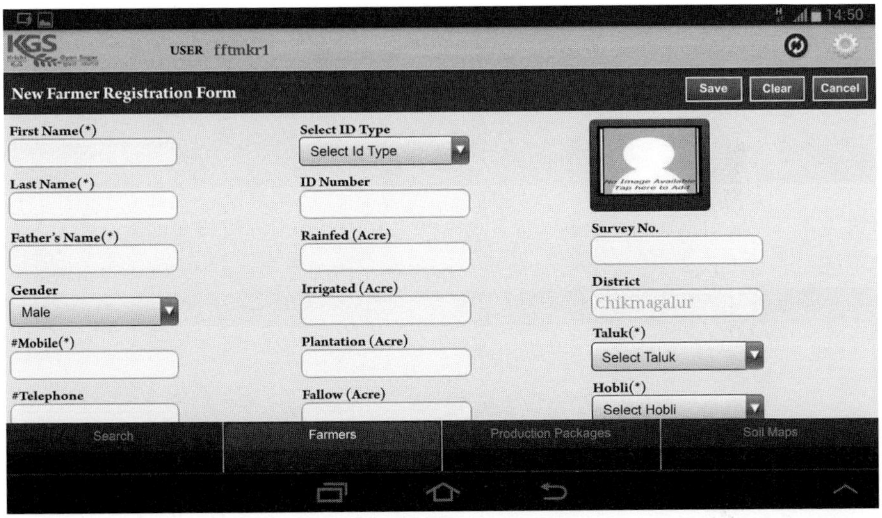

Fig. 5.2. Farmer's registration module in the KGS app.

SOIL-TEST-BASED FERTILIZER RECOMMENDATION One of the important outputs from the Bhoochetana project is the soil fertility atlas of Karnataka state of India. Methodologies adapted for developing the soil fertility atlas are described in Chapter 3 of this book. The published atlas is available in the public domain. These data are adopted in the KGS app in two forms: (i) district-wise soil fertility maps, including the status of organic carbon, phosphorus, potassium, sulfur, boron and zinc, are embedded in the app (Fig. 5.3); and (ii) a site-specific fertilizer recommendation. Based on the user's district, the soil maps will be displayed in the app.

The KGS app is supported by georeferenced soil fertility data and location information recorded during the farmer's registration. The queried data are pre-processed on the basis of location, farm area and crop-specific nutrient requirements to provide a customized fertilizer recommendation (Fig. 5.4). This dynamic customization is not possible with soil health cards or information written on walls.

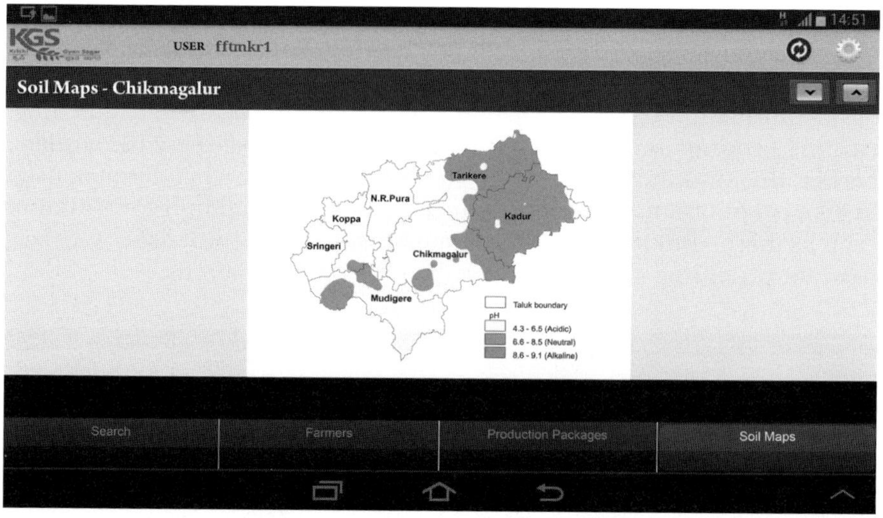

Fig. 5.3. District-wise soil fertility maps embedded in the KGS app.

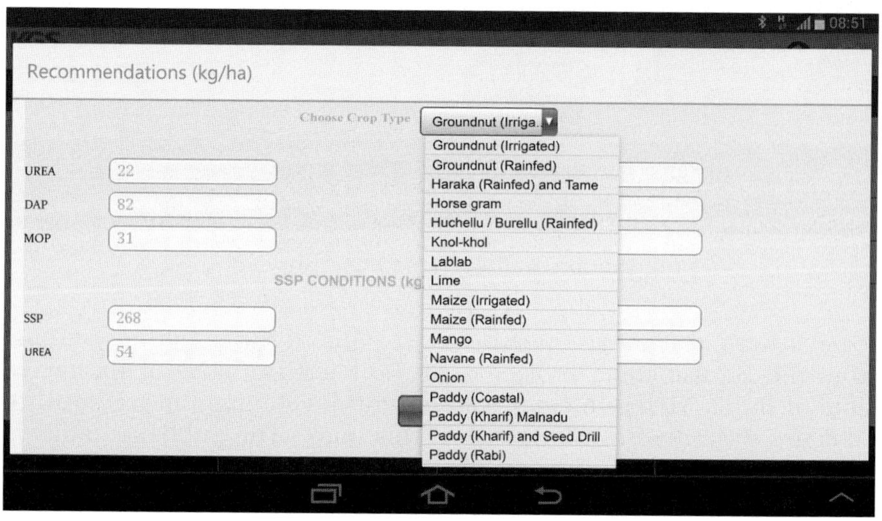

Fig. 5.4. Crop- and location-wise fertilizer recommendation for each farmer.

PACKAGE OF AGRONOMIC PRACTICES This part of the app provides updated information about good agricultural practices with respect to each crop. It contains information on: (i) soil and climate requirements; (ii) land preparation; (iii) available cultivars; (iv) seed treatment; (v) sowing/planting; (vi) fertilizer and water management; (vii) plant protection practices; and (viii) harvesting and postharvest practices. The updated information is provided by the National Agricultural Research and

Extension System and ICRISAT. This information is translated into the local language and supported with pictures so that the FFs can easily understand it (Fig. 5.5).

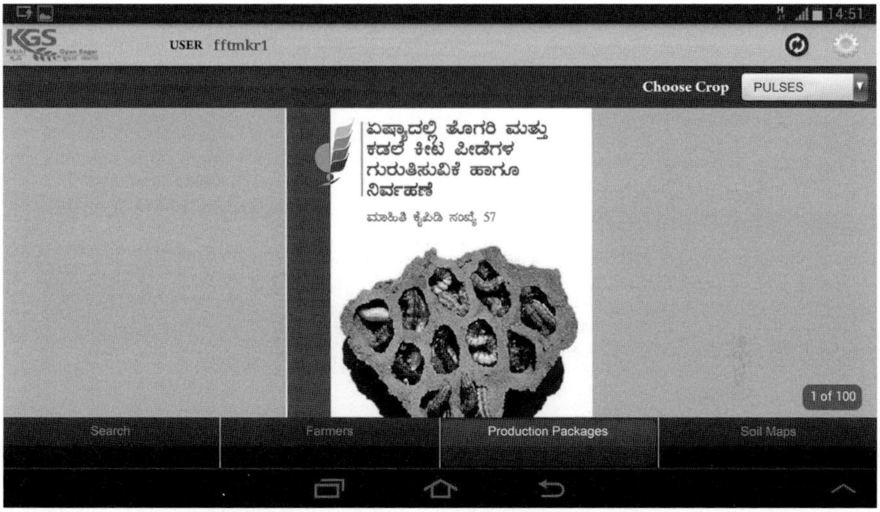

Fig. 5.5. Package of agronomic practices in Kannada.

Web app

The android KGS app is a field tool for information dissemination and data collection, whereas the web app is the website for visualizing the data gathered by the FF using the KGS app (see Box 5.2 and Fig. 5.6). Both android app and web app have a common database server. The web address for this application is www.krishigyansagar.com; however, this application is not accessible to all users. Web administrators, government officials and technical staff from ICRISAT have access to this application. The web app contains all the information available on the android app. The important features of this app are user registration and report generation tools.

5.2.3 Krishi Vani

ICRISAT in collaboration with IKSL and Bharti Airtel initiated the Krishi Vani platform. This initiative has been piloted in 171 villages in Telangana and Karnataka benefiting 40,000 farmers (ICRISAT, 2014). Krishi Vani is a mobile phone/phablet-based application. Through this application generic advisories are delivered to groups of farmers in a location through the mobile phone enabled by Green SIM. IKSL has pioneered the voice-message-based agro-advisory system. To subscribe to Krishi Vani, the user has to buy Green SIM from Airtel. These SIM

Box 5.2. Field operation of the KGS app.

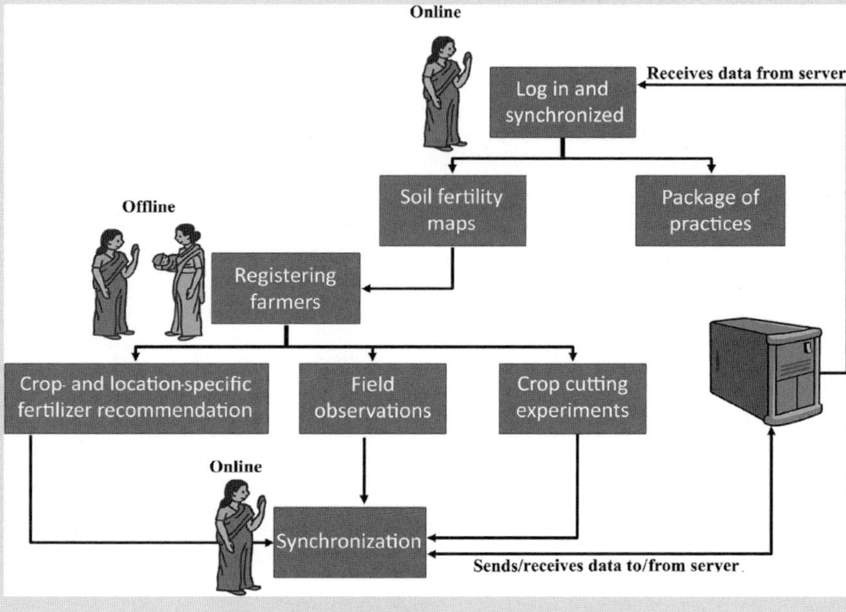

Flow diagram of field operation of the KGS system.

1. A trained FF visits farmers during the registration campaign.
2. The FF signs in on the KGS app and synchronizes the data between the KGS server and the KGS app. The sign-in and synchronization processes require Internet connectivity.
3. Soil fertility maps of the district of the FF are available.
4. The farmer's basic information, including location, unique identity number, contact number and photograph is recorded on the KGS app.
5. Location of farms and crops cultivated are registered under farmers' profiles. According to farm location and crop, the KGS app provides fertilizer recommendations that are narrated by the FF to the farmer.
6. Production packages of various crops and insect and pest management practices, provided in Kannada in the KGS, help the FF to read easily and narrate more efficiently to farmers.
7. During the harvest period, the FF collects data of crop cutting experiments from selected farmers' fields.
8. All the data recording processes are offline, which store the data in the android app. A partial synchronization process sends all stored data in the KGS app to the remote server.

cards are specially configured for receiving voice messages and other agro-advisory services. Every day, four free voice messages are delivered to the subscribers. The contents of voice messages are advised by subject matter specialists and cover diverse areas such as soil management,

Fig. 5.6. A farm facilitator (FF) using the KGS app on Samsung Galaxy Tab 2.

crop management, dairy and animal husbandry management, horticulture and vegetable management, plant protection, market rates, weather forecast information, human and cattle health, employment opportunities and government schemes.

5.2.4 Useful utility tools on the android tablet

Mobile data collection tools

Mobile data collection tools are becoming important resources for development programmes for improving efficiency, accuracy and the pace of assessing the situation on the ground. Setting up the survey questionnaires and getting them filled, real-time data flow to the central server and instant availability of clean data for the analysis are the important features of mobile data collection tools. The faster cellular technology and open source software development option has brought down the cost factor associated with the mobile data collection. Paper forms, camera and a GPS device are replaced by a single smartphone. There are several mobile data collection tools available for the android platform. Open Data Kit (ODK) is one of the open source applications and is easy for quick deployment. The deployments of ODK for various purposes are mentioned on the ODK website (www.opendatakit.org).

ODK is a set of three basic tools: (i) ODK Build for creating the questionnaire (www.build.opendatakit.org); (ii) ODK Collect, a smartphone app available on Google Play Store; and (iii) ODK Aggregate, a data aggregation application (Plate 9). We have used ODK for a baseline household survey in two watershed projects in Karnataka, soil sampling locations, field observations from villages in Karnataka and crop cutting experiments. The forms as per requirements were prepared using ODK Build and uploaded onto the ODK Aggregate server. ODK Aggregate was deployed on Google's free App Engine service. ODK Aggregate can also be deployed locally on a Tomcat server backed with a MySQL or PostgreSQL database server. The forms uploaded on the ODK Aggregate server are available for the ODK Collect application installed on the enumerator's mobile device. Data are recorded using ODK Collect store on the mobile device until the filled form is sent to the Aggregate server. ODK has the option to record text, the date, numbers, geo-coordinates, photographs, videos, multiple choice questions, etc.

Instant messaging apps

Instant messaging apps such as WhatsApp, Hike, WeChat, Line, telegram and Viber have become very convenient tools of mass communication (Fig. 5.7). Social networking tools such as LinkedIn, Twitter and Facebook have already proved their importance in the social

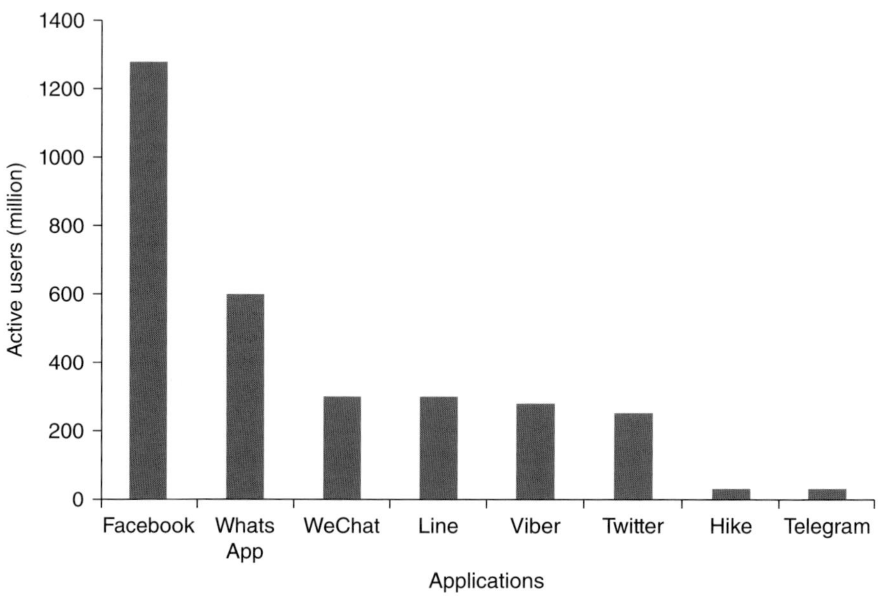

Fig. 5.7. Monthly global active users for top instant messaging applications.

welfare sector. Microblogging tools such as Twitter have been used by many individuals as well as government agencies. The usual use of these applications is to let others know about one's experiences during daily life; however, most of the time users exchange general information, jokes/quotes, interesting pictures and personal/official documents. The target user group for these apps is below the age of 30 years. However, these applications have been installed in most of the mobile devices by default.

WhatsApp has more than 50 million subscribers in India and the 2-year-old app Hike Messenger now exceeds 20 million users. Such social networking tools are also very useful for awareness campaigns. In Bhoochetana, all ICRISAT staff are connected through WhatsApp. The ground staff share picture messages with a short text message of daily activities in the field. All these messages are being archived at ICRISAT. Updates regarding the programmes or issues related to the KGS app are being shared through the WhatsApp group.

5.3 Dissemination from Farmer to Farmer

For effective dissemination of good management practices, a farmer-to-farmer dissemination route is being explored through farmer-centric video documentation. Digital Green (http://www.digitalgreen.org/) is the technology partner for this innovative dissemination route. Digital Green has initiated the participatory video and mediated instruction for agricultural extension. The advantage of a farmer-to-farmer system is the fact that farmers trust fellow farmers to adopt improved management practices. Farmers can easily understand these farming practices as they explain in their languages. The farmer-to-farmer video documentation and dissemination system was piloted in two districts (Dharwad and Hassan) during the Bhoochetana project. This system is being tested in four other districts in Karnataka. This system has two processes: (i) video production; and (ii) video screening.

The entire process of video production to screening and the various personnel involved in this process are analogous to the movie industry (Fig. 5.8). In this process the producer is a research and development agency or scientist from ICRISAT. The producer decides the subjects of the videos based on location, crops and stage of the crop. Since these videos are real life stories, the producer's field staff identifies the progressive farmer who has adopted the improved agricultural practices with respect to various topics. The video usually contains two characters in conversation about the improved technology and a demonstration. The farmer shares his/her experience about the technology on camera, whereas the FF plays a supporting role as an interviewer (Fig. 5.9). The interviews were scripted in such a way that the length of video remains short. However, the message from the video is very clear so that the other farmers can easily adopt the demonstrated technology. The director, camera person and editor for the video are ICRISAT's ground staff. Digital Green has trained ICRISAT staff for the video production process.

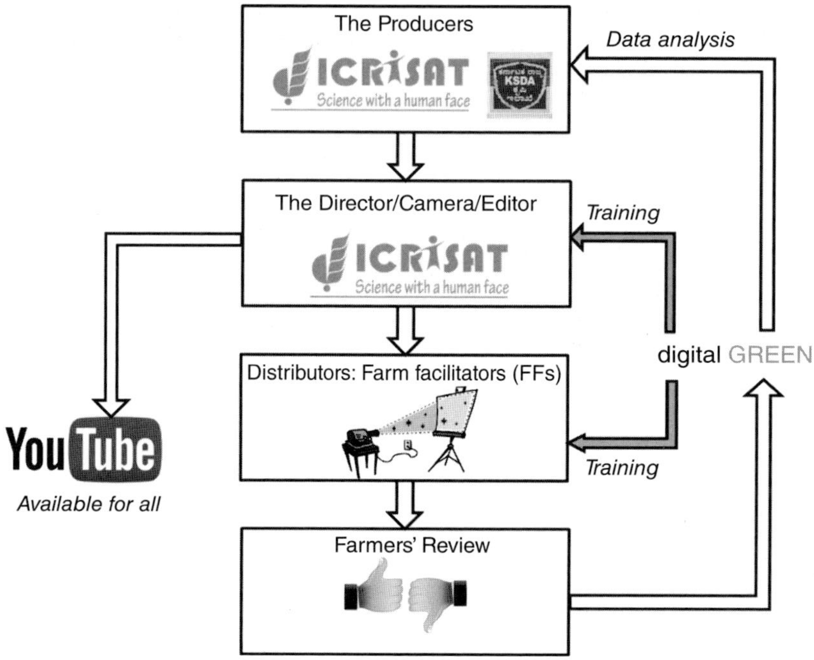

Fig. 5.8. Flow diagram of the various processes in farmer-to-farmer video documentation and screening.

The processed videos are given to the FFs for screening in the villages (Fig. 5.10). The battery-operated portable projectors (Pico projector) along with necessary accessories were provided to the FFs. The FF screens the video to a small gathering (20–30 farmers) in the village. At the end of the video, the FF collects feedback from farmers regarding previous videos. The feedback system also captures the adaptation rate of screened technologies. Digital Green has developed an online/offline data management framework (COCO) that captures data related to the key processes of the Digital Green approach – video production, dissemination and adoption of practices. A total of 48 videos were produced by ICRISAT staff and were screened in target villages by FFs. Based on the feedback from the farmers, one in seven farmers has adopted the screened technology.

Fig. 5.9. Handycam being used for recording videos in the field.

Fig. 5.10. A farm facilitator (FF) is screening the video for a group of women, using a battery-operated portable projector.

5.4 Constraints

The use of technology in AES is changing as per the requirements. Changes in dissemination pathways were driven by innovation in technology and its adaptation. The major constraints we faced during implementation of the digital extension system are as follows:

- Information communication tools are themselves dependent on other technologies: for example, lack of 3G connectivity in remote villages restricts users' ability to update the android application in villages.
- The device is relatively expensive and is not designed to be weather proof. Rugged and weather-proof industrial-grade tablet devices are available in the market, but are very expensive. ICRISAT has asked manufacturers to build a rugged but less-expensive tablet device. This new rugged device will be launched as a green phablet.

5.5 Way Forward: Digital AES as a Public–Private Partnership (PPP) Model

The Department of Agriculture, Government of Karnataka has implemented innovative programmes like Bhoochetana which have benefited millions of farmers by taking science to the doorstep of farmers. Bhoochetana has created an active ground force of para-agronomists (FFs) who are directly involved in information dissemination and data collection. Over the past five decades, in spite of enormous technological developments, current yields in farmers' fields are lower by two- to threefold as compared with the achievable potential yield. In order to bridge the large yield gaps to unlock the potential of agriculture, there is an urgent need for an effective, sustainable and quality extension system to reach millions of smallholder farmers as well as large farm holders in a consortium mode through public–private partnerships (PPPs). To resolve these issues, pilot testing of various digital extension tools were carried out in Karnataka. ICRISAT has brought private partners such as IKSL, Airtel, NUNC Systems and Digital Green for pilot testing of digital technologies for strengthening the existing extension system.

The experiences from the pilot studies have led to a policy decision to plan a comprehensive holistic PPP extension service. The PPP initiative envisages a timely quality agricultural extension service by utilizing the ICT network. These services are designed to find solutions for all farming issues by establishing a one-stop shop at reasonable cost by bundling knowledge with inputs services and value addition and linking these with markets. The self-sustainable PPP model will consist of agriculture and allied government departments, national agricultural research and extension agencies and private companies. The real-time information exchange and update of improved agricultural practices strengthens the farmers to obtain overall improved farm production or crop productivity and thus enhance profit margins.

References

Government of India (GoI) (2009) *Agriculture Statistics at a Glance 2009*. Department of Agriculture and Cooperation, Directorate of Economics and Statistics, New Delhi.

Government of India (GoI) (2013) Ministry of Agriculture, Rajya Sabha unstarred question number 2175, answered on 30 August 2013. Available at: http://rajyasabha.nic.in/ (accessed 17 February 2016).

International Crops Research Institute for the Semi-Arid Tropics (ICRISAT) (2014) *Annual Report 2013*. ICRISAT, Patancheru, Telangana, India.

Singh, P., Aggarwal, P.K., Bhatia, V.S., Murty, M.V.R., Pala, M., Oweis, T., Benli, B., Rao, K.P.C. and Wani, S.P. (2009) Yield gap analysis: modelling of achievable yields at farm level. In: Wani, S.P., Rockström, J. and Oweis, T. (eds) *Rainfed Agriculture: Unlocking the Potential*. CAB International, Wallingford, UK, pp. 81–23.

United Nations Development Programme (UNDP) (2009) *Human Development Report 2009. Overcoming Barriers: Human Mobility and Development*. UNDP, New York.

6 Institutional Arrangements and Innovations

K. Krishnappa,* K.H. Anantha, Suhas P. Wani and K.V. Raju

International Crops Research Institute for the Semi-Arid Tropics (ICRISAT), Patancheru, India

6.1 Introduction

Modernization of agriculture is a key element in the overall development of developing countries like India. The agriculture sector employs on an average about 55% of the population in India and has significant forward and backward linkages to the rest of the economy. Raising agricultural productivity is the key factor in making agriculture more competitive in rapidly liberalizing world markets as well as increasing the incomes of rural populations. Agriculture is receiving increasing attention as an instrument for growth and in which institutional innovations are seen as key to achieve not only agricultural growth, but also to include poor smallholders in this growth (World Bank, 2008). These institutional innovations are expected to be able to overcome various market failures, including missing or incomplete input and output markets. The *World Development Report – 2008* sees a particularly important role for the communities, collective action and non-governmental organizations (NGOs) in overcoming some of the market and state failures (World Bank, 2008). Reducing hunger through increasing agricultural production to keep up with population growth while increasing access to food for marginalized populations requires a better understanding of the dynamic and complex relationships between the socio-political, economic, scientific and environmental factors that

*k.kamma@cgiar.org

underpin food security. This requires strong institutional and policy support to match knowledge integration.

Agricultural research and extension services play a crucial role in raising agricultural productivity, but the performance of these services has for long been considered problematic in India as they failed to adapt to changing circumstances (Glendenning *et al.*, 2010; Suresh Babu *et al.*, 2013). To revitalize the performance of the agricultural research and extension systems, federal governments throughout India started to experiment with a series of institutional reforms since the 1990s (Suresh Babu *et al.*, 2013). These institutional reforms – in particular with respect to the sustainability of funding for research and technology transfer – have in common a set of principles with the aim of empowerment of local communities.

The Government of Karnataka introduced a unique model through institutional reforms for the Bhoochetana programme, which was operationalized through a structure composed of state- and grass-root-level institutions. The highest level of institutions that was created through the Bhoochetana programme was the separate Bhoochetana cell at the state level to facilitate input delivery and monitoring aspects and the concept of farm facilitators (FFs) (para-extension workers), with representation from the farming community, formed the village-level institutional mechanism for upscaling the Bhoochetana model. The innovativeness of Bhoochetana includes convergence of central and state-supported programmes/schemes to increase financial efficiency. This chapter aims to review these institutional reforms on the basis of Bhoochetana experiences in the state of Karnataka in South India.

6.2 Bhoochetana: A Mission Programme

The Bhoochetana project idea was first discussed by scientists of the International Crops Research Institute for the Semi-Arid Tropics (ICRISAT), Patancheru, India, and presented to the Government of Karnataka. In 2009, ICRISAT was invited to form a consortium with state agricultural universities (SAUs), line departments such as the Department of Agriculture (DoA), the Watershed Development Department (WDD), the Directorate of Economics and Statistics (DES), community-based organizations and farmers for the project. The consortium collectively took part in planning and promoting the project activities. The project activities were designed for six districts during the first year. It was upscaled to 16 districts (including the original six districts) during the second year. After realizing the success, the project activities were upscaled to the entire state in the third year. The project Bhoochetana then became a mission project for the government and the role of the consortium was changed to an advisory board. A high-level coordination committee was formed at the state level for regular monitoring and evaluation of the project activities.

The Bhoochetana project systematically involved knowledge-generating institutes such as universities and an international agricultural research organization (ICRISAT). Its purpose was to reach millions of smallholders in the rainfed region to adopt the soil-test-based fertility management for improving soil health and crop productivity. The farmers participated in the project on an individual basis and adopted the land rejuvenation interventions, while the DoA provided extension services and inputs (seeds, secondary and micronutrients) with 50% subsidy. The ICRISAT-led consortium along with the DoA prepared the plans to increase agricultural production by 5% annually. Major constraints/bottlenecks for increasing agricultural productivity were identified as: (i) imbalanced use of fertilizers; and (ii) lack of information about new knowledge (technologies, cultivars and seeds, water and land management, pest and disease management adaptation strategies to overcome adversities such as dry spells, waterlogging, frost or high temperature, and market information). The ICRISAT-led consortium suggested the creation of trained para-agricultural workers by training local farmers as FFs and lead farmers during peak demand time. Master trainers were trained and in turn master trainers trained FFs at district and *hobli* (cluster of villages) levels. Innovative agricultural extension services provided through the concept of FFs was the contributing factor for knowledge dissemination among farmers for increasing technology adoption in the state. There was no charge for the service. Handholding support was provided by ICRISAT staff located in each district along with DoA and SAU staff.

The innovativeness of this project includes identification of soil nutrient deficiency status and taluk-wise (block level) nutrient recommendations based on nutrient status supported by making available the necessary inputs at 50% subsidy. For timely and easy availability of inputs, these were delivered in advance to each cluster of villages as well as to the *Raithu Samparka Kendras* (RSKs) at every *hobli*. These inputs (seeds, seed treatment chemicals, gypsum, micronutrients and biofertilizers) had been supplied as a package. To operationalize the project, the services of FFs to take technologies to the farmers' doorsteps were utilized. Village meetings, wide publicity through mass media and wall writings in each of the selected villages were organized. Above all, the innovative strategy was to converge all the existing schemes of the DoA and channel them through Bhoochetana. The DoA is formally responsible for the project, while the consortium role gradually changed from technical support to advisory role. The programme provided farmers with technical advice on production processes, especially on the use of variable inputs, with the stated objectives of increasing the efficiency of production and improving the quality of products. The services were promoted by the DoA at village level and monitored regularly by the staff of the DoA.

6.3 Building Institutional Partnerships

Past experience showed that enhancing partnerships and institutional innovations is the key factor for harnessing the potential of community watershed management to reduce poverty and environmental degradation (Shambu Prasad *et al.*, 2006). Therefore, the consortium approach was introduced to harness the expertise of different areas to expand the effectiveness of various watershed initiatives and interventions. The underlying element of the consortium approach is to engage a range of actors to harness their strengths and synergies, with the local community as the primary implementing unit. Through the consortium approach, complex issues can effectively be addressed by the joint efforts of key partners. In this context, the Bhoochetana initiative started with pooling together both knowledge-generating institutions and knowledge-disseminating institutions to cater for the needs of different stakeholders in the state. As a result, the consortium of the University of Agricultural Sciences, ICRISAT and line departments such as the DoA, WDD and *Krishi Vigyan Kendras* (KVKs) was formed with the aim of imparting knowledge, skills and technologies to the farming community (Fig. 6.1). In this initiative, the role of consortium partners was to enhance the awareness of stakeholders to adopt improved technologies to achieve sustainable crop production. Each consortium partner had a specific responsibility for enhancing the capacity of different stakeholders in the project. This was to avoid duplication and delay in implementation as well as delivering the outputs.

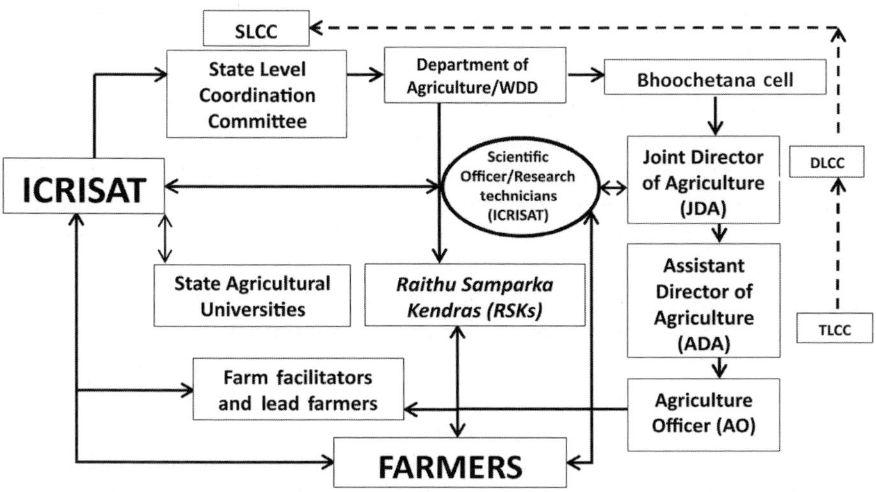

Fig. 6.1. Institutional partnership in Bhoochetana in Karnataka. DLCC, District Level Coordination Committee; SLCC, State Level Coordination Committee; TLCC, Taluk Level Coordination Committee; WDD, Watershed Development Department.

The DoA as a nodal agency has the responsibility of overall coordination of the programme for smooth implementation. The responsibility of organizing, planning, monitoring and review workshops/meetings rests with the DoA; it also issues timely guidelines and circulars with new information to enable field personnel to effectively implement the programme. In addition, the DoA organizes brainstorming sessions with experts from research and development agencies such as ICRISAT, SAUs, lead NGOs and line department staff in order to ensure interventions are tailor made to suit the needs of each respective district. It also facilitates the planning process of crop seasons and undertakes large-scale awareness programmes/campaigns and communication initiatives to spread the message to the larger community at different levels, namely, state, district, taluk and village.

The University of Agricultural Sciences of Bengaluru, Dharwad, Raichur and Shivamogga shared the responsibilities of capacity development of various actors involved in the implementation of the Bhoochetana programme. The major responsibility is to train master trainers periodically, with the main objective of dissemination of new knowledge uniformly across the actors at various levels. The master trainers were trained through training of trainers by expert scientists of ICRISAT and partner organizations, covering various issues of agricultural operations. The purpose of training master trainers so that they were on the same wavelength was to ensure that the messages conveyed down the line were the same and they were conveyed in simple language that was easy to understand. Further, KVKs volunteered to impart training along with conducting demonstrations. This has been the strength of the consortium which generally complements strengths and overcomes the weaknesses of partners. In the Bhoochetana initiative, KVKs acted as a technical support agency and contributed to awareness building, training programmes and sharing of new technologies with the farming community through demonstrations. The KVKs have enhanced the capacity of FFs through organizing training programmes on integrated nutrient management, integrated pest management, improved varieties, good agricultural practices, etc. However, it was observed that capacity building is essential even for KVK scientists and subject matter specialists on new technologies, strategies and methods.

Along with the DoA, ICRISAT had a major role in facilitating and handholding for the smooth implementation of the initiative. ICRISAT, as a technical support provider, ensures new knowledge generation, sharing of knowledge and capacity development of the various stakeholders, including policy makers, researchers, line departments and farmers. Regular monitoring and evaluation, generation of reports and feedback on improvements in the initiative is also the responsibility of ICRISAT. Demonstration of new crop varieties across different rainfall and soil types is a high priority area in order to suggest new varieties in the changing climate scenario. Similarly, development of contingency plans, expert advice on different issues, organizing annual progress review meetings and facilitating field exposure visits to policy makers, technocrats and farmers are other high priority areas.

The National Commission on farmers has drawn attention to the knowledge deficit, which constrains agricultural productivity and to overcome this, farmers need to have an effective linkage with the relevant institutions to access best practices (GoI, 2006). A good extension system is the means for achieving this linkage, but for the present this has virtually collapsed in most parts of the country. In the context of meeting the holistic needs of increasing agricultural production in a sustainable manner, agricultural extension has a key role to perform. Recognizing these challenges, the Bhoochetana programme in Karnataka has adopted an effective coordination mechanism between the Government of Karnataka and ICRISAT (Fig. 6.2). The Government of Karnataka had utilized the services of FFs and lead farmers as extension agents to reach millions of farmers with innovative technologies. Periodic capacity-building training sessions and hands-on exercises put them in a strong position to disseminate the required information to farmers. To spread the new information on the performance of different

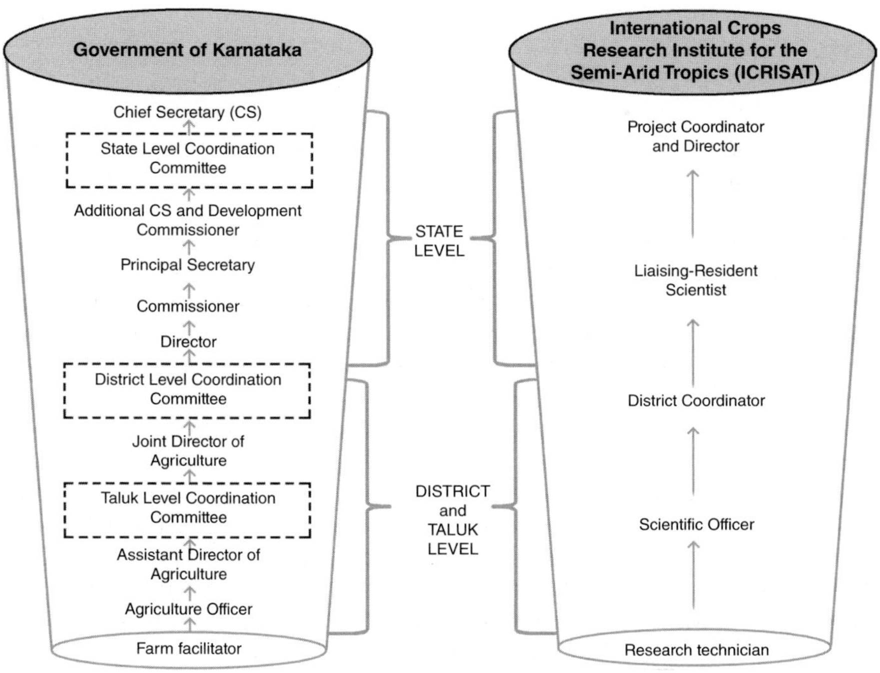

Fig. 6.2. Bhoochetana coordination mechanism between the Government of Karnataka and ICRISAT at state, district and taluk levels.

technologies, interactive farmers' days were organized so that farmers could visit the plots where such technologies have been used, to make them aware of these technologies and to show them the impact.

6.4 Process of Conceptualization, Implementation and Scaling up of Bhoochetana

The conceptualization of Bhoochetana started with organizing a brainstorming session in 2009 by inviting major stakeholders, including policy makers, SAUs, line departments and ICRISAT. During the deliberation, the good performing technologies were very much appreciated and it was decided to demonstrate the innovative technologies through a programme with technical assistance from ICRISAT. The strategy was prepared and discussed with high-level policy makers and technocrats. The programme was launched within a short time and implementation started with soil-test-based fertilizer recommendation as an entry point activity in six districts in the 2009 *kharif* (rainy season). The modalities of input procurement, supply and storage were worked out and crop sampling, nutrient analysis and capacity building programmes were organized to understand the process of implementation. The programme was based on the concept of generating funds using state resources as well as central resources. The proposal was developed based on this concept and funds were received by converging schemes like *Rashtriya Krishi Vikas Yojana* (RKVY) and other programmes. It is an innovative approach to converge with all the ongoing programmes to address the larger issues in a scientific manner. To implement this ambitious programme, the state government made provision for 25% of the total programme cost to come from the state budget and the remaining 75% to come from central funding. The budget was created using funds of different schemes such as the Integrated Scheme of Oilseeds, Pulses, Oilpalm and Maize (ISOPOM), the National Food Security Mission (NFSM) and the Accelerated Pulse Production Programme (A3P), and additional funds required were met under the integrated agricultural extension system (IAES) and RKVY (Table 6.1).

For better monitoring, a three-tier decentralized system was adopted; namely, State Level Coordination Committee (SLCC), District Level Coordination Committee (DLCC) and Taluk Level Coordination Committee (TLCC). These committees meet regularly, review the progress and address the issues/concerns appropriately. Every Wednesday, a video conference is organized at the state level to review the programme, and higher authorities attend this conference regularly and take swift decisions to facilitate the programme.

Bhoochetana was implemented strategically in different phases over a period to make essential gains in the struggle for improved agricultural productivity, rural incomes and nutrition. The scaling-up process

Table 6.1. Source of funding, release and expenditure for the implementation of Bhoochetana.

Year	Source of funding[a]	Grant released (₹ in million)	Expenditure (₹ in million)
2009/10	RKVY	129.1	123.2
2010/11	RKVY	255.8	252.9
	IAES	50.0	46.4
2011/12	RKVY	450.0	428.6
	IAES	100.0	98.9
2012/13	RKVY	400.0	392.5
	IAES	675.0	674.0

[a]IAES, Integrated agricultural extension system; RKVY, *Rashtriya Krishi Vikas Yojana*.

in this particular project adopted a multi-level 'refinement strategy' to increase the effectiveness of technologies and reach a greater number of farmers. It is part of a broader process of innovation and learning. With effective monitoring and evaluation processes, the knowledge acquired from the initial year was used to scale up the model to create larger impacts in the entire state. The process occurred in an iterative and interactive cycle, as the experience from scaling up feeds back into new ideas and learning. The process adopted in Bhoochetana involves harnessing the potential of good partnership, political will, administrative support, science-backed capacity building, and regular and effective monitoring mechanisms (Fig. 6.3). This initiative also adopted innovative ideas of support services such as extension services to reach more farmers to create awareness about improved agricultural methods. It is a systematic approach followed to maximize the impact on the ground in Karnataka. This clearly brings out that Bhoochetana is a holistic approach adopted with the support from all stakeholders for benefiting the state as a whole and smallholders in particular.

6.5 High-level Policy Support

The project adopted a mission approach through convergence of various government programmes and a scheme implemented by a consortium consisting of different line departments of the Government of Karnataka along with academic institutions like the University of Agricultural Sciences located in Bengaluru, Dharwad, Raichur and Shivamogga and the international institution working in the area of dryland agriculture worldwide. There are several programmes within the DoA financed by both state and central sectors which were functioning independently. Under the umbrella of Bhoochetana, financial and human resources were converged and officers were allowed to utilize

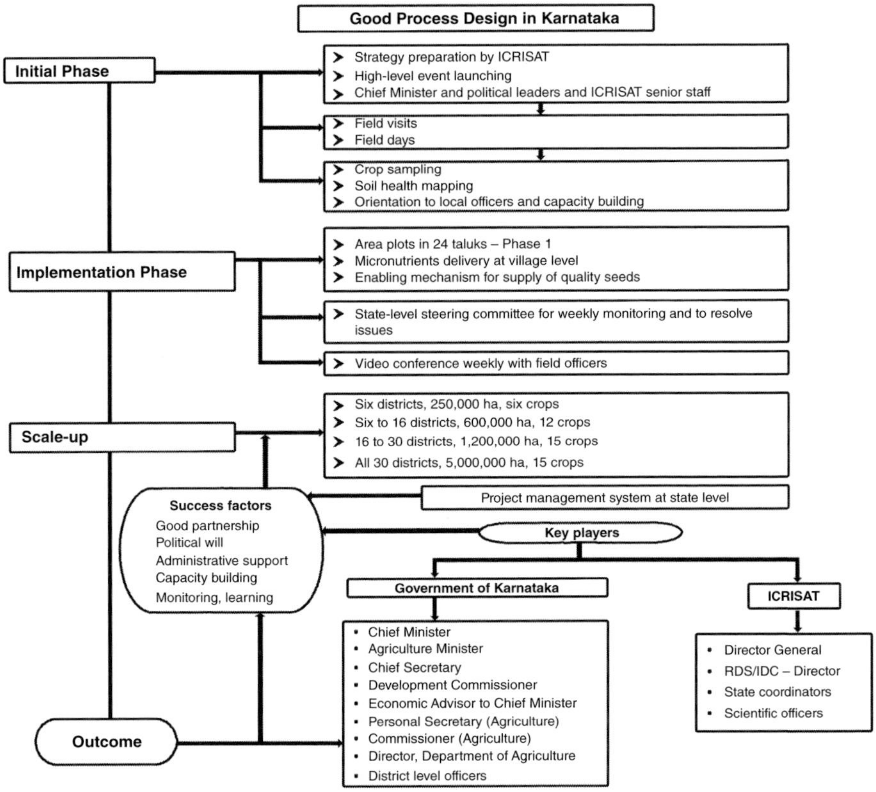

Fig. 6.3. Process of conceptualization, implementation and scaling up of Bhoochetana in Karnataka.

funds and human resources available from all programmes and meet the expenditure for implementing Bhoochetana as a package (Fig. 6.4). For better planning, execution and monitoring, the Government of Karnataka constituted a high-powered committee (SLCC) chaired by the Additional Chief Secretary and Development Commissioner. The committee reviews the performance of the project every fortnight. The project has been implemented in a mission mode and coordination at different levels starting with a cluster of villages in each taluk linking up with TLCCs, DLCCs and the SLCC. Communication was very regular and shared electronically (i.e. through e-mails) to speed up the processes at field level.

At the initial stage, policy makers as well as executing authorities at different levels contributed to the development, execution and implementation of the Bhoochetana strategy. At the policy-making and execution level, the Chief Minister, Chief Secretary, Additional Chief Secretary and Development Commissioner took interest and monitored

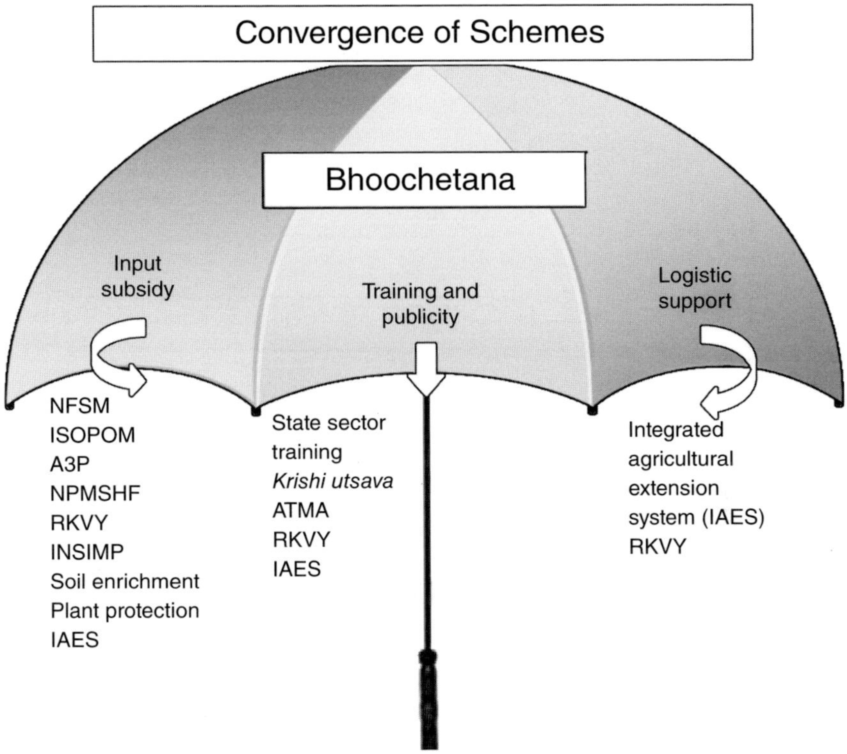

Fig. 6.4. Convergence of human and financial resources in Bhoochetana. A3P, Accelerated Pulse Production Programme; ATMA, Agriculture Technology Management Agency; IAES, integrated agricultural extension system; INSIMP, Initiatives for Nutritional Security through Intensive Millets Promotion; ISOPOM, Integrated Scheme of Oilseeds, Pulses, Oilpalm and Maize; NFSM, National Food Security Mission; NPMSHF, National Project on Management of Soil Health and Fertility; RKVY, *Rashtriya Krishi Vikas Yojana*.

the programme regularly for its success. Bhoochetana featured in the Chief Minister's budget speech continuously from 2010/11 to 2013/14 and is still continuing (Fig. 6.5). This kind of ownership at different levels has helped to mainstream the process of Bhoochetana implementation in the state. The high-level policy support enabled technocrats at different levels and ground-level implementing officers to take swift action to implement new ideas to benefit farmers in the state.

6.6 Effective Planning and Monitoring

As a strong monitoring mechanism provides strength to the system to upgrade and update changes required, it was thought that agencies

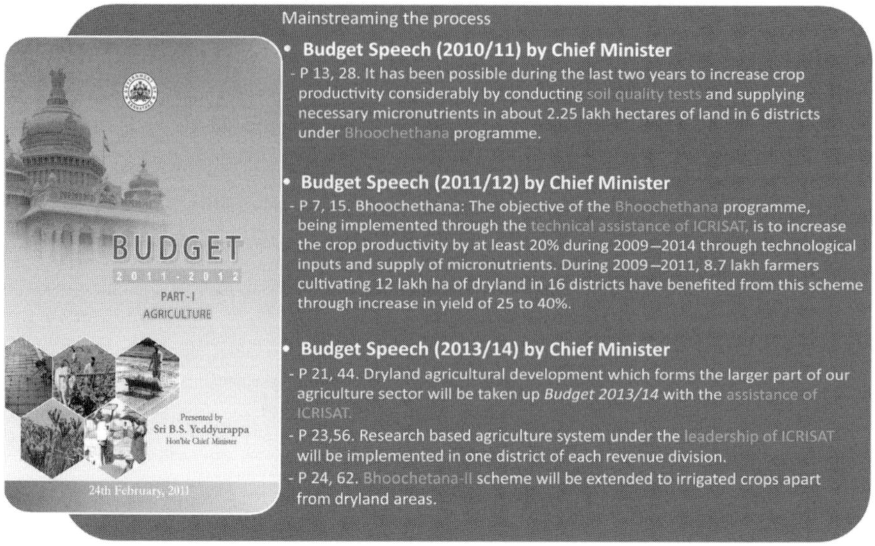

Fig. 6.5. Bhoochetana budget highlights of the Chief Minister of Karnataka.

need to use creative ways of monitoring in areas where no or limited access exists. One idea was to introduce video conferencing and ensure that all the concerned officials participated with updated information. Implementation of this kind of information communication technology-based monitoring mechanism is extensive and creative so as to triangulate information. Although these monitoring challenges exist in all parts of the state, the introduction of a large-scale productivity enhancement programme hopefully increased the minimum standards for monitoring and introducing new mechanisms that can be applied, and are required, for all service delivery functions in Karnataka. A performance-based ranking mechanism was adopted in Bhoochetana and every district is given a rank every month. It is a system for developing a healthy competitive environment in the programme and within the line department to achieve the goal of food security. This helps in maintaining the pace of work at ground level.

The SLCC reviews the progress of project activities and interacts with district-level officials instantaneously through the video conference and takes stock of solutions to address problems arising in the field and issue directives for each district (Fig. 6.6). The SLCC members attended DLCC meetings and visited fields along with the Joint Director of Agriculture (JDA) of the district to monitor and provide on-the-ground guidance to tackle problems in the district. High-level monitoring by senior officials helped in arranging inputs. This close monitoring also made this programme roll out at field level.

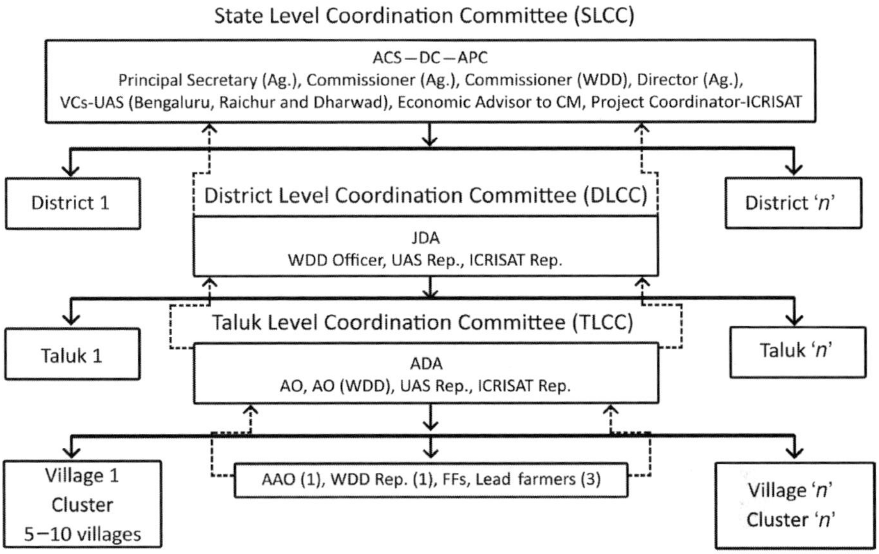

Fig. 6.6. Planning and monitoring mechanism adopted in Bhoochetana in Karnataka. AAO, Assistant Agriculture Officer; ACS, Additional Chief Secretary; ADA, Assistant Director of Agriculture; Ag., Agriculture; AO, Agriculture Officer; APC, Agriculture Production Commissioner; CM, Chief Minister; DC, Development Commissioner; FFs, farm facilitators; JDA, Joint Director of Agriculture; Rep., representative; UAS, University of Agricultural Sciences; VCs, Vice-chancellors; WDD, Watershed Development Department.

Monthly programme implementation calendar (MPIC) guidelines were developed for better monitoring of the programme. The MPIC aims at apportioning the budget allocation for a scheme based on physical activities during the 12 months of the financial year and thereby achieving satisfactory implementation of the state plan and non-plan schemes. Spending the allocation in the last month of the financial year can be prevented and the developmental objectives of the schemes can be achieved. The MPIC includes state-, district-, taluk- and Gram Panchayat-level activities.[1] It also provides space for monthly planning of the grant and its timely release. The document has defined output and outcomes linked with the plan. It also helps in review of the programme implementation at district level to the concerned Chief Executive Officer of Zilla Panchayat.[2]

The Bhoochetana programme was reviewed in special meetings in its initial stage. These review meetings were mainly to rectify field problems, taking policy decisions to implement the concept in a positive mode. The review mechanism also helped in building relationships between all the consortium partners. Review meetings were initially conducted fortnightly but later weekly meetings were required, so a nodal officers' meeting was organized on a Tuesday and video conferences on Wednesday. In these review meetings, other departmental programmes were also reviewed.

Regular monitoring through weekly review meetings, progress reports and field-level review helped to achieve impact on the ground. To reduce the transaction cost of administrative procedures as well as to fast track the progress and communicate the decisions to ground-level authorities, a weekly video conference on every Wednesday was organized and it is still continuing as an effective monitoring strategy. This review meeting aims to critically review administrative, financial, logistical and scientific issues with inputs from various stakeholders, including RSK-level agriculture officers (Fig. 6.7). With inputs from all stakeholders, issues are discussed and decisions are taken by high-level authorities such as the Additional Chief Secretary and Development Commissioner, Principal Secretary and Commissioner. It is important to mention that the Minister of Agriculture also attends these meetings periodically to update any development on the policy front and review the progress of the programme. It helps to boost the morale of the ground-level staff to undertake the implementation of activities that benefit the farmers. It is heartening to note that major policy decisions which have been taken in the weekly review meetings and SLCCs have

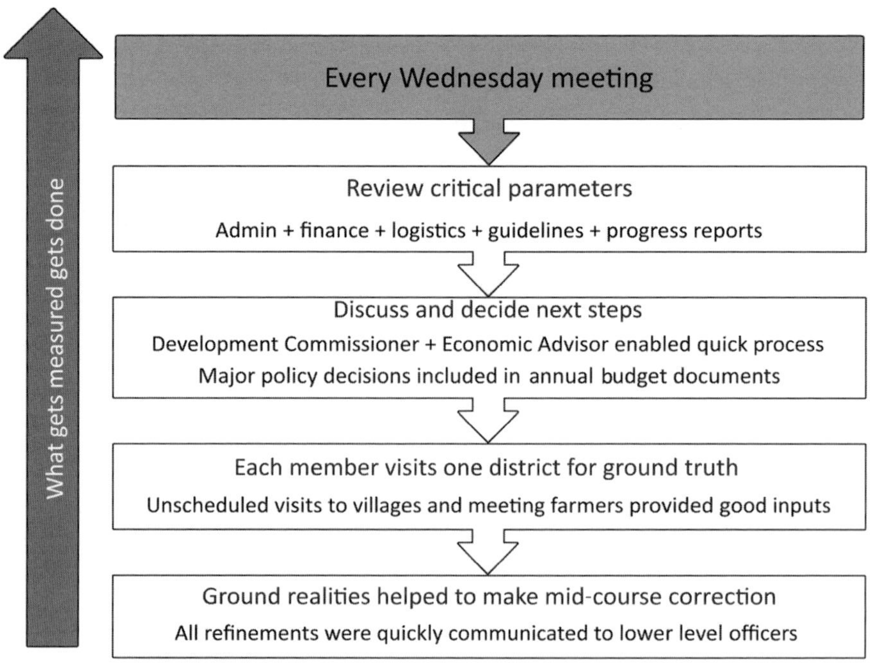

Fig. 6.7. Performance monitoring at different levels in Bhoochetana.

been included in the annual budget speech. This highlights the importance and quality of these meetings. Besides, unscheduled visits to villages by high-level authorities helped those in authority to understand the issues and farmers provided good inputs for modification of the programme. Therefore, discussion with farmers and different stakeholders helped those in authority to understand the realities on the ground which again facilitated the mid-course corrections and these changes have been communicated to lower level officers. Therefore, there is a proper communication channel established to link with high-level authorities.

6.7 Ensuring Availability of Micronutrients at Village Level

A total of 92,904 soil samples were collected from farmers' fields across all districts in Karnataka through a stratified soil sampling technique (Sahrawat *et al.*, 2008). This technique considered all factors (e.g. topography, soil colour, cropping system) that may lead to variable soil fertility and ensures representative sampling. But this brings in the scales of economy by sampling 20–25% villages in a target region and then the same percentage of farms within to represent a village. To inculcate among farmers a sense of ownership, we have adopted a participatory approach right from the beginning (more details in Chapter 2 of this book).

Based on analysis of a large number of soil samples, ICRISAT developed crop-wise nutrient recommendations at taluk level to guide farmers towards effective nutrient management. In order to meet increased demand for inputs, a systematic assessment is done twice a year, well before the commencement of the seasons. At the district level, the JDA assesses the changes in the cropping pattern and plans for inputs requirement based on the consumption pattern of different crops over the previous years. This helps in developing a realistic assessment of required inputs at district level. After assessment for every season, the demand of inputs is sent to the Government of India. In addition to supply from central government, the state government also tries to get additional inputs through private suppliers and cooperative institutions. About 50–60% of input supply in the state is made through private suppliers and the rest is through cooperative institutions (Wani *et al.*, 2013).

Proper and timely procurement is essential to ensure timely implementation of agriculture interventions at field level (Fig. 6.8). In Karnataka, procurement of inputs is part of an ongoing process. A tendering process is adopted for procurement of inputs. The DoA follows the norms of the Karnataka Transparency in Public Procurement Act, 1999. It is mainly to ensure a transparent procurement process within the government system. Requirement of inputs is based on the calculation of total demand submitted by the various wings of the department. A central rate contract is issued for all programmes. According to the proposed area to be adopted under Bhoochetana, the DoA made the arrangements for input supply. ICRISAT did the soil analysis and submitted the recommendations for application of nutrients, including micronutrients in different districts. This became the basis for planning of inputs. During the process of planning of inputs, achievable targets are set up to 50–60% of the total planned area demand (Wani *et al.*, 2013).

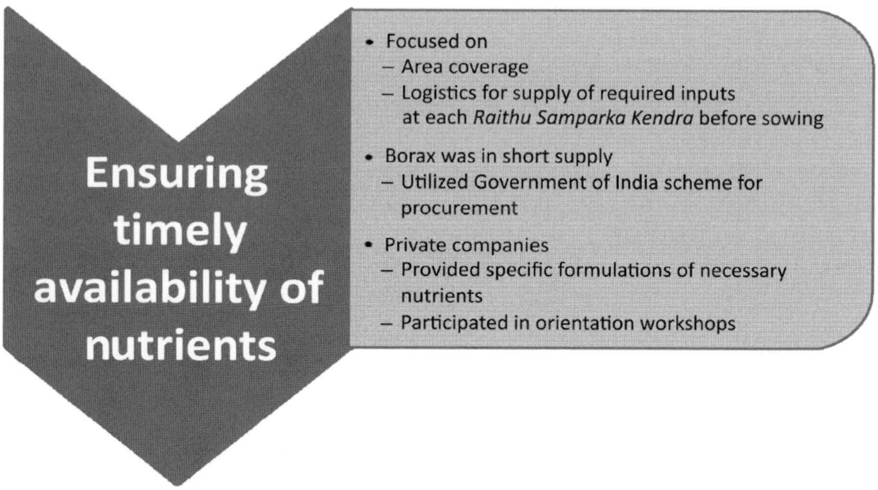

Fig. 6.8. Process of ensuring timely supply of inputs to farmers.

The Director (Agriculture) is the tender-inviting authority whereas the Commissioner is the tender-accepting authority in the state. A committee for issuing the tender is constituted, which looks into this matter at state level. After approval for tendering, the tender document is prepared and issued in the public domain such as a newspaper and a website. The peak period of input requirements is clearly mentioned in the document. Approval is taken from the government before issuing the tender. E-tendering is adopted in the state. The tender process is in December every year and rates are finalized in March.

Requirement for different inputs is compiled at district level according to the defined procedure and compiled at state level. Based on the demand, a pool of four to five suppliers is allocated for each district. Indents are given by the JDA. The Office of the JDA is responsible for placing the order to the supplier to ensure timely supply of micronutrients. This is mainly for major sources of micronutrient formulations such as gypsum, zinc sulfate and borax.

At cluster-village level, a unique arrangement has been made to store nutrients and the responsibility for this was assigned to FFs. Storage rooms were hired and this proved successful in getting the required nutrients to a large number of farmers. For monitoring quality of inputs supply, a vigilance committee was constituted by the department at district level. The Assistant Director of Agriculture heads the vigilance team and is supported by two agriculture officers. This team works with the JDA. This vigilance team ensures that all inputs are as per the prescribed quality. At times, the vigilance team takes samples and tests them in the laboratory. Most of the samples are of good quality. If any sample is not of good quality or as per the standard, the particular supplier has to replace the substandard material with good quality material.

6.8 Outcome of Good Governance: Institutional Building

The Bhoochetana initiative originated with the proposition of exploring new ways of thinking about interventions by enabling a better partnership process to promote agricultural development. Arising from good governance, the Bhoochetana programme had the following positive outcomes:

- Bhoochetana provided a platform for better resource allocation with added responsibility among partners. The convergence of programmes/schemes and knowledge was useful in allocating human as well as financial resources. All major programmes in the DoA have been converged and treated as a 'single file system' in Bhoochetana.
- The major proportion of the financial resource is from central government (75%) and the remaining share is from the state government (25%).
- A major lesson was research and development practitioners; line departments and non-state actors must be willing to work with emerging concepts and must recognize that interventions planned by them evolve as a part of the learning process.
- The partnership concept provides a framework for inclusive, knowledge-intensive agricultural development. But more focused, committed efforts are required to accomplish the goal of disseminating pro-poor technologies for small and marginal farmers for better impact.
- The soil diagnosis as an entry point activity showed the potential of rainfed agriculture in Karnataka. The state government has shown a keen interest in scaling up this initiative to the entire state to improve soil health and crop yield levels and income to millions of farmers.
- Dynamic and coordinated interaction among partners in a partnership system often was seen as the main driver of innovation for technology dissemination. The ability to respond quickly to

change is an increasingly important element of strong partnerships. Hence, capacity-strengthening interventions require a major focus on measures that foster strong patterns of interaction and build coordinated action to respond to continuously changing competitive and other challenges.

- New types of skills must be developed and partners need to learn from their own and others' experiences of coping with change in a highly uncertain environment, such as a contingency plan during drought situations. This effort may involve new initiatives and organizational processes that can promote knowledge management, and sharing and learning to respond to change effectively. In Bhoochetana, new ways of learning loop were developed and benefited a large number of stakeholders in the state (Fig. 6.9).
- The partnership helped in dealing with agricultural extension services by creating new institutional arrangements such as FFs, lead farmers and convergence and creation of the Bhoochetana cell in the state and this helped millions of small and marginal farmers. Since the inception of this initiative, FFs and lead farmers are the new extension agents effectively disseminating knowledge to the community, thereby creating a huge impact on the state's agricultural scenario.
- The importance of FFs in the extension system was realized. Thus, the partnership provided time for each partner to find their own place in the programme. As a result, this concept was adopted by other departments of the state government such as the Department of Horticulture and the Department of Sericulture to implement schemes such as *Suvarna Bhoomi Yojane* in the state.
- The Bhoochetana cell contributed to effective management of financial and administrative problems as well as smooth implementation, monitoring and evaluation.
- This innovative partnership has its own merits in terms of reaching out to poor farmers and helping them increase productivity through technology and institutional innovations. It has generated a stimulated discussion among policy makers and researchers to follow the process of formation of similar partnerships to improve the rural livelihoods of poor and vulnerable sections of society. However, more transparent and clear communication needs to be developed for a vibrant partnership, even at grass-root levels such as district, taluk and village levels.
- The success of Bhoochetana is the culmination of a variety of factors. First of all, the coordination and cooperation of diverse institutions under the umbrella of the consortium provided a strong foundation for the partnership, towards its implementation which aimed at bridging the large yield gaps that existed in the state. This has been accompanied by specific roles clarified and agreed by each institution to strengthen the process and to build the commitment towards a set goal.

- One of the features of effective partnership systems is the way organizations beyond the state play a proactive role in compartmentalized and rigidly defined roles. Often flexibility leads to innovations. However, some degree of monitoring is necessary to avoid overconfidence among partners.
- The monitoring and evaluation framework being used is one of the more thorough monitoring systems in place for agriculture service delivery in Karnataka. This needs further strengthening. An area that requires increased support is further professionalization of FFs, maybe through public–private partnerships.
- The impact of regular review meetings conducted by the DoA is visible and the sincere efforts of top-level officials has contributed largely to the success, as is evident from the results. To sustain this, planning and monitoring mechanisms need to be continued and further good practices need to be evolved and adopted.
- The partnership has explored new ways of operating an extension system, which is unique in its composition and functioning. It is essential that traditional extension systems are exchanged for this model, where research supports innovation at the local level.
- The important thing that has been learnt from the programme was that support for research systems needs to focus more on developing an interface with other sectors to achieve the desired growth in the agriculture sector. The mechanism governing the research system needs major attention, focusing on the ability and attitude of stakeholders required for engaging in partnerships. Attention needs to be given towards implementation of public awareness strategies through print and mass media, along with training and field-exposure activities. These types of changes are not necessarily expensive, but they are preconditions for effective investment in research and can contribute towards innovation.
- Similarly, extension investments should create the capacity to identify new, promising alternatives at the farm level and ensure that they are supported in the right way through engaging potential partners.

6.9 Conclusion

The success achieved within a short period is largely because of strong institutional arrangements and policy support received from higher level authorities, coupled with intense monitoring and guidance by the senior policy makers in the state (Fig. 6.10). Sincere efforts of the staff of the DoA throughout the state ensured the convergence of various government schemes through Bhoochetana. The capacity-building initiative has played an important role and institutional building was

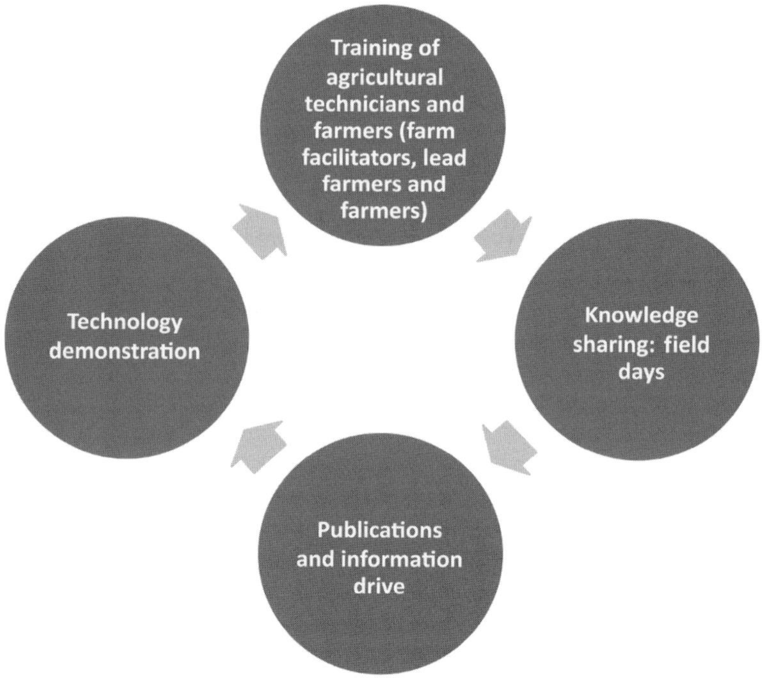

Fig. 6.9. Learning loop developed in Bhoochetana.

the keystone for the success of Bhoochetana. The concept of FFs and lead farmers in the programme was the first of its kind to meet the required extension help for a large number of farm families in rural areas. It is important to note that despite the existence of private and community-based extension services, public extension services remain important and are not likely to be replaced entirely. Therefore, it is recommended that the public extension system should be strengthened either through regular extension services or through a contract mode of extension services with control over its operations for enhancing the efficiency. Agricultural extension services ought to be upgraded to provide the education needed to modernize production practices and change past ways and perceptions of agriculture from being a means of subsistence into being a feasible business opportunity. The DLCCs and SLCC monitor the work progress periodically through video conferencing which keeps the pace of the programme and helps in executing the plans with the best team effort. Bhoochetana demonstrated that agricultural ventures are attractive, profitable and sustainable to induce economic growth. These major attributes can be achieved through scientific research to develop quality inputs, improve agronomic practices and develop good management skills with the support of appropriate institutional arrangements and policy guidelines.

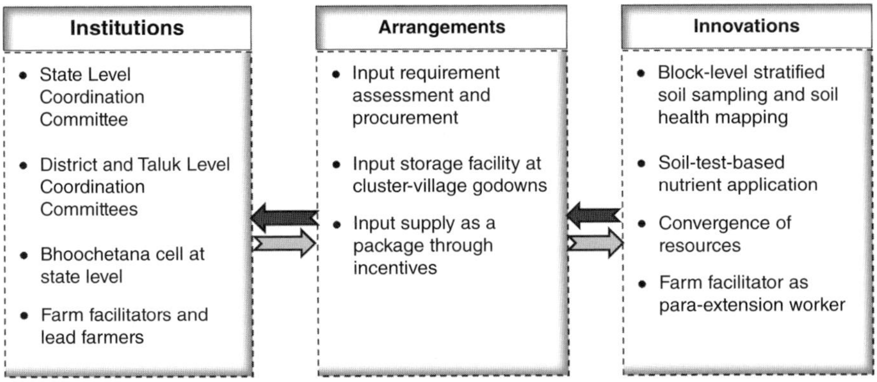

Fig. 6.10. Summary of institutional arrangements and innovations in Bhoochetana.

Notes

[1] Gram Panchayat means village council.
[2] Zilla Panchayat means district council.

References

Glendenning, C.J., Suresh Babu, C. and Asenso-Okyere, K. (2010) Review of agricultural extension in India: are farmers' information needs being met? *IFPRI Discussion Paper 01048*. International Food Policy Research Institute (IFPRI), Washington, DC.

Government of India (GoI) (2006) *National Commission on Farmers – Serving Farmers and Saving Farming – Towards Faster and More Inclusive Growth of Farmers' Welfare*. Ministry of Agriculture, GoI , New Delhi.

Sahrawat, K.L., Rego, T.J., Wani, S.P. and Pardhasaradhi, G. (2008) Stretching soil sampling to watersheds: evaluation of soil-test parameters in a semi-arid tropical watershed. *Communications in Soil Science and Plant Analysis* 39, 2950–2960.

Shambu Prasad, Laxmi, C.T. and Wani, S.P. (2006) *Institutional Learning and Change (ILAC) at ICRISAT: A Case Study of the Tata–ICRISAT Project*. Global Theme on Agroecosystems Report no. 19. International Crops Research Institute for the Semi-Arid Tropics, Patancheru, Andhra Pradesh, India.

Suresh Babu, C., Joshi, P.K., Glendenning, C.J., Asenso-Okyere, K. and Sulaiman, V.R. (2013) The state of agricultural extension reforms in India: strategic priorities and policy options. *Agricultural Economics Research Review* 26(2), 159–172.

Wani, S.P., Sandeep Khanwalkar, Krishnappa, K., Raju, K.V. and Sarvesh, K.V. (2013) *Bhoochetana: Process Documentation – A Program that Helped Farmers in Coming Out From Low Productivity and Poor Economic Status*. Resilient Dryland Systems Report no. 60. International Crops Research Institute for the Semi-Arid Tropics, Patancheru, Andhra Pradesh, India.

World Bank (2008) *World Development Report – 2008: Agriculture for Development*. World Bank, Washington, DC.

Climate Variability and Agriculture

7

A.V.R. Kesava Rao,* Suhas P. Wani and K. Srinivas

International Crops Research Institute for the Semi-Arid Tropics (ICRISAT), Patancheru, India

7.1 Introduction

Evidence over the past few decades has shown that significant changes in climate are taking place all over the world as a result of enhanced human activities in deforestation, emission of various greenhouse gases and indiscriminate use of fossil fuels. The results of climate change research indicate that climate variability and change may lead to more frequent weather-related disasters in the form of floods, droughts, landslides and sea level rise. Many countries, including India, are making efforts to undertake adaptation measures as well as to mitigate the challenges posed by global warming and climate change. There is an urgent need to develop a climate change network for Indian agriculture that will go a long way to build the resilience of the community to cope with the impacts of climate change, particularly in rainfed areas (Wani *et al.*, 2012).

In a state like Karnataka in South India, which has a spectrum of climates ranging from perhumid type in the coastal and Malnad region to arid type in the Ballari-Vijayapura (Bellary-Bijapur) region, the south-west monsoon rainfall is likely to be more uncertain, with both increasing and decreasing trends in different parts of the state. Surface air temperature and diurnal temperature ranges are likely

*k.rao@cgiar.org

to increase along the high ranges of the Western Ghats and under such conditions there is a threat to thermo-sensitive crops such as black pepper (*Piper nigrum*), cardamom (*Elettaria cardamomum*), tea (*Camellia sinensis*), coffee (*Coffea* spp.), cashew (*Anacardium occidentale*) and other plantation crops. A study indicated shifts in rainfall peaks by 2–3 weeks in some parts of Karnataka influencing the time of sowing and subsequent crop growth, necessitating shifts in crops and cropping patterns to match the modified rainfall regime. There are shifts in areas under different climates in the state. Frequent occurrence of droughts led to development of drought-tolerant cultivars to sustain agricultural production over the state. Unlike in other states, forest cover in Karnataka is increasing, which is a positive sign. The Karnataka State Action Plan on Climate Change was developed in 2012; it discusses climate trends, projected vulnerabilities along with adaptation and mitigation priorities for various sectors (EMPRI and TERI, 2012).

Karnataka has about 8.2 million ha of rainfed area out of the total cultivated area of 12.3 million ha; thus it has the second largest area under rainfed agriculture in the country after Rajasthan. Crop yields in dryland areas are quite low (1–1.5 t/ha) and are two- to five-fold lower than the yield from researchers' managed plots (Bhatia *et al.*, 2006). Current rainwater use efficiency in dryland agriculture varies between 35% and 45% and the vast potential of rainfed agriculture could be unlocked by using available scientific technologies, including improved cultivars. The vast opportunities existing in dryland areas can be harnessed for improving rural livelihoods (Wani *et al.*, 2009). The project 'Bhoochetana: Mission to Enhance Productivity of Rainfed Crops in Karnataka' was implemented during 2009–2012 by a consortium led by the International Crops Research Institute for the Semi-Arid Tropics (ICRISAT) with partners including government line departments, state agricultural universities, community-based organizations and others. The mission of the Bhoochetana project was to increase the average productivity of selected crops in the selected districts by 20% in 4 years. The Results-Framework Document for the Department of Agriculture (2013–2014) describes the vision of Karnataka as to achieve the targeted growth rate of 4.5% in the agriculture sector by enhancing agricultural production and productivity (Government of Karnataka, 2013).

An attempt is made here to understand climate variability and change occurring over Karnataka with particular reference to the 3-year period 2011–2013, during which large changes in rainfall occurred in the state. These results would help formulate climate-change risk management strategies for bringing in resilience to agriculture in the state.

7.2 Agroecology of Karnataka

7.2.1 Agroclimate

Karnataka lies between 11.5° and 18.5° N latitudes and 74.0° and 78.5° E longitudes. The state has three principal geographical zones: (i) the coastal region; (ii) the hilly region known as Malnad; and (iii) the Bayaluseeme region comprising the plains of the Deccan plateau. Meteorologically the state is divided into Coastal Karnataka, North Interior Karnataka and South Interior Karnataka. According to the Government of Karnataka, the state comprises ten agroclimatic zones (Government of Karnataka, 2014) (Fig. 7.1 and Table 7.1) based on rainfall amount and distribution, soil type, texture, depth and physicochemical properties, elevation, topography, major crops and type of vegetation.

7.2.2 Agroecological subregions

Knowledge of the date of the onset of rains will help in better planning agricultural operations, particularly land preparation and sowing. The length of the rainy season is the duration between the onset and end of agriculturally significant rains. The length of growing period (LGP) is defined as the length of the rainy season plus the period for which the soil moisture storage at the end of the rainy season and the post-rainy season and winter rainfall can meet the crop water needs. Therefore, the LGP depends on not only the rainfall distribution but also the soil type, soil depth, water retention and release characteristics of the soil. Thus, the LGP can be used as a parameter for delineating agroecological regions.

The National Bureau of Soil Science and Land Use Planning (NBSS&LUP), Nagpur has delineated India into 20 agroecological regions by superimposing bioclimate and LGP onto the soil-scape (Sehgal *et al.*, 1992). The LGP is an indicator of crop production because the soil–water balance, which is a direct function of moisture availability in a landform, is considered rather than total rainfall. In addition, the LGP allows for computing a modification in the quantum of residual moisture when the soil depth is less than 100 cm.

Realizing the importance of a narrow LGP interval of 30 days for diverse crop suitability and also the need for further subdivisions of bioclimate and some important soil quality parameters such as depth and available water capacity (AWC), the NBSS&LUP divided the 20 agroecological regions into 60 agroecological subregions (AESRs) (Velayutham *et al.*, 1999).

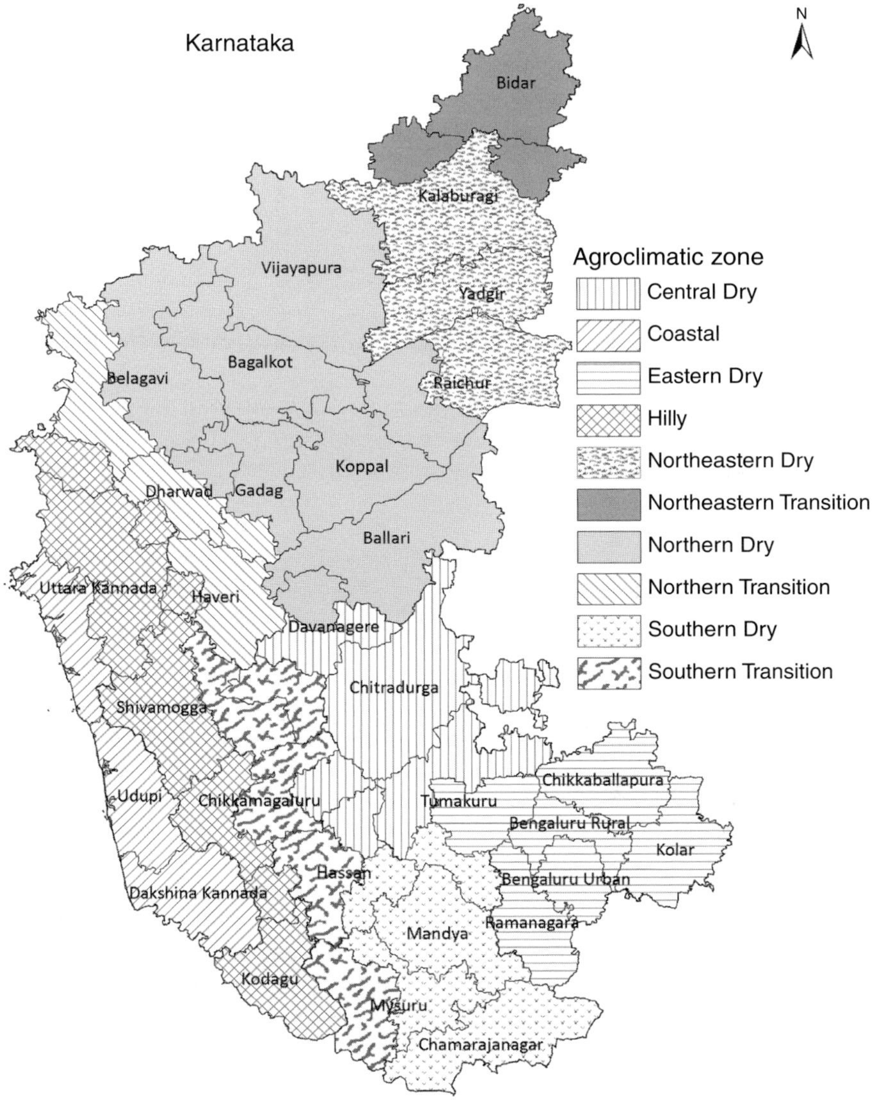

Fig. 7.1. Agroclimatic zones of Karnataka.

Karnataka is divided into ten AESRs (Fig. 7.2) considering physiographic features, soil characteristics, bioclimatic types (rainfall, potential evapotranspiration, soil storage) and LGP. These are as follows:

AESR 3: Deccan plateau region with arid (typic) climate – Mixed red and black soils having shallow depth; low to medium AWC of 50–150 mm; and LGP ranges from 60 to 90 days.

AESR 6.1: Deccan plateau region with semi-arid (dry) climate – Shallow black soils (with medium and deep black soils as inclusion) having

Table 7.1. Agroclimatic zones of Karnataka.

Climate zone	Soil	Rainfall (mm)	Area (million ha)	No. of taluks	Major crops
Northeastern Transition Zone	Shallow to medium black	830–890	0.871	7	Pulses, sorghum, oilseeds, pearl millet, cotton, sugarcane
Northeastern Dry Zone	Deep to very deep black clay	630–810	1.762	11	Pearl millet, pulses, oilseeds, cotton and post-rainy sorghum
Northern Dry Zone	Shallow to deep black clay	460–790	4.78	35	Maize, pearl millet, groundnut, cotton, wheat, sugarcane, tobacco and post-rainy sorghum
Central Dry Zone	Shallow to deep black soils	450–720	1.943	17	Finger millet, sorghum, pulses, oilseeds
Eastern Dry Zone	Loamy	680–890	1.808	24	Finger millet, rice, pulses, maize, oilseeds
Southern Dry Zone	Red sandy loam	670–890	1.739	18	Rice, finger millet, pulses, sorghum, tobacco
Southern Transition Zone	Sandy loam	610–1050	1.218	14	Rice, finger millet, pulses, sorghum, tobacco
Northern Transition Zone	Shallow to medium black clay and red sandy	620–1300	1.194	14	Rice, sorghum, groundnut, pulses, sugarcane, tobacco
Hilly Zone	Red sandy loam	900–3700	2.56	22	Rice, pulses
Coastal Zone	Red lateritic and coastal alluvial	3010–4700	1.167	13	Rice, pulses, sugarcane

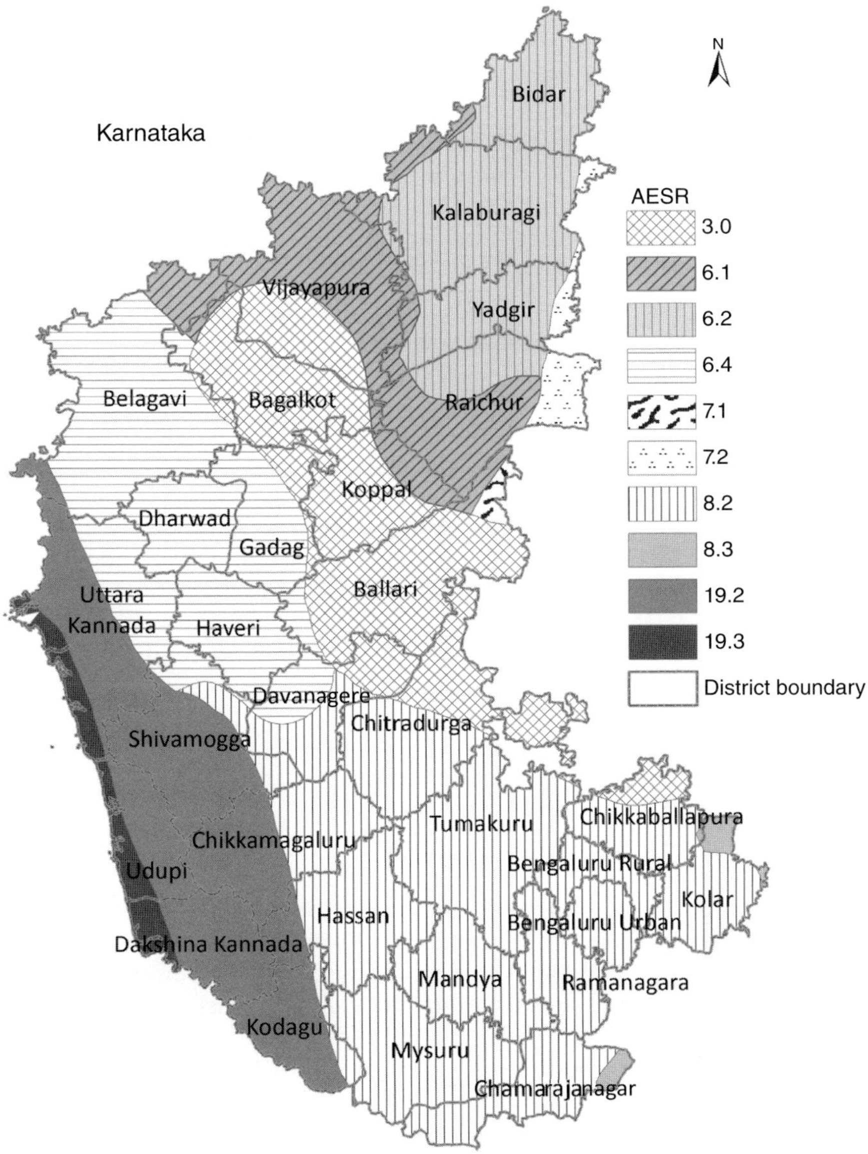

Fig. 7.2. Agroecological subregions (AESRs) in Karnataka.

shallow to medium depth; LGP is 90–120 days; and AWC is medium to high (100–200 mm).

AESR 6.2: Deccan plateau region with semi-arid (moist) climate – Shallow black soils (with medium and deep black soils as inclusion) having shallow to medium depth; LGP is 120–150 days; and AWC is medium to high (100–200 mm).

AESR 6.4: Deccan plateau region with subhumid (dry) climate – Shallow black soils (with medium and deep black soils as inclusion) having shallow depth; LGP is 150–180 days; and AWC is medium to high (100–200 mm).
AESR 7.1: Deccan plateau region with semi-arid (dry) climate – Mixed red and black soils having shallow to medium depth; LGP is 90–120 days; and AWC is medium (100–150 mm).
AESR 7.2: Deccan plateau region with semi-arid (moist) climate – Mixed red and black soils having deep depth; LGP is 120–150 days; and AWC is medium to very high (100 mm to > 200 mm).
AESR 8.2: Deccan plateau region with semi-arid (moist) climate – Soils are red loamy having medium to deep depth; LGP is 120–150 days; and AWC is low (50–100 mm).
AESR 8.3: Eastern Ghats and Tamil Nadu Uplands region with semi-arid (moist) climate – Soils are red loamy having deep depth; LGP is 120–150 days; and AWC is low (50–100 mm).
AESR 19.2: Western Ghats with subhumid (moist) climate – Soils are red and lateritic having deep depth; LGP is 210–240 days; and AWC is low to medium (50–150 mm).
AESR 19.3: West Coastal Plain region with perhumid (humid) climate – Soils are coastal and deltaic alluvium-derived having deep depth; LGP is 240–270 days; and AWC is low to medium (50–150 mm).

7.3 Climate Variability and Change in Karnataka

7.3.1 Rainfall and temperature variability

Several studies reveal that there is a shift in rainfall over Karnataka during the main crop season. The shift in rainfall and its variability over a period of time led to a change in cropping systems in different regions of Karnataka. Unlike in other states, forest cover is increasing in Karnataka. An increase in temperature and a decline in rainfall were noticed over different regions of the state. Several climate projections indicate a general increase in temperature and rainfall over India; however, in parts of Karnataka there is an increase in temperature and a decline in rainfall.

Devappa and Khageshan (2011) reported a decreasing trend in the annual rainfall at 3.44 mm/year for Kalaburagi (previously known as Gulbarga) district in Karnataka, based on data for 1961–2008. Variations in rainfall amount and distribution, increased temperatures, depleting soil productivity and disturbing water balance are affecting pigeon pea (*Cajanus cajan*) productivity in Kalaburagi. The climate of Kalaburagi was almost stable in the semi-arid type except for a few years when it changed to dry subhumid and arid types of climate (Kesava Rao *et al.*, 2013b). Analysis of seasonal rainfall indicated that no significant trend exists in south-west monsoon rainfall during the period 1969–2009. However, there is an increasing trend in temperature, particularly in winter. The maximum temperature during the winter (November–February) in Kalaburagi shows a statistically significant increasing trend. During the period 1950–1990, the average rise in annual temperature was 1.3°C in Karnataka (Rajegowda, 1992).

In Karnataka as a whole, annual rainfall decreased during 1951–2000. The mean annual rainfall of the state for the period from 1901 to 2000 indicated a definite declining trend (Panduranga *et al.,* 2006). A time series of the annual rainfall of Karnataka shows a cycle of 16 years in which the first half of the cycle received less than the normal rainfall for the period from 1950 to 1958 and the second half of the cycle received more than normal for the period from 1959 to 1964. A few districts (Bengaluru, Kolar and Tumakuru (Tumkur)) in Karnataka show an increasing trend in the annual rainfall. The mean annual rainfall values for the period from 1901 to 1950 for Bengaluru, Kolar and Tumakuru are 867 mm, 745 mm and 688 mm, respectively, and for the period from 1951 to 2006 are 883 mm, 767 mm and 730 mm, respectively.

The India Meteorological Department (IMD) has identified three meteorological subdivisions in Karnataka as: (i) Coastal Karnataka; (ii) North Interior Karnataka; and (iii) South Interior Karnataka. Long-period monthly rainfall data for 142 years (1871–2012) was downloaded from the Indian Institute of Tropical Meteorology, Pune (Indian Institute of Tropical Meteorology, 2013). Annual values were compiled and decadal averages worked out to study the variability in decadal rainfall (Fig. 7.3). There is an increasing trend in rainfall from 1911–1920 to 1971–1980 in Coastal Karnataka meteorological subdivision. In the past three decades, however, a slight decreasing trend is observed. In North Interior Karnataka, an increasing trend is seen during 1921–1930 to 1951–1960; in the later decades the annual rainfall decreased from about 950 mm to 825 mm. No such trends are seen for South Interior Karnataka.

The Eastern Dry agroclimatic zone consists of Bengaluru and Kolar districts and parts of Tumakuru district, which is also known as the tank-fed region. Rajegowda *et al.* (2000) have shown that there is a predominant shift in the initiation and termination of rainfall to supply adequate moisture for the crop growing period. This shift has been observed after 1990 and their mean monthly values also have changed. Before 1990, the annual rainfall ranged from 619 mm to 1119 mm with a mean of 869 mm. After 1990, the annual rainfall ranged between 611 mm and 1311 mm with a mean of 1011 mm. Before 1990, on average, peaks were observed during May, July and September while after 1990, peaks were observed during May, August and October. There is a perceptible shift in rainfall pattern from July to August and also from September to October in this agroclimatic zone.

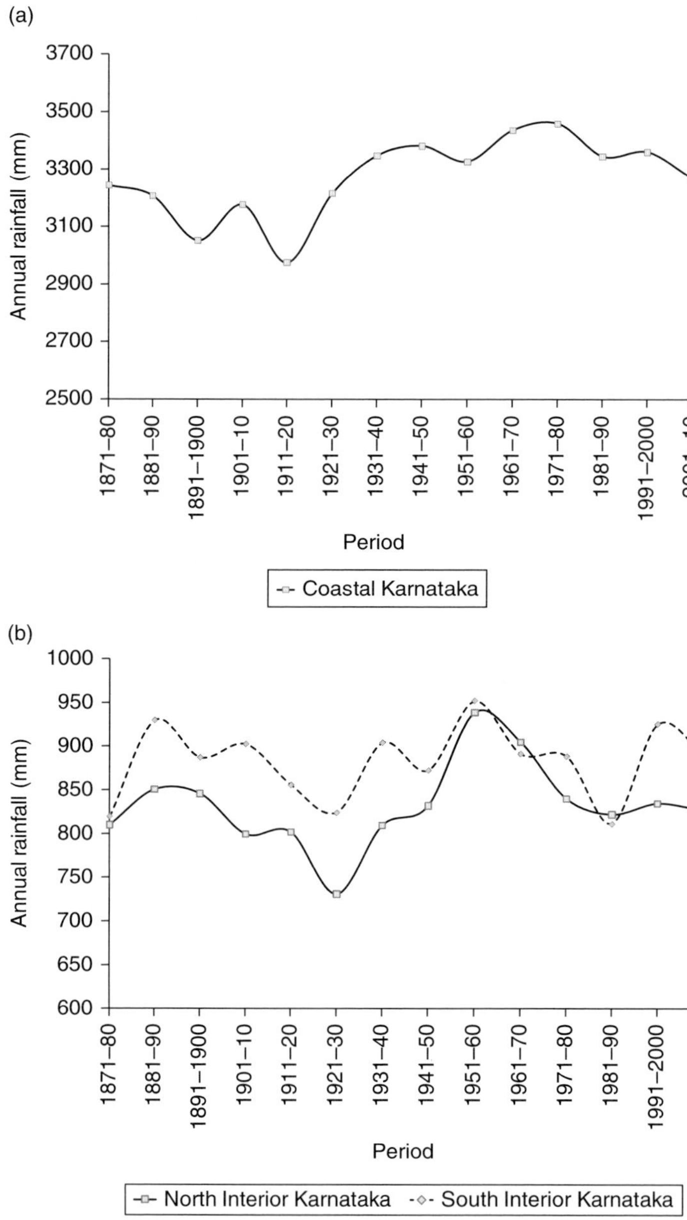

Fig. 7.3. Variability in decadal rainfall in three regions of Karnataka. (a) Coastal Karnataka; (b) North Interior Karnataka and South Interior Karnataka.

Crops sown during the July rains reached the grand-growth period, namely, flowering to grain formation stage (long-duration varieties of about 115 days) during September with the highest rainfall until 1990, so there was no moisture stress during the grand-growth period. After 1990, as a result of reduced rainfall in July and September, the crops could not be sown during July, though June rainfall enabled land preparation. Even with scanty rains, if sowing is done during July, the crop would suffer from moisture stress due to the reduction in rainfall during September and also the crop grown would be caught in the October rains, causing considerable loss in grain yield. This analysis reveals that in the Eastern Dry agroclimatic zone, sowing of crops (long-duration varieties of about 115 days) could be done during August, preparing the land using the June and July rains. In the years of early onset of the south-west monsoon, sowing can be recommended also during the last week of July. Crops sown during August would reach the grand-growth period in October. As October receives higher rainfall the crop in its grand-growth period would not suffer for want of moisture and higher crop yields are expected. Crops sown after August may not be able to complete the life cycle as a result of inadequate moisture availability after the second fortnight of November (in the event of the intensity of the north-east monsoon being low) as crop maturity coincides with this period. Under such circumstances, the short-duration crop varieties have to be preferred.

In the northern districts of Karnataka, annual rainfall variability analysis based on 100 years' data (1901–2000) indicated a periodicity of a 13–17 years' cycle. There has been a general decrease in rainfall in September and a large increase in October over Northern Karnataka during the 20th century. A corresponding decrease in maximum temperature during October was noticed. Venkatesh *et al.* (2008) observed increased temperatures in November and December, which indicated availability of higher thermal energy for better vegetative growth during November, while there was greater thermal stress during the flowering period in December for the post-rainy season sorghum (*Sorghum bicolor*) crop.

Rainfall analysis of the Sirsi region of Karnataka based on data of 48 years (1955–2002) indicated that the monsoon season accounted for 88% (2170 mm) of annual rainfall with a coefficient of variation of 27% (Manjappa and Kelaginamani, 2005). There were 11 deficit rainfall years (23%), 28 normal rainfall years (58%), 8 excess rainfall years (17%) and 1 wet year (2%) with a coefficient of variation of 24%.

Singh *et al.* (2014) studied climate change variability in Coastal Karnataka by using the IMD daily rainfall data for the period of 1971–2005 and Climatic Research Unit (CRU), UK data for temperature for the period 1901–2002. Coastal Karnataka districts showed a trend of decreasing precipitation, with an average rainfall of > 25 mm/day. Rainfall has decreased by 17.7, 1.9 and 22.4 mm/day/100 years in the districts Mangaluru (Mangalore), Uttara Kannada and Udupi, respectively. The increase in minimum temperature is 0.14, 0.15 and 0.11°C/100 years, whereas the increase in maximum temperature is 0.15, 0.15 and 0.11°C/100 years in the districts Mangaluru, Uttara Kannada and Udupi, respectively.

Biradar *et al.* (2012) studied the annual and seasonal rainfall variability of Aurad taluk in Bidar district by using daily rainfall data of a 35-year period from 1976 to 2010. Pre-monsoon season rains were at a deficit in 16 years, in excess in 13 years and normal in 6 years, as compared with the normal rainfall of 69 mm. The south-west monsoon rainfall was normal in 4 years, in excess in 13 years and at a deficit in 18 years, while post-monsoon season rainfall was only normal in 2 years, at a deficit in 18 years and in excess in 15 years.

Aavudai Anandhi (2010) reported uncertainty over the season length in Karnataka due to the choice of scenarios, season type and number of seasons. Based on the type of season, the monthly sequences of variables (predictors) were selected from datasets from the National Centers for Environmental Prediction (NCEP) and Canadian General Circulation Model (CGCM3). Results of cluster analysis revealed an increase in the length of the average wet season in A2, A1B and B1 scenarios towards the end of the 21st century. No change in average warm and cold season length was observed across the four scenarios considered.

In the Upper Krishna Project area of Karnataka, the temperature trend line for the period of 1990–2003 showed that the mean annual maximum temperature has increased by 0.0640°C/year and the mean annual minimum temperature has increased by 0.010°C/year (Shashidhara and Reddy, 2012).

7.3.2 Changes in areas under climate types

It is evident from the various studies that climate change in India is real and it is one of the major challenges faced by Indian agriculture, more so in the semi-arid tropics (SAT) of the country. India ranks first among the countries that practise rainfed agriculture in terms of both extent and value of production. The rainfed agroecologies cover about 60% of the net sown area of 141 million ha and are widely distributed in the country (DOAC, 2011). Karnataka has the second largest area under rainfed agriculture after Rajasthan in the country and a reduction in yields due to climate change is likely to be more prominent in rainfed agriculture and under limited water availability. Changes in arid and semi-arid areas in India with associated shifts during 1971–2004 were reported earlier (Kesava Rao *et al.*, 2013a). Climate change is a major issue for sustainable agriculture and thus there is a need to review the areas under the different climatic zones in Karnataka to understand the changing rainfall and temperature patterns over the past few decades.

A new daily gridded rainfall data set (IMD 4) at a high spatial resolution (0.25° × 0.25°, latitude × longitude) covering a long period of 110 years (1901–2010) over the Indian mainland was developed (Pai *et al.*, 2014). For preparing the new gridded data, daily rainfall records from 6995 rain-gauge stations in India were used, the highest number of stations ever used in any study for such a purpose. Various standard quality control tests were applied on the data before the interpolation of the station rainfall data on to fixed spatial grid points. The comparison of IMD 4

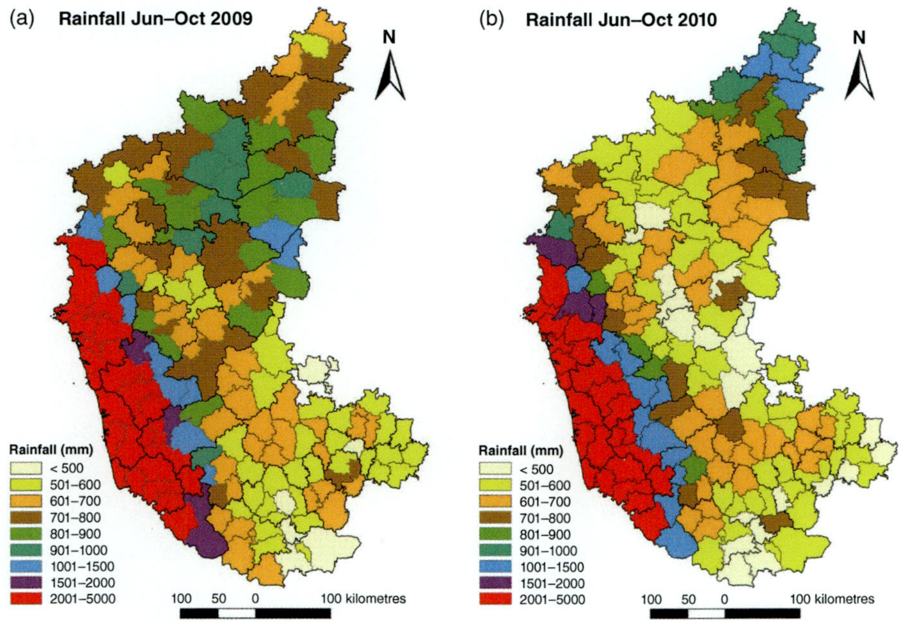

Plate 1. Spatial and temporal variability of rainfall during the monsoon period between 2009 and 2012 in different taluks across Karnataka. Rainfall received between June and October in (a) 2009; (b) 2010; (c) 2011; and (d) 2012.

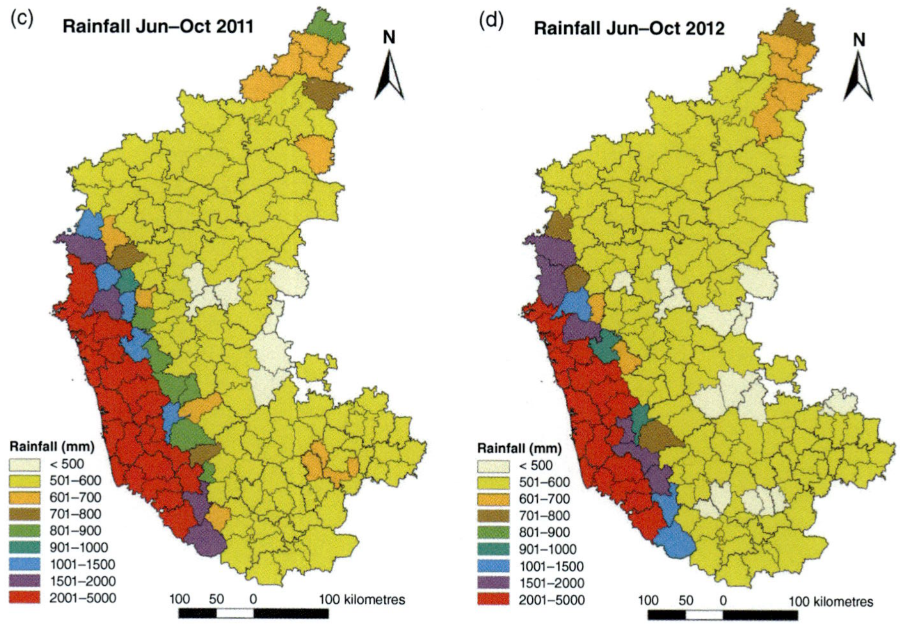

Plate 1. Continued.

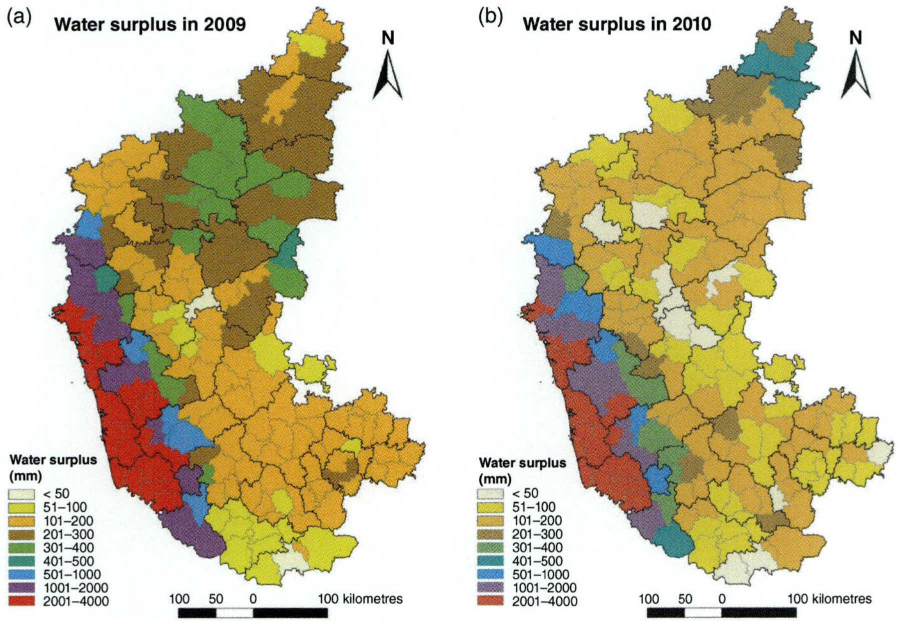

Plate 2. Spatial and temporal variability of surplus water (a–d), evapotranspiration (ET) (e–h) and available soil moisture content (SMC) (i–l) at the end of the monsoon period during 2009–2012 in different taluks across Karnataka.

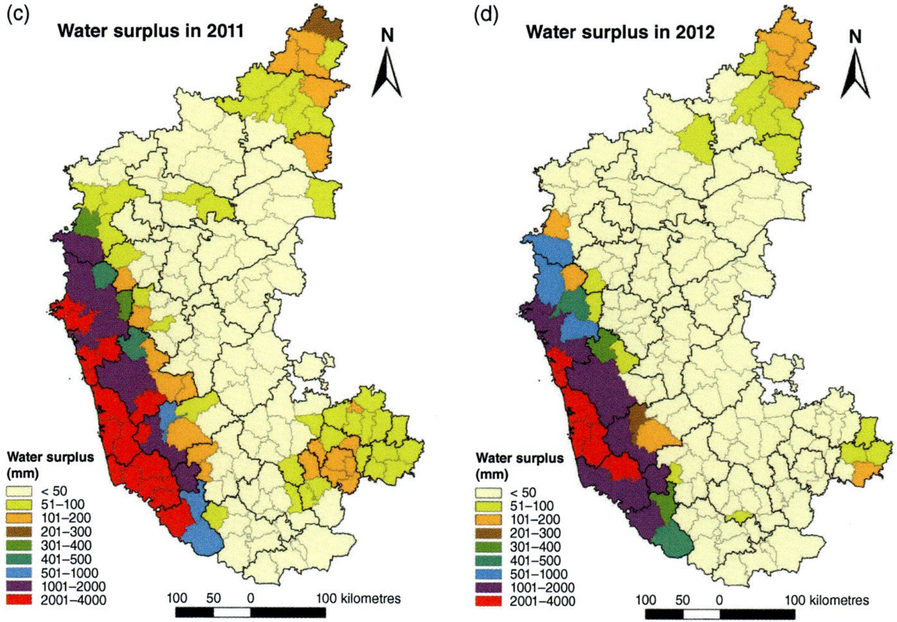

Plate 2. Continued.

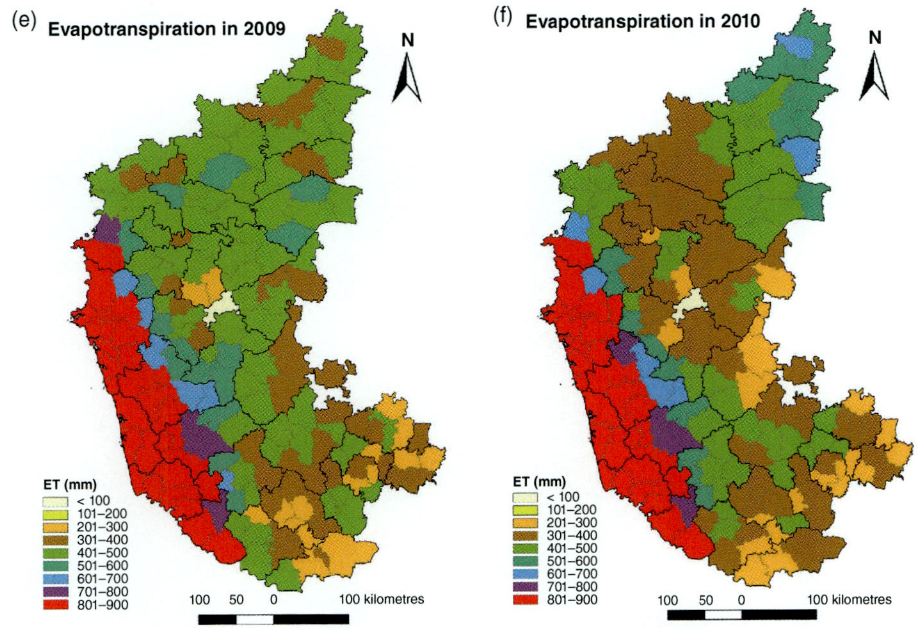
(e) Evapotranspiration in 2009
N
ET (mm)
< 100
101–200
201–300
301–400
401–500
501–600
601–700
701–800
801–900
100 50 0 100 kilometres
(f) Evapotranspiration in 2010
N
ET (mm)
< 100
101–200
201–300
301–400
401–500
501–600
601–700
701–800
801–900
100 50 0 100 kilometres

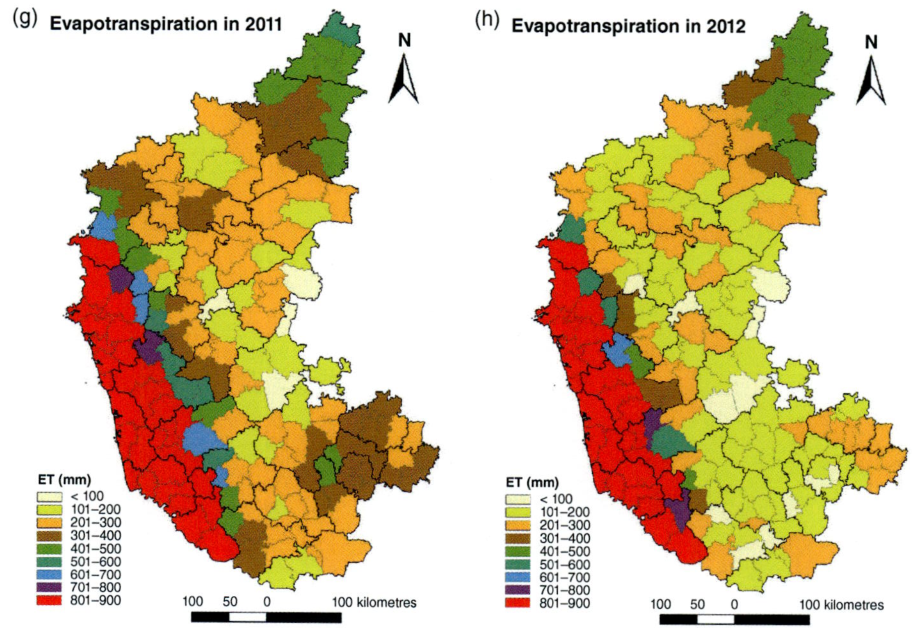

Plate 2. Continued.

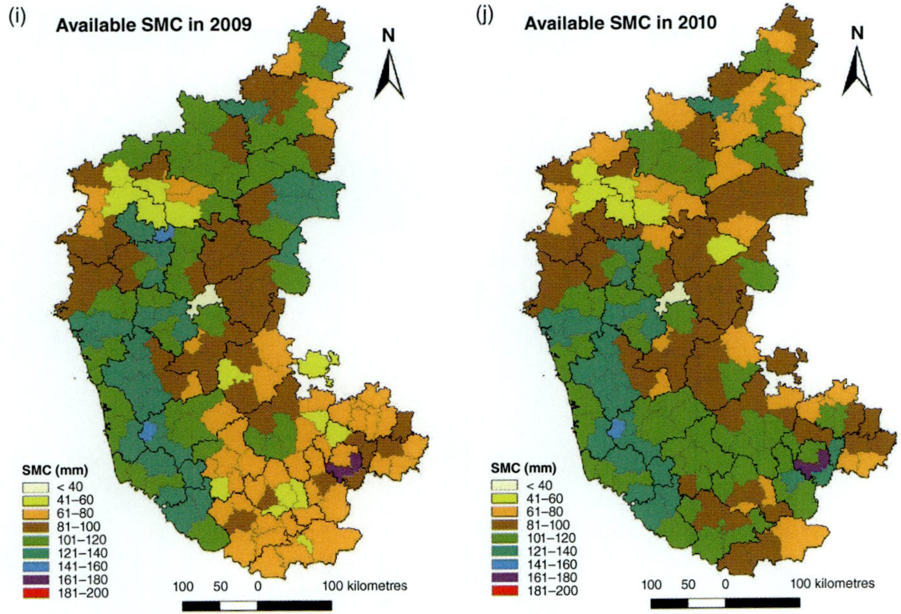
(i)
Available SMC in 2009
(j)
Available SMC in 2010
N
SMC (mm)
< 40
41–60
61–80
81–100
101–120
121–140
141–160
161–180
181–200
100 50 0 100 kilometres

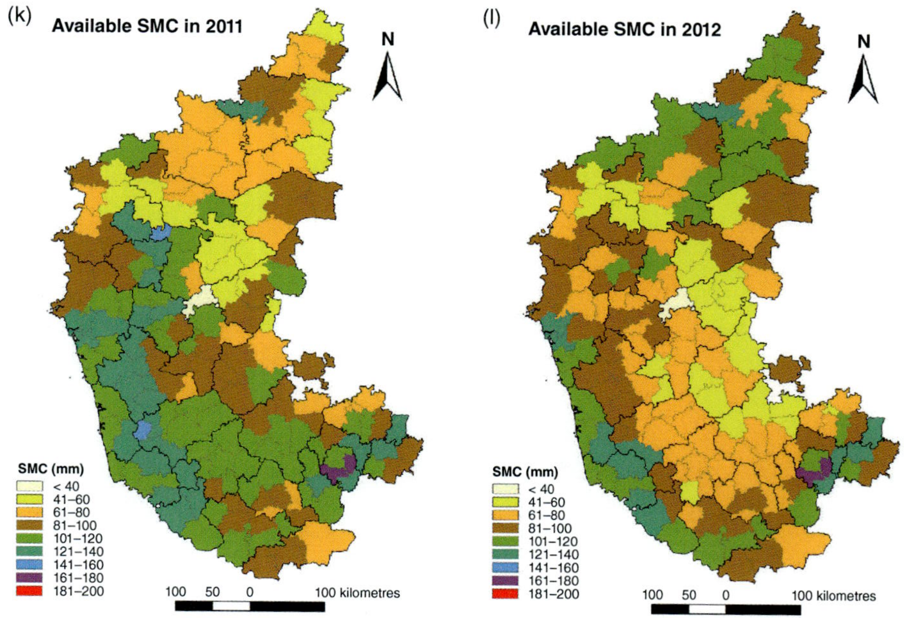

Plate 2. Continued.

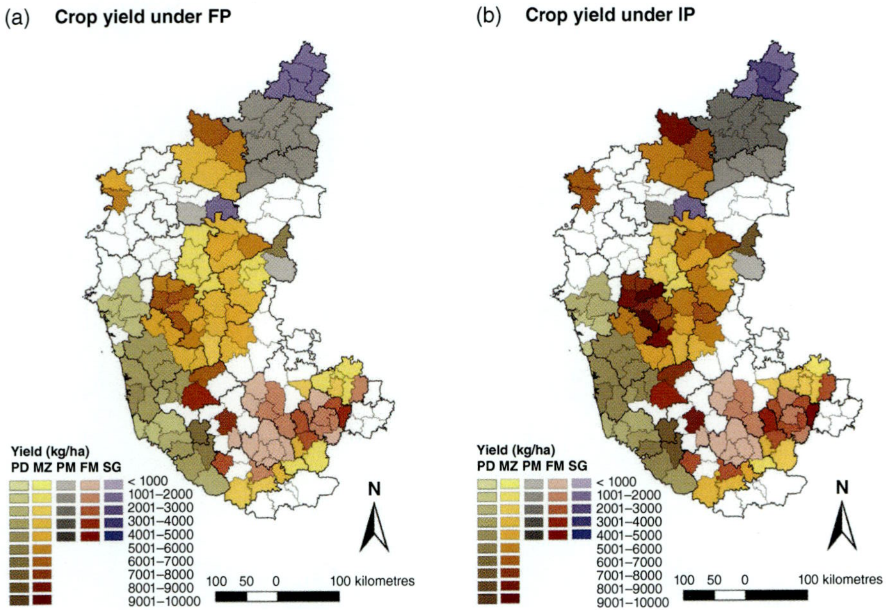

Plate 3. Spatial variability in: (a) crop yield under farmers' practice (FP); (b) crop yield under improved practice (IP); (c) additional net income due to IP; and (d) benefit–cost (BC) ratio of IP for selected cereals across taluks in Karnataka during 2012. Cereals: PD, paddy; MZ, maize; PM, pearl millet; FM, finger millet; SG, sorghum.

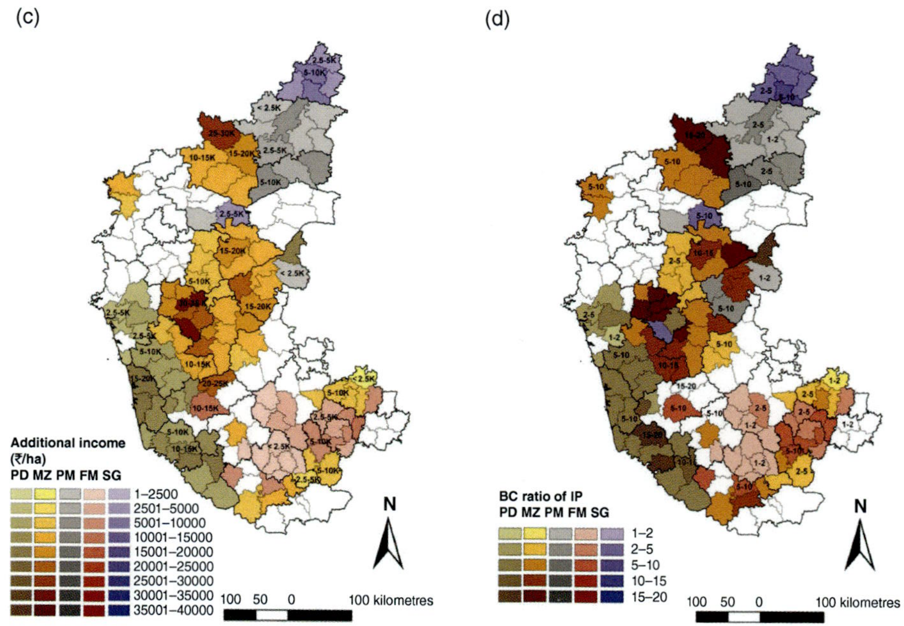

Plate 3. Continued.

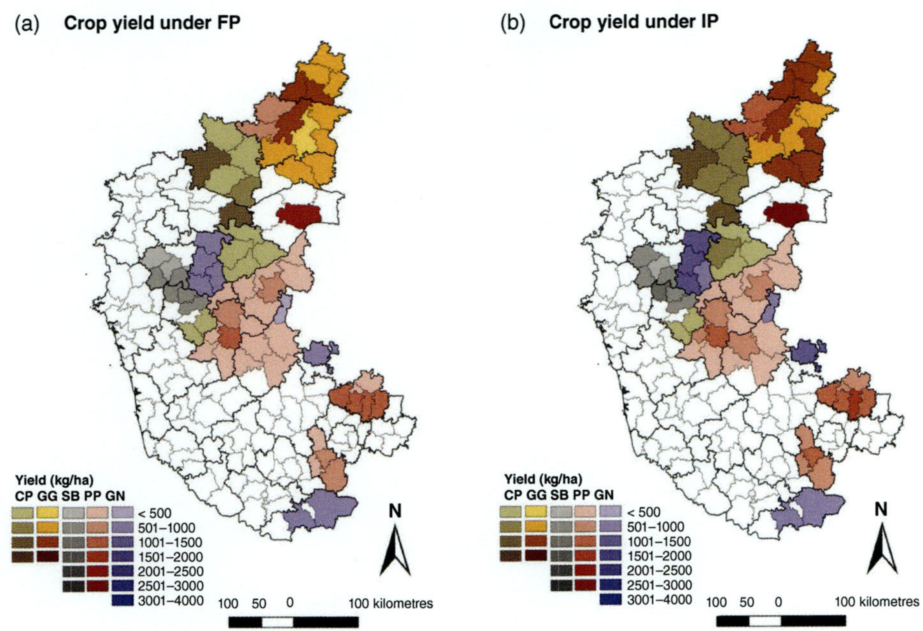

Plate 4. Spatial variability in: (a) crop yields under farmers' practice (FP); (b) crop yield under improved practice (IP); (c) additional net income due to IP; and (d) benefit–cost (BC) ratio of IP for selected pulses and oilseeds across taluks in Karnataka during 2012. Pulses: CP, chickpea; GG, green gram; PP, pigeonpea. Oilseeds: SB, soybean; GN, groundnut.

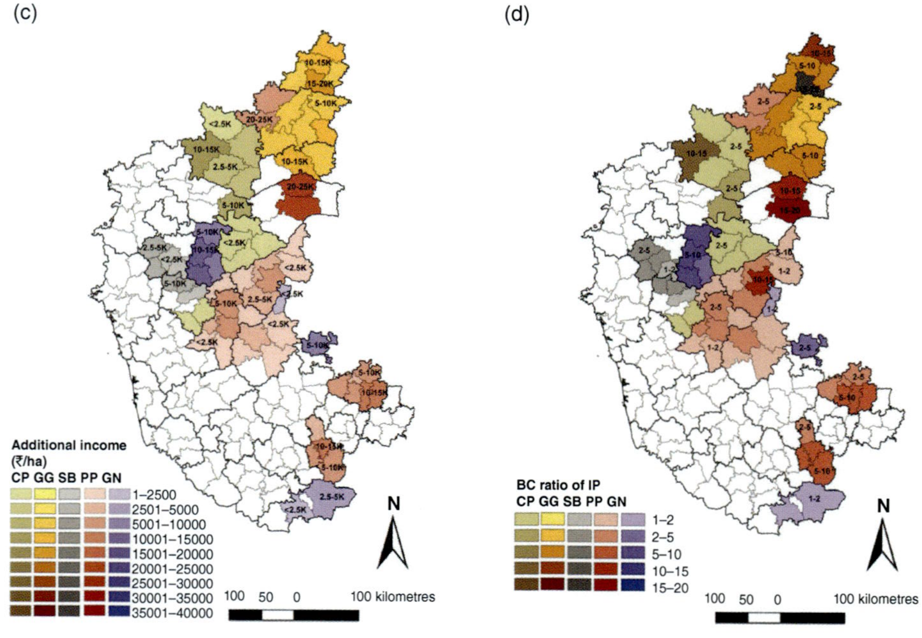

Plate 4. Continued.

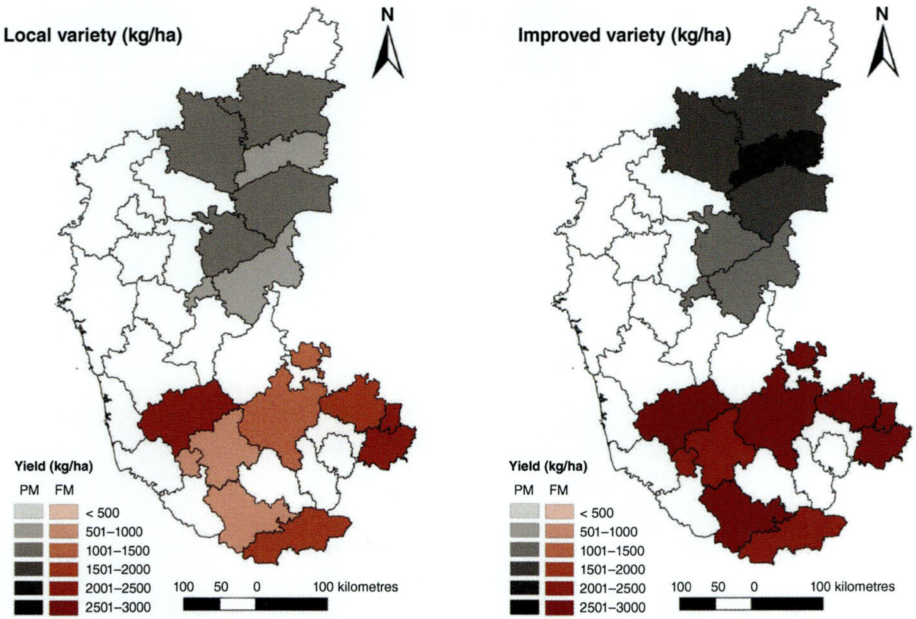

Plate 5. Spatial variability in crop yield of local and improved varieties of selected cereals, pulses and oilseed across the districts of Karnataka during 2012. Cereals: PM, pearl millet; FM, finger millet. Pulses: CP, chickpea; PP, pigeonpea. Oilseed: GN, groundnut.

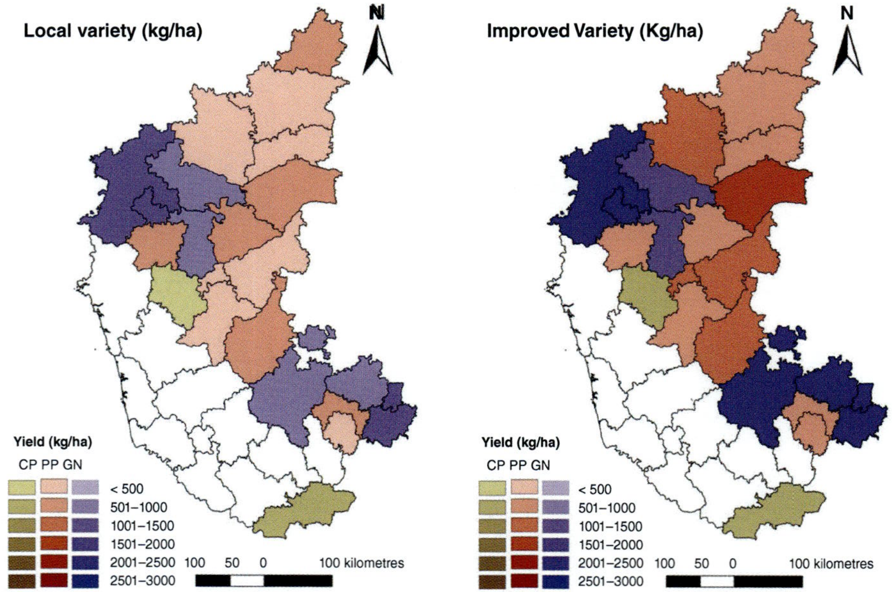

Plate 5. Continued.

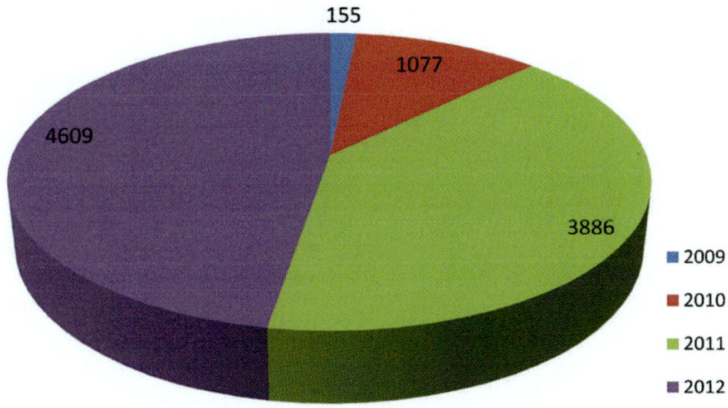

Plate 6. Net returns (₹ million) due to improved management at the state level in Karnataka from 2009 to 2012.

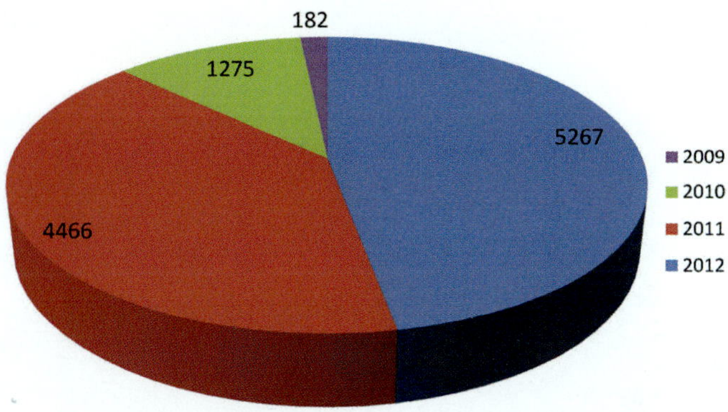

Plate 7. Value (₹ million) of additional production due to Bhoochetana in Karnataka from 2009 to 2012.

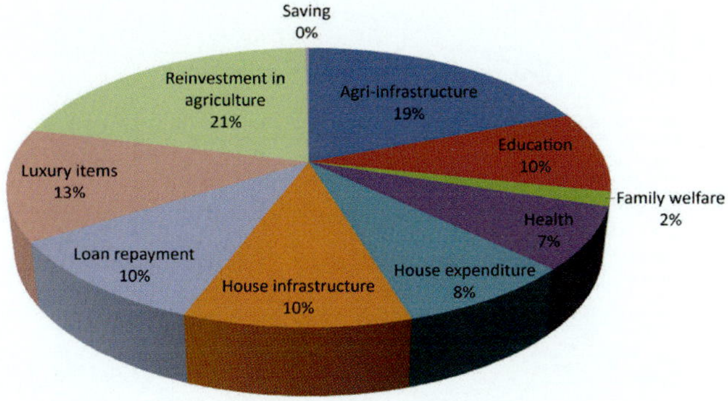

Plate 8. Contribution of Bhoochetana to household assets formation, reinvestment in agriculture, health and education due to increased income in Karnataka.

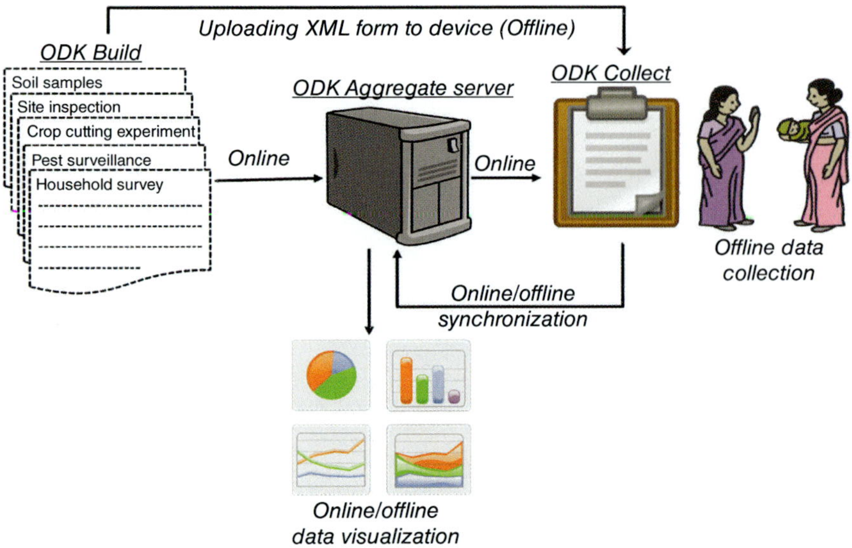

Plate 9. Flow diagram of various applications and operations in Open Data Kit (ODK) for mobile data collection.

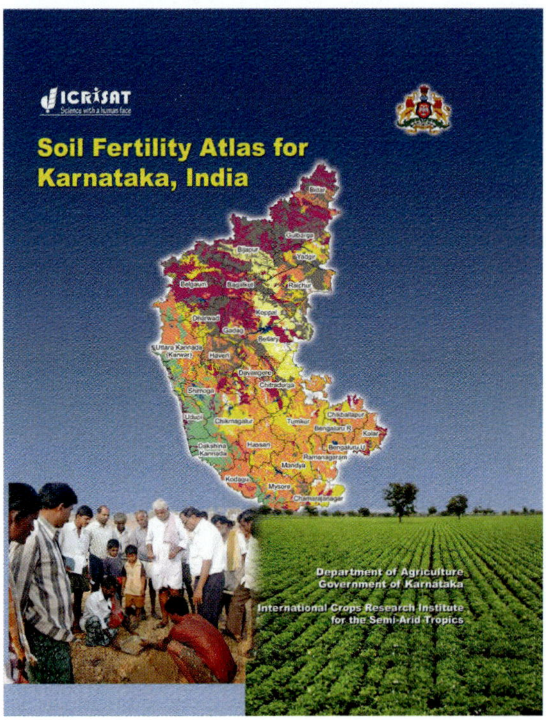

Plate 10. Soil fertility atlas for all nutrients for the entire state of Karnataka.

with other data sets suggested that the climatological and variability features of rainfall over India derived from IMD 4 were comparable with the existing gridded daily rainfall data sets. In addition, the spatial rainfall distribution in, for example, heavy rainfall areas in the orographic regions of the west coast and over the north-east, and low rainfall in the leeward side of the Western Ghats, were more realistic and better presented in IMD 4 due to its higher spatial resolution and to the higher density of rainfall stations used for its development. The IMD 4 data were updated by IMD to include data up to the year 2013; these updated data were procured from the IMD and daily gridded rainfall climate data of 4964 pixels in India for 113 years (1901–2013) were retrieved.

The CRU Time-Series (TS) Version 3.22 of High Resolution Gridded Potential Evapotranspiration (PET) data (January 1901 to December 2013) was downloaded for India (http://badc.nerc.ac.uk/browse/badc/cru/data/cru_ts/cru_ts_3.22/data/, accessed 5 August 2014). These are at 0.5° resolution and using a suitable method, PET values at 0.25° resolution for the 4964 pixels in India were derived for 113 years. The CRU TS 3.22 data are produced by the CRU at the University of East Anglia, UK and are updated from the previous datasets (Harris *et al.*, 2014).

Soil water-holding capacities for the 4964 pixels in India were estimated based on the soil map of NBSS&LUP (1985). Pixel-wise water balances and climate indices for 113 years were computed based on the revised water budgeting approach of Thornthwaite and Mather (1955). Climates for each year were classified based on the annual moisture index (Table 7.2) according to the classification of Thornthwaite and Mather (1955).

Table 7.2. Climatic classification based on moisture index.

Type	Moisture index	Symbol
Perhumid	100 and above	A
Humid	20–100	B
Moist subhumid	0–20	C_2
Dry subhumid	–33.3 to 0	C_1
Semi-arid	–66.6 to –33.3	D
Arid	Less than –66.6	E

While assessing climate change, it is an accepted method to find deviations from a base period. According to the World Meteorological Organization (WMO) guidelines, 30-year continuous data are required to compute climatic norms. Standard periods for climatic norms are 1901–1930, 1931–1960 and 1961–1990. In the present case, the gridded data available were for the period 1901–2013, hence 1901–1990 is considered as the base period or period 1 and 1991–2013 is considered as period 2. Average climates are classified into six types for both periods 1 and 2 (Fig. 7.4).

(a)

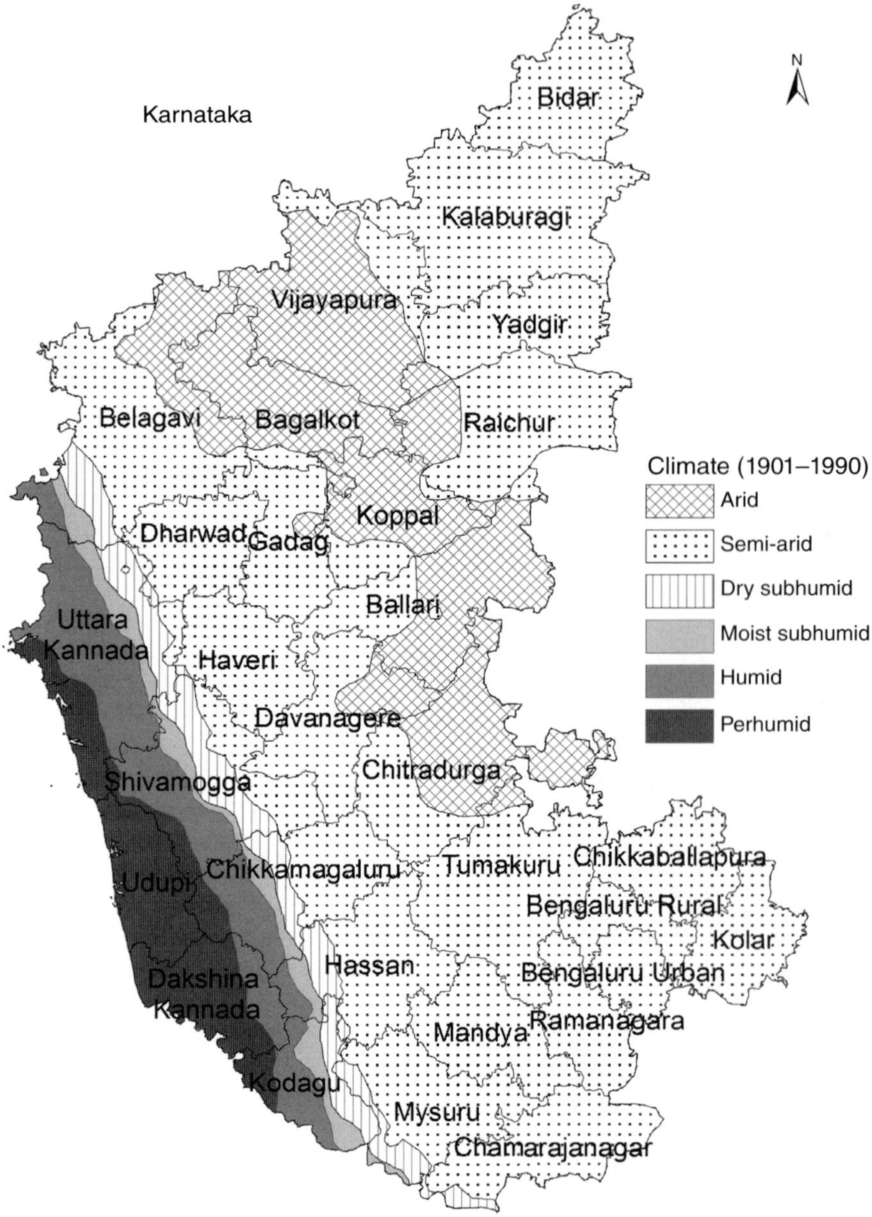

Fig. 7.4. Spatial distribution of climates in Karnataka during two periods. (a) Period 1 (1901–1990); and (b) period 2 (1991–2013).

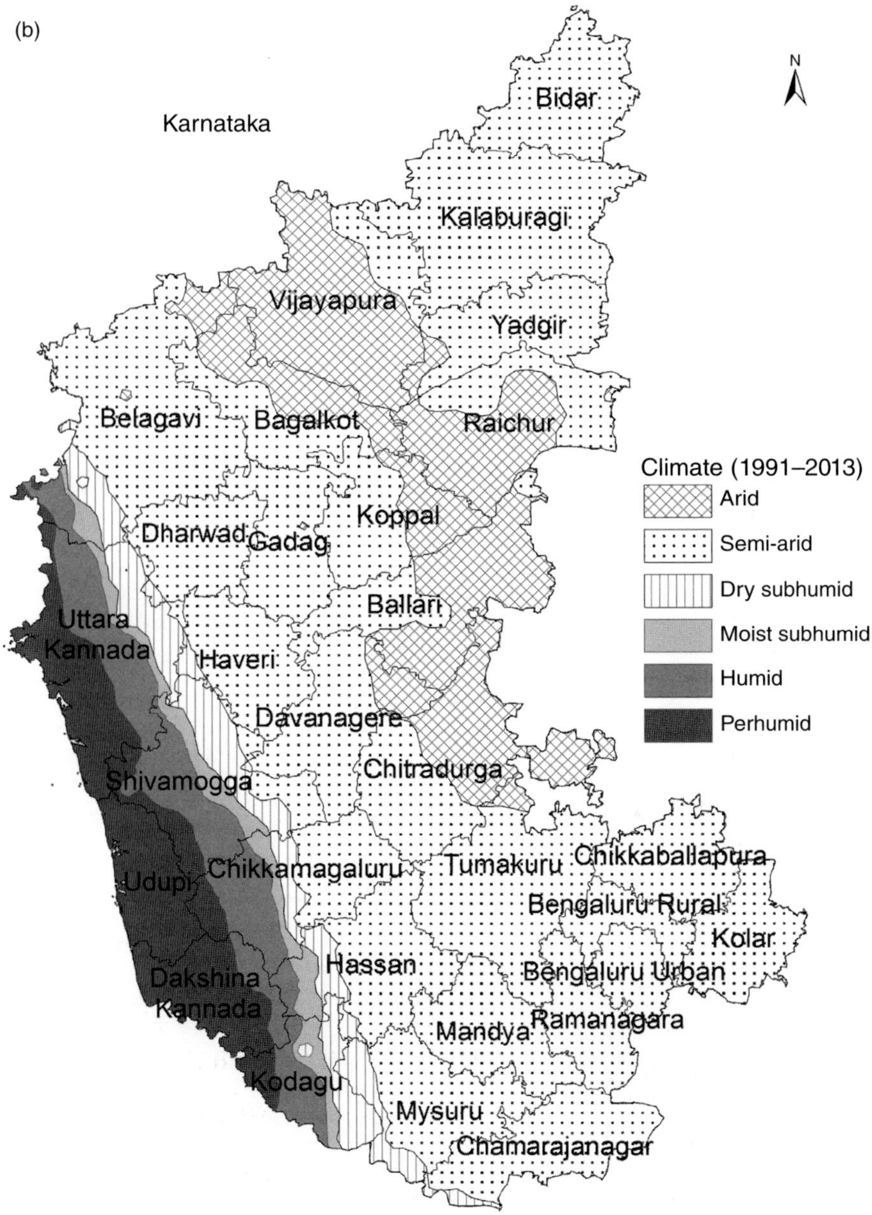

Fig. 7.4. Continued.

Climate classification results of 4964 pixels were converted to points and re-interpolated using ArcGIS 10.0 since a 0.25° × 0.25° pixel is coarse and patchy to show the climatic zones clearly. The inverse distance weighted (IDW) method is used to interpolate the point data with an exponent of distance as 2 and the search radius fixed to 0.5 minutes and the number of points around the estimated value limited to 6. This was achieved after exploring different combinations of input variables which can be changed within the set limits. The resolution of the output grid is fixed at approximately 1 km. This method was used because IDW is an exact interpolator and estimated values do not cross the range of values in the total dataset. The area under each climate is the number of pixels multiplied by the area of each pixel, which is fixed at 1 km. Minor aberrations in the area estimated by IDW and the area estimated by the conventional method were adjusted to remove ambiguity and areas under each climate were quantified for both periods in Karnataka.

Considerable changes in climate are observed between the two periods, 1901–1990 and 1991–2013 (Table 7.3). The arid areas have decreased by about 0.574 million ha and much of it has shifted to the semi-arid climates. The interesting feature is that there is an eastward shift in the arid areas in the districts of Belagavi (Belgaum), Bagalkot, Koppal and Raichur. Areas under semi-arid and perhumid climates have increased by about 0.334 million ha and 0.316 million ha, respectively. Dry subhumid areas have increased by about 0.12 million ha and humid areas are reduced by about 0.161 million ha. A very small decrease of 0.035 million ha was observed in the areas with moist subhumid type of climate. Overall, it is seen that perhumid, dry subhumid, moist subhumid and semi-arid areas have increased by about 27%, 13%, 8% and 3%, respectively, compared with their normal 90-year climate as seen in the period 1 (1901–1990). Arid and humid climate areas have decreased by about 15% and 4%, respectively.

Table 7.3. Area (million ha) under different climate types in Karnataka.

Period	Arid	Semi-arid	Dry subhumid	Moist subhumid	Humid	Perhumid	Total
1901–1990	3.775	11.688	0.929	0.452	1.149	1.186	19.179
1991–2013	3.201	12.022	1.049	0.417	0.988	1.502	19.179
Difference	–0.574	0.334	0.12	–0.035	–0.161	0.316	0
Change (%)	–15	3	13	8	–4	27	0

7.3.3 ICRISAT's hypothesis of hope to address climate variability and change

ICRISAT's research findings showed that integrated genetic and natural resource management (IGNRM) through participatory watershed management is the key for improving rural livelihoods in the SAT (Wani *et al.*, 2002, 2003). Even under a climate change regime, crop yield gaps can still be significantly narrowed down with improved management practices and by using germplasm adapted to warmer temperatures (Wani *et al.*, 2003, 2009; Cooper *et al.*, 2009). Some of the climate-resilient crops are short-duration chickpea (*Cicer arietinum*) cultivars ICC 96029 (super early), ICCV 2 (extra-early) and KAK 2 (early maturing); wilt-tolerant pigeon pea hybrid (ICPH 2671) with a potential to give 80% higher yields than traditional varieties; and short-duration groundnut (*Arachis hypogaea*) variety ICGV 91114 that escapes terminal drought.

Integrated watershed management comprises improvement of land and water management, integrated nutrient management including application of micronutrients, improved varieties and integrated pest and disease management for substantial productivity gains and economic returns by farmers (Wani *et al.*, 2003). The goal of watershed management is to improve livelihood security by mitigating the negative effects of climatic variability while protecting or enhancing the sustainability of the environment and the agricultural resource base. Greater resilience of crop income in Kothapally (Andhra Pradesh) during the drought year 2002 was indeed due to watershed interventions. While the share of crops in household income declined from 44% to 12% in the non-watershed project villages, crop income remained largely unchanged from 36% to 37% in the watershed village (Wani *et al.*, 2009).

7.4 Rainfall Situation During 2001–2013

Great variability in rainfall is observed in the past decade in Karnataka. Since 2000, the state has experienced drought during 2001, 2002, 2003 and 2004 consecutively. During 2005, the state was under heavy floods. During 2006, it experienced both flood and drought situations. During 2007, it repeatedly faced floods four times. During 2008 and 2009, there were both drought and floods in the state. During September 2011, rainfall was the second worst event in South Interior Karnataka since 1971, and the third worst event since 1971 in North Interior Karnataka. Failure of the monsoon during September caused a late season drought of rare severity. The dry spell in Interior Karnataka during September continued until October 2011 in many districts; 77 taluks recorded a deficit in rainfall during the period 1–14 October 2011. The Government of Karnataka declared 99 taluks as drought affected (National Institute of Disaster Management, 2012). Very low rainfall in the south-west monsoon season was recorded in the year 2012, which was the second lowest in the previous 42 years (1971–2012); the lowest rainfall during this period was recorded in 2002.

Karnataka has the distinction of being the first state in the country to establish a Drought Monitoring Cell (DMC) in 1988 as an institutional mechanism to monitor drought. Activities broadened to also include monitoring other natural disasters and the DMC was renamed as Karnataka

State Natural Disaster Monitoring Centre (KSNDMC) in 2007. KSNDMC provides inputs to the farming community as well as to agriculture, horticulture and various other sectors.

The centre operates a number of telemetric rain-gauges, satellite-linked automatic weather stations and conventional rain-gauges. KSNDMC publishes weekly state weather and crop reports, which include rainfall data. These reports were accessed (https://www.ksndmc.org/ReportHomePage.aspx) regularly and the rainfall data as given in these reports for the 5 years 2009–2013 is used in the present study.

7.4.1 Rainfall situation during 2009 and 2010

In 2009, the south-west monsoon which set on 25 May covered the entire state of Karnataka by 25 June. During the period from 1 June to 30 September 2009, the state as a whole recorded 954 mm rainfall as against the normal rainfall of 820 mm. During the period from 1 June to 1 July 2009, the state as a whole recorded 160 mm rainfall as against the normal rainfall of 199 mm and was classified under the deficit category. By the third week of October 2009, the south-west monsoon had completely withdrawn from the state. The north-east monsoon became active from the first week of November and widespread rainfall was received in the state.

In 2010, the south-west monsoon set in South Interior Karnataka on 6 June and advanced further on 7 June to Coastal Karnataka and it remained stationary up to 9 June 2010. The monsoon covered the entire state by 11 June 2010. During the period from 1 June to 30 September 2010, the state as a whole recorded actual rainfall of 890 mm as against the normal rainfall of 833 mm. The departure from normal was 7% and Karnataka was classified under the normal rainfall category. If the departure of actual rainfall received from normal is more than 20%, then rainfall is classified as 'excess'; if it is between –19% and +19% it is classified as 'normal'; if it is between –20% and –59% it is classified as 'deficit'; and if it is between –60% and –99% then it is classified as 'scanty'.

7.4.2 Rainfall situation during 2011 and 2012

In 2011, the south-west monsoon entered Coastal Karnataka on 2 June and the entire state was covered by 6 June 2011. During June 2011, the rainfall was widespread over Coastal Karnataka and Malnad and scattered over Interior Karnataka; and the whole state received 204 mm of rainfall as against the normal rainfall of 192 mm. In July and August, the monsoon was normal for Karnataka. During September 2011, the monsoon was normal in Malnad and Coastal Karnataka and scanty in South Interior Karnataka and North Interior Karnataka. During the period from 1 June to 30 September 2011, the state as a whole recorded 817 mm of rainfall as against the normal rainfall of 826 mm with departure from normal being –3% and rainfall was classified under the normal category. The monsoon had withdrawn on 24 October 2011 and simultaneously the north-east monsoon had set in over the state. In the post-monsoon season the state as a whole received 151 mm of rainfall as against the normal rainfall of 191 mm; and rainfall was classified under the normal category. Region-wise rainfall in both the seasons is provided in Table 7.4. Thus, Karnataka received normal rainfall during the south-west monsoon and post-monsoon periods.

Table 7.4. Region-wise rainfall in Karnataka during 2011.

	South-west monsoon (June–September)			Post-monsoon (October–December)		
Region	Normal (mm)	Actual (mm)	Percentage departure	Normal (mm)	Actual (mm)	Percentage departure
South Interior Karnataka	371	274	–24	216	193	–9
North Interior Karnataka	504	420	–17	141	77	–44
Malnad	1343	1417	4	223	227	2
Coastal Karnataka	2825	3106	11	247	258	5
State	826	817	–3	191	151	–18

In 2012, Karnataka state as a whole received only 618 mm of rainfall during the south-west monsoon period (June–September) compared with the normal of 835 mm. Thus, the departure from normal was –26% and rainfall was classified under the deficit category. This kind of deficit rainfall was the second lowest in the last 42 years (1971–2012); in 2002, the departure from normal was –33%. Out of the 30 districts, only seven districts received normal rainfall while the remaining 23 districts received deficit rainfall. Out of 176 taluks, only one taluk received excess rainfall, 32 taluks received normal rainfall (departure between –19% and +19%), 139 taluks received deficit rainfall (departure between –20% and –59%) and four taluks received only scanty rainfall (departure less than –60%). During October to December 2012, Karnataka as a whole received 146 mm rainfall as against the normal rainfall of 189 mm. Departure from normal was –23% and rainfall was classified under the deficit category. Region-wise rainfall in Karnataka for 2012 is shown in Table 7.5. Thus, in both the seasons, Karnataka received deficit rainfall impacting the availability of water resources for agriculture, livestock, human consumption, electricity generation and industry.

Table 7.5. Region-wise rainfall in Karnataka during 2012.

	South-west monsoon (June–September)			Post-monsoon (October–December)		
Region	Normal (mm)	Actual (mm)	Percentage departure	Normal (mm)	Actual (mm)	Percentage departure
South Interior Karnataka	357	232	–35	211	162	–23
North Interior Karnataka	493	327	–34	146	125	–14
Malnad	1469	1146	–22	230	148	–36
Coastal Karnataka	3048	2449	–20	267	197	–26
State	835	618	–26	189	146	–23

Weekly rainfall departure from normal during both the monsoon periods in 2012 is presented in Fig. 7.5. During the 30-week period from June to December 2012, rainfall was always lower than normal except during 4 weeks. Low rainfall affected various agricultural operations and crop performance during the year.

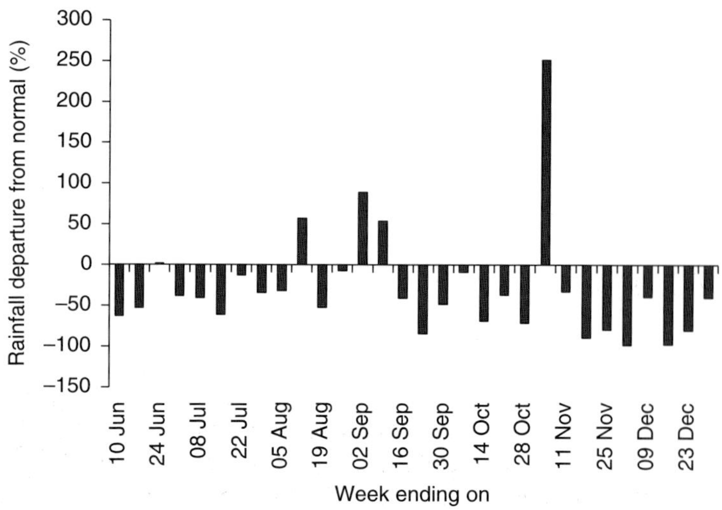

Fig. 7.5. Weekly rainfall departure in Karnataka during rainy season 2012.

7.4.3 Rainfall situation during 2013

In 2013, the south-west monsoon set in over parts of South Interior Karnataka and coastal regions on 1 June and covered the entire state by 7 June. During June 2013, the monsoon was active and the state as a whole received 216 mm rainfall as against the normal rainfall of 195 mm.

In July, the monsoon was active over Karnataka except for some parts of South Interior Karnataka and North Interior Karnataka; and the state as a whole received 350 mm rainfall as against the normal 278 mm. In August 2013, the monsoon was relatively weak over Karnataka and the state as a whole received only 163 mm rainfall as against the normal 205 mm. In September 2013, the monsoon was active and the state as a whole received 205 mm rainfall compared with the normal 158 mm. In summary, it is observed that 2013 was a good year for Karnataka as the state received about 934 mm rainfall in the south-west monsoon period as against the normal 835 mm; and the departure from normal was +12%, classified as 'normal'. Rainfall received was normal in June, excess in July and September and deficit in August.

During the post-monsoon period (October–December), the state as a whole recorded 128 mm rainfall as against the normal 188 mm; the departure was –32% and this was classified as in the 'deficit' category. Region-wise rainfall for 2013 is shown in Fig. 7.6. Thus, Karnataka received normal rainfall during the south-west monsoon period and deficit rainfall during the post-monsoon period.

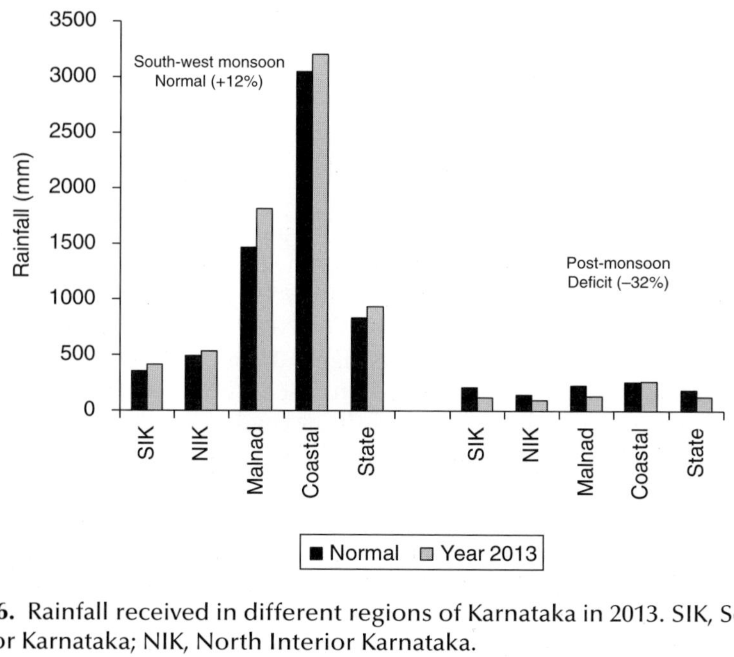

Fig. 7.6. Rainfall received in different regions of Karnataka in 2013. SIK, South Interior Karnataka; NIK, North Interior Karnataka.

Weekly rainfall departure from normal during June–December 2013 shows that rainfall was either normal or above normal in the south-west monsoon period, except for 2 weeks in August (Fig. 7.7). However, during the post-monsoon period, rainfall departure was always negative

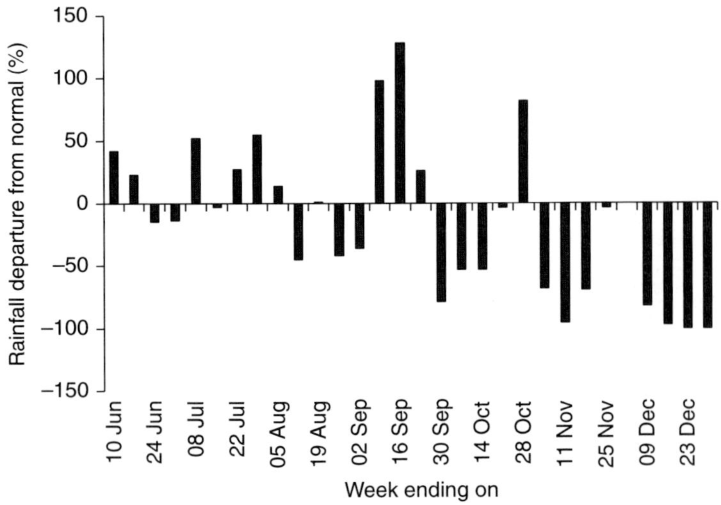

Fig. 7.7. Weekly rainfall departure in Karnataka in 2013.

except during 1 week ending on 28 October. The deficit rainfall situation in the post-monsoon period in 2013 impacted water availability to crops to some extent.

7.4.4 Rainfall situation during 2009–2013

Normal annual rainfall of Karnataka is about 1150 mm of which about 80% is received in the south-west monsoon season. Rainfall during June–September determines the crop growth and productivity. Taluk level rainfall analyses for 5 years from 2009 to 2013 (Fig. 7.8) show that 2012 was a drought year and out of 176 taluks, 139 taluks (78%) received deficit rainfall and four taluks received only scanty rainfall. Only one taluk received excess rainfall in 2012 while normal rainfall was received in 32 taluks. However, 115 taluks received normal rainfall in 2013. Excess rainfall was received in 112 taluks in 2009. The year 2010 was unique as about 96% of all the taluks received either excess or normal rainfall.

Seasonal rainfall departure during the 5-year period 2009–2013 for the whole Karnataka state (Fig. 7.9) indicates that in 2009 and 2010, rainfall was above normal during both south-west and post-monsoon seasons, while in the following consecutive 2 years, 2011 and 2012, it was below normal. In 2013, above normal rainfall was received in the south-west monsoon period and very low rainfall was received in the post-monsoon period. Taluk-wise rainfall departure during the south-west monsoon and post-monsoon periods for 3 years during 2011–2013 is shown in Figs 7.10 and 7.11, respectively.

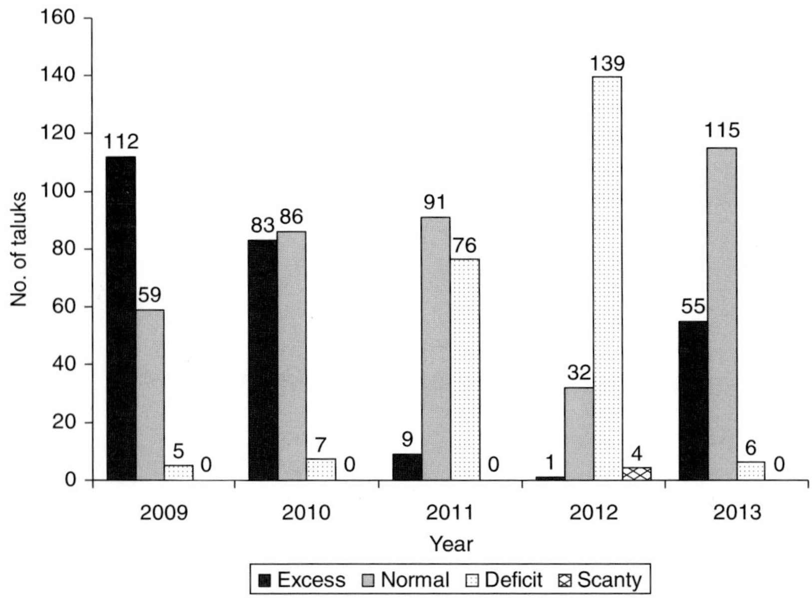

Fig. 7.8. Categorization of south-west monsoon rainfall in Karnataka.

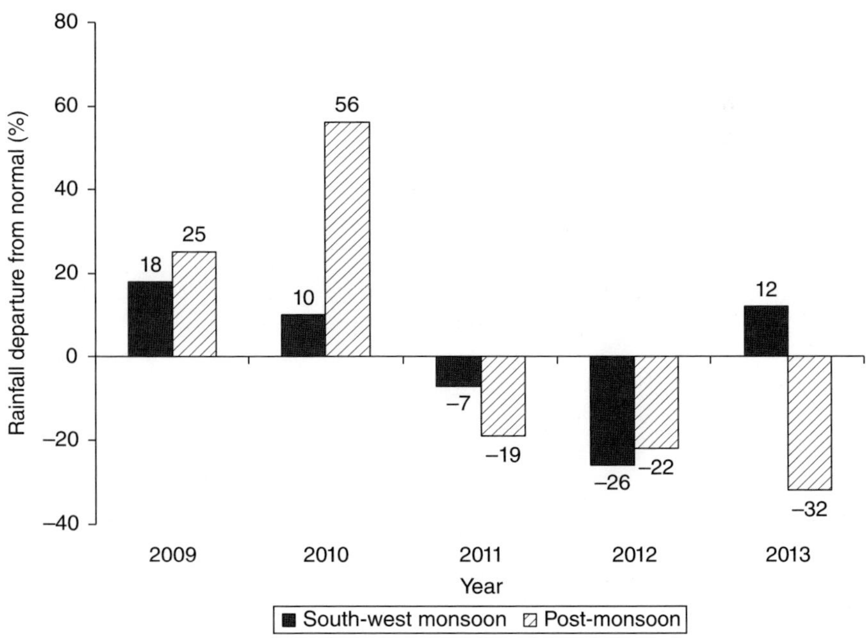

Fig. 7.9. Seasonal rainfall departure in Karnataka during 2009–2013.

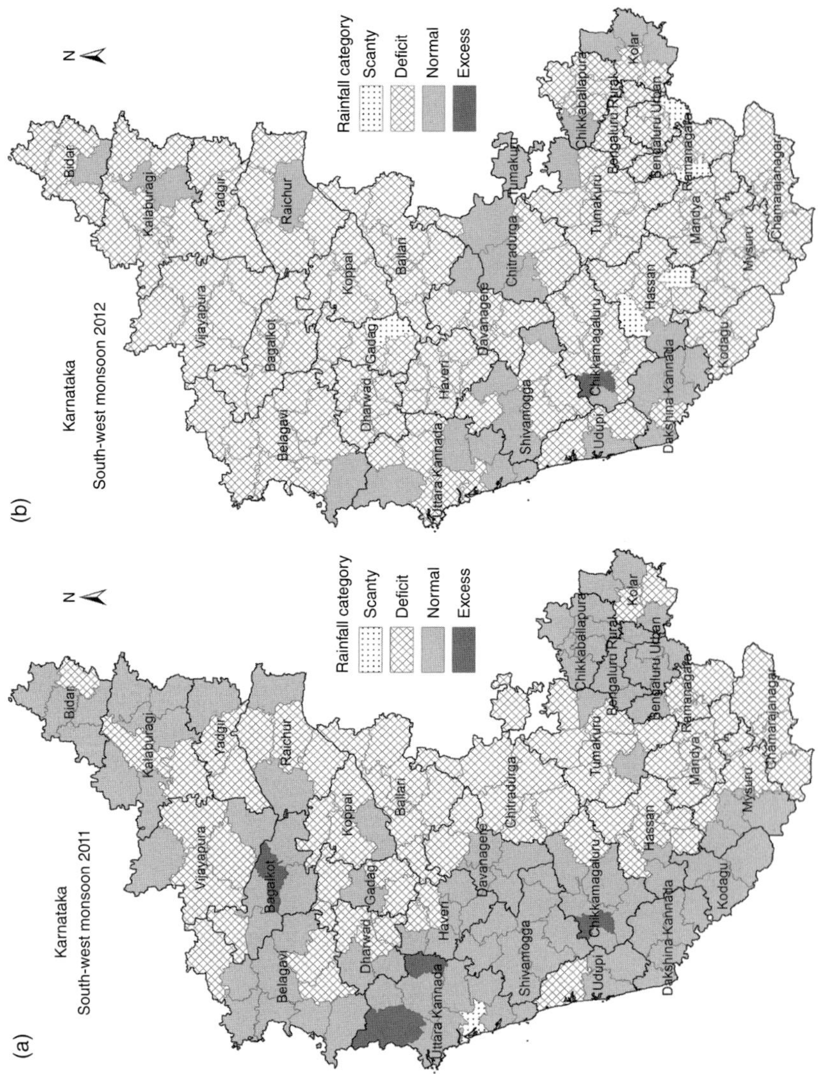

Fig. 7.10. Taluk-wise south-west monsoon rainfall in Karnataka. (a) 2011; (b) 2012; and (c) 2013.

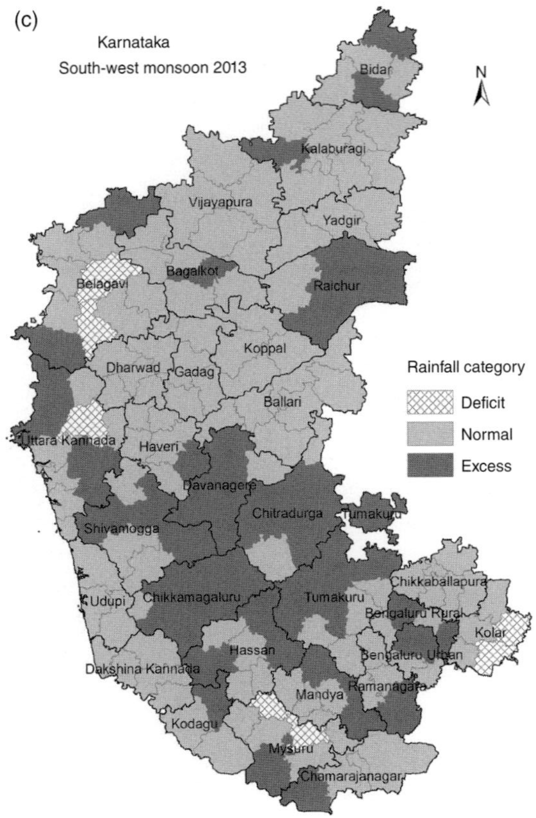

South-west monsoon rainfall over Karnataka

- Karnataka receives 80% of its annual rainfall during the south-west monsoon period.
- During 2011–2013 the south-west monsoon rainfall in Karnataka was a mixture of normal and deficit rainfall.
- Rainfall was normal in 2011 and 2013, while the year 2012 witnessed a rainfall departure of –26%, which was the second lowest in the past 42 years.
- In 2012, 79% of the taluks received deficit rainfall.
- The year 2013 received normal rainfall and only six taluks had deficit rainfall, while 115 taluks were under normal rainfall and 55 taluks under excess rainfall.
- Almost 162,000 ha of cultivated crops were affected by the dry spell in July.

Fig. 7.10. Continued.

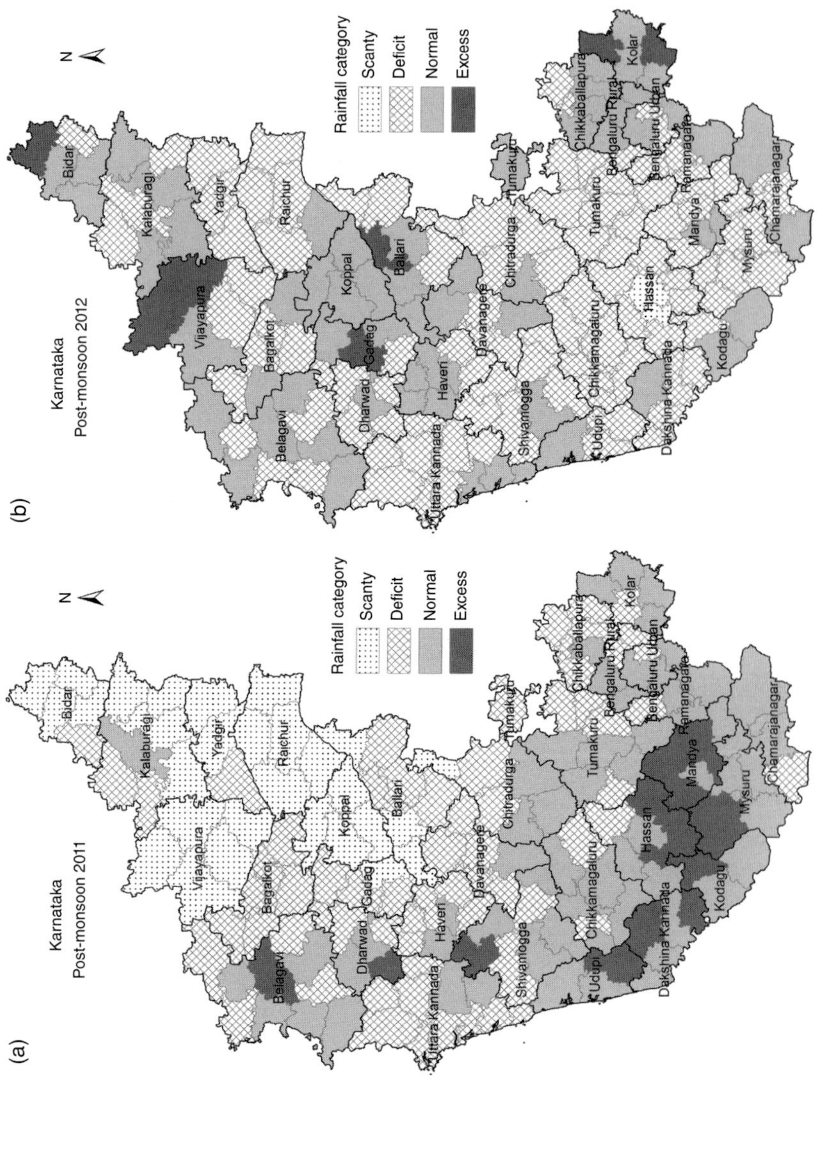

Fig. 7.11. Taluk-wise post-monsoon rainfall in Karnataka. (a) 2011; (b) 2012; and (c) 2013.

(c)

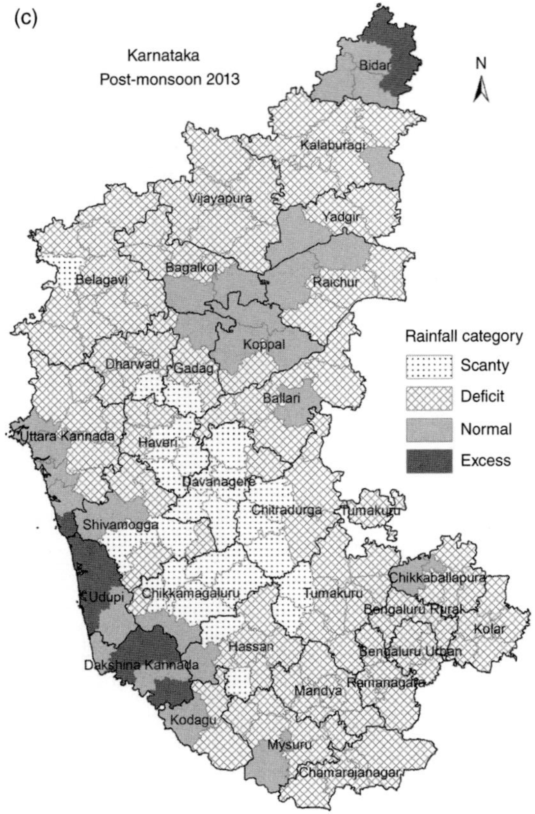

Post-monsoon rainfall over Karnataka

- Karnataka receives about 13% of its annual rainfall during the post-monsoon period.
- Post-monsoon normal rainfall is about 188 mm.
- Deficit rainfall was received in 2012 and 2013 while it was normal in the year 2011.
- In 2011, 121 taluks received either no rainfall or scanty rainfall. Only 38 taluks received normal to excess rainfall.
- In 2012, though 101 taluks had deficit rainfall, 73 taluks received normal to excess rainfall.
- Among the 3 years, 2013 received the lowest rainfall of 128 mm; about 136 taluks received either no rainfall or scanty rainfall.
- Although there was deficit south-west monsoon and post-monsoon rainfall in 2012, almost 335,000 ha land was cultivated under various *rabi* crops, which is higher compared with 2011 and 2013.

Fig. 7.11. Continued.

7.5 Crop and Climate Interactions

7.5.1 Performance of crops during 2009–2013

The Bhoochetana initiative covered six districts in 2009 and 14 districts in 2010. Later in 2011, 2012 and 2013, Bhoochetana was extended to all 30 districts in Karnataka. Major cereal crops such as paddy (*Oryza sativa*), sorghum, finger millet (*Eleusine coracana*), maize (*Zea mays*), pearl millet (*Pennisetum glaucum*) and others; legumes and oilseed crops such as groundnut, pigeon pea, chickpea, black gram (*Vigna mungo*), green gram (*Vigna radiata*; mung bean), soybean (*Glycine max*), sunflower (*Helianthus annuus*) and safflower (*Carthamus tinctorius*); and other commercial crops such as sugarcane (*Saccharum officinarum*) and cotton (*Gossypium* sp.) were covered under Bhoochetana. Using a stratified sampling method, soil samples from farmers' fields were collected and analysed for macro- and micronutrients and balanced fertilizer recommendations were developed and promoted. Results of the on-farm balanced nutrition management trials (2010–2012) in Karnataka to evaluate the performance of pearl millet in contrasting seasons with improved agronomic management have shown that grain yield and aboveground dry matter improved significantly with balanced nutrient application of nitrogen, phosphorus, potassium, sulfur, zinc and boron in farmers' fields, which were critically deficient in soil nutrients. Even in comparatively drier years, application of balanced nutrients significantly increased grain yield and aboveground dry matter, which provides resilience against drought through enhanced water productivity (Uppal *et al.*, 2015). Pearl millet can be an important component of climate resilient agriculture in low production environments when managed with improved agronomic practices.

In situ soil moisture conservation techniques such as contour cultivation, conservation furrows and broad-bed and furrow systems were implemented. Many improved practices like seed treatment with biofertilizers, planting *Gliricidia* on field bunds, promotion of high-yielding and short-duration varieties, capacity enhancement of farmers in farm machinery, integrated pest management practices and others were taken up under the Bhoochetana initiative. Participatory crop cutting experiments to record the yields obtained in the fields under farmers' practices and Bhoochetana improved practices were conducted and the yield data were compiled at farmer, crop and taluk level. A summary of increase in yield for major crops is presented in Tables 7.6 and 7.7. Taluk-wise yield increase during 2011, 2012 and 2013 is depicted in Fig. 7.12. Bhoochetana interventions have helped to enhance the yields of several crops in different regions of the state by 21–66% (Wani *et al.*, 2013).

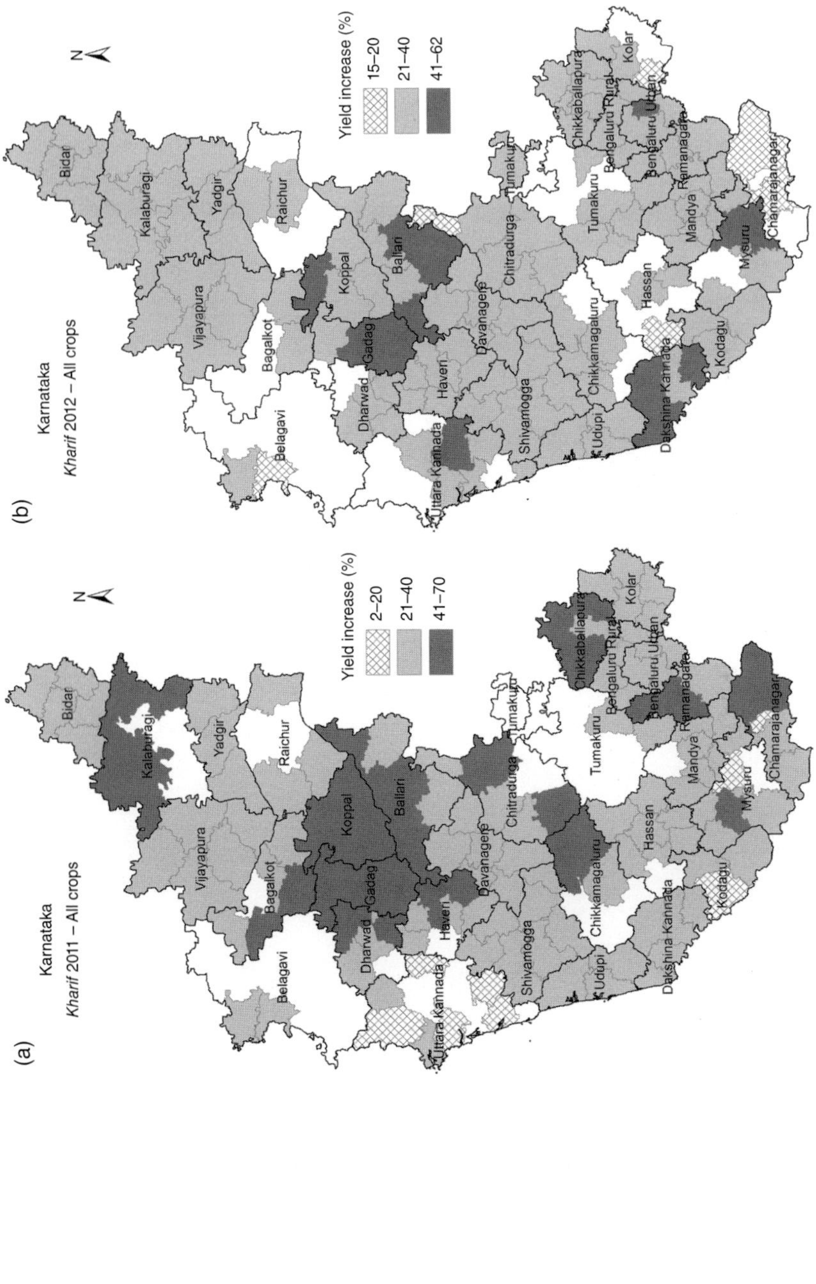

Fig. 7.12. Taluk-wise crop yield increase (for all crops) through Bhoochetana in Karnataka. (a) 2011; (b) 2012; and (c) 2013.

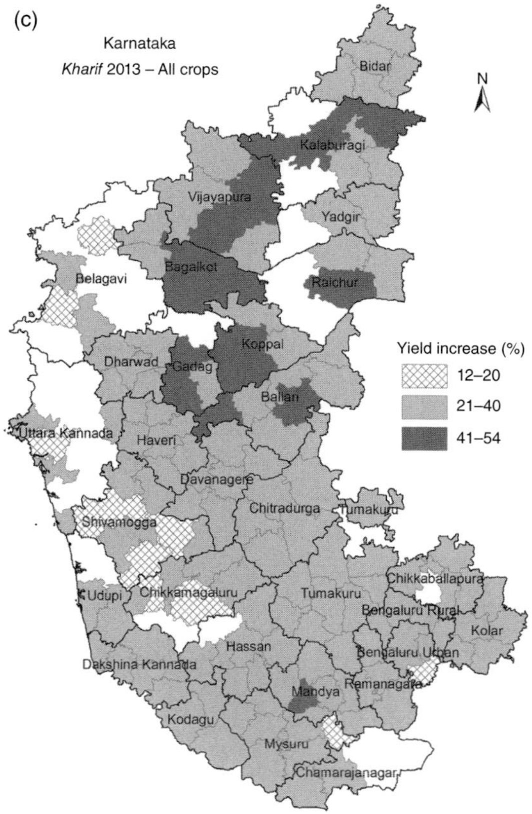

Bhoochetana enhanced crop yields

- Major cereal crops such as paddy, sorghum, finger millet, maize, pearl millet and others; legumes and oilseed crops such as groundnut, pigeon pea, chickpea, black gram, green gram, soybean, sunflower and safflower; and other commercial crops such as sugarcane and cotton were covered under Bhoochetana.
- Improved management practices such as balanced fertilizer application, *in situ* soil moisture conservation, seed treatment with biofertilizers, *Gliricidia* planting on field bunds, cultivating high-yielding and short-duration varieties, and integrated pest management were promoted.
- These interventions have helped to enhance the yields of several crops in different regions of the state by 20–60% and also to bring resilience to climate aberrations (extremely low rainfall situation in 2012).

Fig. 7.12. Continued.

Table 7.6. Increased yields of cereal crops under Bhoochetana during 2009–2013.

		Year				
Crop	Details	2009	2010	2011	2012	2013
Paddy	Improved yield (kg/ha)			5580	4780	4950
	Farmers' yield (kg/ha)			4480	3760	3980
	Increase in yield (%)			25	27	24
	No. of districts			5	12	14
Maize	Improved yield (kg/ha)	7600	7250	5200	5080	5310
	Farmers' yield (kg/ha)	5470	5410	3920	3850	4090
	Increase in yield (%)	39	34	33	32	30
	No. of districts	2	8	19	18	18
Finger millet	Improved yield (kg/ha)	2680	2320	2040	1680	1850
	Farmers' yield (kg/ha)	1740	1700	1550	1260	1450
	Increase in yield (%)	54	36	32	33	28
	No. of districts	2	7	10	12	11
Sorghum	Improved yield (kg/ha)		2410	3020	2510	2790
	Farmers' yield (kg/ha)		1780	2150	1940	2090
	Increase in yield (%)		35	40	29	33
	No. of districts		2	4	2	7
Pearl millet	Improved yield (kg/ha)		2390	2370	1980	1960
	Farmers' yield (kg/ha)		1710	1710	1490	1390
	Increase in yield (%)		40	39	33	41
	No. of districts		3	6	6	6
Rabi sorghum	Improved yield (kg/ha)	1880	1830	1780	1630	1770
	Farmers' yield (kg/ha)	1290	1290	1330	1230	1500
	Increase in yield (%)	46	42	34	33	18
	No. of districts	2	2	9	7	11

Table 7.7. Increased yields of legumes and oilseeds under Bhoochetana during 2009–2013.

		Year				
Crops	Details	2009	2010	2011	2012	2013
Green gram	Improved yield (kg/ha)		760	830	1090	650
	Farmers' yield (kg/ha)		550	580	820	510
	Increase in yield (%)		38	43	33	27
	No. of districts		5	7	3	5
Black gram	Improved yield (kg/ha)		1260	750	1630	830
	Farmers' yield (kg/ha)		930	550	1210	620
	Increase in yield (%)		35	36	35	34
	No. of districts		1	2	1	3
Pigeon pea	Improved yield (kg/ha)		1630	1230	1040	910
	Farmers' yield (kg/ha)		1210	890	810	670

Continued

Table 7.7. Continued.

		Year				
Crops	Details	2009	2010	2011	2012	2013
	Increase in yield (%)		35	38	28	36
	No. of districts		4	9	9	11
Chickpea	Improved yield (kg/ha)	1410	1820	1120	780	1040
	Farmers' yield (kg/ha)	1050	1350	830	600	780
	Increase in yield (%)	34	35	35	30	33
	No. of districts	2	2	8	7	11
Groundnut	Improved yield (kg/ha)	1760	1810	1890	1030	1740
	Farmers' yield (kg/ha)	1280	1300	1340	780	1330
	Increase in yield (%)	38	39	41	32	31
	No. of districts	4	6	12	12	14
Soybean	Improved yield (kg/ha)	2630	2650	1920	1570	1950
	Farmers' yield (kg/ha)	1770	2010	1400	1170	1540
	Increase in yield (%)	49	32	37	34	27
	No. of districts	1	3	5	3	5
Rabi	Improved yield (kg/ha)			660	560	950
safflower	Farmers' yield (kg/ha)			520	420	730
	Increase in yield (%)			27	33	30
	No. of districts			1	2	1
Kharif	Improved yield (kg/ha)	1120	780	1370	1100	1160
sunflower	Farmers' yield (kg/ha)	810	640	1010	830	840
	Increase in yield (%)	38	22	36	33	38
	No. of districts	1	2	6	5	7
Rabi	Improved yield (kg/ha)			2240	780	730
sunflower	Farmers' yield (kg/ha)			1640	580	540
	Increase in yield (%)			37	34	35
	No. of districts			2	3	3
Cotton	Improved yield (kg/ha)			3050	2370	1410
	Farmers' yield (kg/ha)			2510	1800	1170
	Increase in yield (%)			22	32	21
	No. of districts			2	3	6

7.5.2 Crop and rainfall interactions during 2011–2013

Weekly water balances were computed for all the 176 taluks in Karnataka for the south-west monsoon and post-monsoon seasons of 2011, 2012 and 2013. One of the outputs of water balance analysis is soil moisture index (SMI), which is the ratio of actual soil moisture to the available water capacity of the soil, expressed as a percentage. The SMI varies from 1% to 100% depending on the rainfall, evapotranspiration requirement and residual soil moisture. An SMI above 75% for two or more consecutive weeks is considered as a no stress condition, while an SMI between 50% and 75% is considered as low stress. An SMI between 25% and 50% is considered as a moderate stress condition, and severe stress conditions are experienced when the SMI is below 25% for more than 2 weeks. Data on areas under different crops in districts of Karnataka were collected from the Ministry of Agriculture, Government of India (DACNET, 2014). Crop, rainfall and SMI interactions in the 3 years 2011, 2012 and 2013 for selected crops and districts were chosen as case studies. The impact of Bhoochetana interventions on crop area is shown in Fig. 7.13.

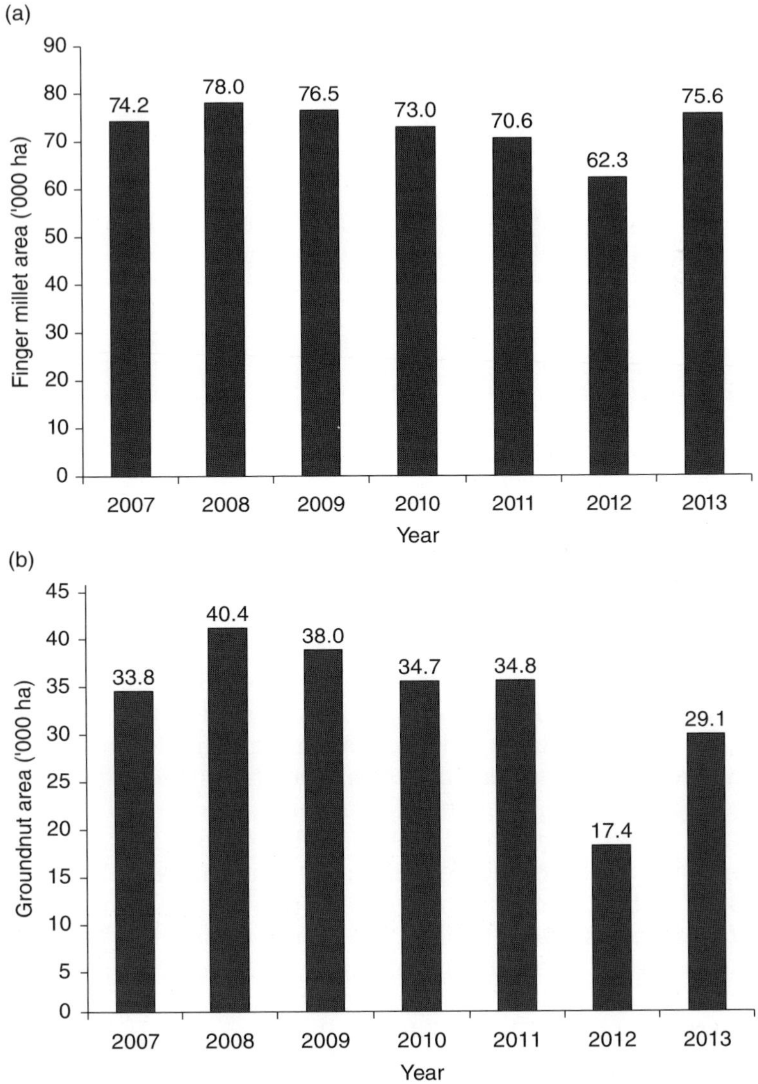

Fig. 7.13. Area under finger millet, groundnut and pigeon pea in three districts of Karnataka during 2007–2013. (a) Finger millet grown in Ramanagara; (b) groundnut grown in Dharwad; and (c) pigeon pea grown in Kalaburagi.

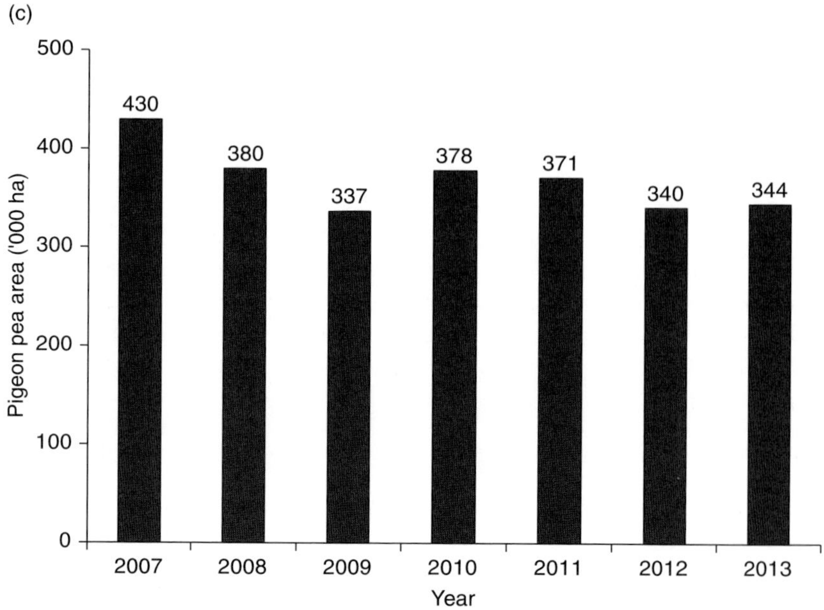

Fig. 7.13. Continued.

Finger millet in Ramanagara

Finger millet is an important rainfed cereal crop in Karnataka and is grown mostly in *kharif* (the rainy season) (June–October) in the Eastern, Central and Southern Dry Zones and also in the Southern Transition Zone. Tumakuru district has more than 175,000 ha under finger millet cultivation, followed by Ramanagara, Kolar, Hassan, Mandya, Chikkamagaluru (Chikkamagalur), Chitradurga, Mysuru (Mysore), Bengaluru (Rural) and Chikkaballapura (Chikkaballapur). As a case study, Bhoochetana field experimental data for finger millet in Ramanagara district were examined with respect to weekly water balance parameters. Ramanagara district has four taluks, namely, Channapatna, Kanakapura, Magadi and Ramanagara. Fig. 7.13(a) shows the area under finger millet during 2007–2013.

Ramanagara district receives about 835 mm of annual rainfall and September and October are the rainiest months, with about 160 mm of rainfall. Good rainfall of about 120 mm is received in May through summer showers. During the 3 years 2011, 2012 and 2013, Ramanagara district received contrasting rainfall quantities. In 2011, the district received about 334 mm of rainfall during the south-west monsoon period (June–September) with a deficit of 18%. The year 2012 was a dry year for Karnataka and Ramanagara district received only 200 mm of south-west monsoon rainfall and the departure from normal was –54%. On the contrary, the following year 2013 was a normal year with a monsoon rainfall of about 456 mm and the departure from normal was +5%. Thus, these consecutive 3 years had contrasting rainfall regimes. Rainfall received in August is critical for finger millet, as it coincides with flowering and grain formation stages.

The SMI for the four meteorological standard weeks (32–35) corresponding to 6 August to 2 September was selected for assessing the impacts on finger millet productivity (Table 7.8). Rainfall during this critical period was low in 2012 and the SMI was less than half of that in 2011 and 2013. Bhoochetana interventions included application of balanced nutrients such as micronutrients (boron, sulfur and zinc) and better crop management practices. Thus, even in the very dry year (2012), finger millet yields did not reduce drastically and the change compared with the proximate years is only 13%.

Table 7.8. Water balance and yield of finger millet in Ramanagara district during 2011–2013.

	Year		
Parameter[a]	2011	2012	2013
Yield under improved practice (t/ha)	3.1	2.6	3.0
Yield under farmers' practice (t/ha)	2.2	2.0	2.3
Yield advantage (%)	41	30	30
Rainfall (mm) during 32–35 MW	125	90	118
Soil moisture index (%) during 32–35 MW	71	22	57

[a]MW, Meteorological week.

Groundnut in Dharwad

Groundnut is an important rainfed oilseed crop in Karnataka and is grown mostly in *kharif* (June–October) in the Central, Eastern and Northern Dry Zones. Chitradurga and Tumakuru districts together contribute about 250,000 ha under groundnut cultivation. Other major groundnut districts are Ballari, Gadag, Belagavi, Vijayapura, Dharwad and Chikkaballapura.

Bhoochetana field experimental data for groundnut in Dharwad district were examined with respect to weekly water balance parameters. Dharwad district has five taluks, namely, Dharwad, Hubli, Kalghatgi, Kundgol and Navalgund. The area under groundnut during 2007–2013 in Dharwad district is shown in Fig. 7.13(b). On average, groundnut is cultivated on about 35,000 ha in the district. Due to low and erratic rainfall the area declined in 2012 to 17,400 ha, which is almost half of the normal area.

Dharwad district receives about 765 mm of annual rainfall and about 77% of the annual rainfall is received in the 5-month period

June–October. May receives good rainfall of about 75 mm, facilitating land preparation and even initiating sowing operations of rainfed crops.

Of the 3 years, Dharwad district received deficit rainfall during the south-west monsoon period in 2011 and 2012 while in 2013 the rainfall was normal. The year 2012 was a dry year for Karnataka and the Dharwad district received only 266 mm of south-west monsoon rainfall and the departure from normal was –47%. In the 3-month period June–August in 2012, rainfall was 212 mm while in the same period rainfall was 384 mm in 2011 and 323 mm in 2013. Higher rainfall in 2011 resulted in higher groundnut yields with farmers' practice as well as improved practice, compared with 2013.

Rainfall received in August is critical for groundnut as it coincides with the flowering and pod formation stages. The SMI for the four meteorological standard weeks (32–35) corresponding to 6 August to 2 September was selected for assessing the impacts on groundnut productivity (Table 7.9). Due to low rainfall received in this critical period, the average SMI was only 16% in 2012 while it was more than 30% in the other 2 years. Bhoochetana interventions included application of balanced nutrients such as micronutrients (boron, sulfur and zinc) and better crop management practices such as seed treatment and use of biofertilizers. Due to these interventions, the yield advantage compared with the farmers' practice was about 30% (Table 7.9) even in the dry year 2012.

Table 7.9. Water balance and yield of groundnut in Dharwad district during 2011–2013.

Parameter[a]	Year		
	2011	2012	2013
Yield under improved practice (t/ha)	3.4	1.2	2.0
Yield under farmers' practice (t/ha)	2.5	0.9	1.5
Yield advantage (%)	36	30	32
Rainfall (mm) during June–August	384	212	323
Soil moisture index (%) during 32–35 MW	43	16	34

[a]MW, Meteorological week.

Pigeon pea in Kalaburagi

Pigeon pea is an important rainfed pulse crop in Karnataka and is grown during June/July to December/January in the Northeastern Dry, Northeastern Transition and Northern Dry Zones. Kalaburagi district is considered as the '*pulse bowl*' of Karnataka and contributes to about 52% of total pigeon pea area in the state. Vijayapura, Yadgir, Bidar and Raichur districts also have a sizeable area under pigeon pea.

As another case study, Bhoochetana field experimental data for pigeon pea in Kalaburagi district were examined with respect to weekly and seasonal water balance parameters. Kalaburagi district has seven taluks, namely, Afzalpur, Aland, Chincholi, Chittapur, Kalaburagi (previously known as Gulbarga), Jevargi and Sedam. Pigeon pea area during 2007–2013 in Kalaburagi district is shown in Fig. 7.13(c). On average, pigeon pea is cultivated in about 368,000 ha in the district. There was a slight reduction in the area in 2012 due to deficient and erratic rainfall.

Kalaburagi district receives about 745 mm of annual rainfall and about 75% of the annual rainfall is received during the south-west monsoon period. Onset of the south-west monsoon is around the second week of June; sowing of rainfed pigeon pea generally starts by the first week of July. September is the rainiest month with about 180 mm of rainfall, followed by July and August. Rainfall in 2011 was 514 mm, which was 15% lower than the normal. During 2012 the district received only 437 mm of rainfall as compared with the normal 608 mm (Table 7.10) and rainfall is considered under the 'deficit' category, while in 2013 the rainfall was 659 mm, which is 8% above normal. Higher rainfall in 2013 resulted in better pigeon pea yield of 1.9 t/ha compared with the low yield of 1.3 t/ha in 2012.

Table 7.10. Water balance and yield of pigeon pea in Kalaburagi district during 2011–2013.

	Year		
Parameter[a]	2011	2012	2013
Yield under improved practice (t/ha)	1.8	1.3	1.9
Yield under farmers' practice (t/ha)	1.2	0.9	1.3
Yield advantage (%)	55	33	49
Rainfall (mm) during south-west monsoon	514	437	659
Soil moisture index (%) during September–October (36–43 MW)	53	45	64

[a]MW, Meteorological week.

Rainfall received in August is important for pigeon pea for establishment and vegetative growth, while rainfall in September–October is critical for the flowering and pod formation stages. SMI values for the 2 months September and October in 2011–2013 shown in Table 7.10 indicate that in 2012, soil moisture was able to satisfy only 45% of the crop requirement, while it was more than 50% in the other 2 years. Soil moisture availability during September–October appears to have a close relationship with the harvested yields both with farmers' practice as well as with Bhoochetana improved practice.

7.6 The Way Ahead

In Karnataka, rainfed agriculture has a predominant role and thus change in rainfall amount and distribution will impact the state's agricultural production in future. Reports indicate an increase in semi-arid area by 8.45 million ha in recent years in five states of India, namely, Madhya Pradesh, Bihar, Uttar Pradesh, Karnataka and Punjab. Dryness and wetness are increasing in different parts of the country in areas where moderate climate existed earlier. There is a need to understand the shift in climate in different agroecoregions of Karnataka in terms of anticipated shift in the crop growing periods and water availability and to use the results of agroclimatic trend analysis to develop suitable adaptation strategies. Increased dependence on groundwater irrigation and indiscriminate usage of chemical fertilizers for increasing crop productivity in Karnataka will not be sustainable in the future. In the likely climate change scenario, this situation would worsen, jeopardizing the long-term perspective of livelihoods of the majority of the rural population.

Resilience to climate change depends on identifying climate smart crops and management practices and a degree of awareness of the community. Knowledge dissemination to farmers about climate change is critical. The present weakest link of knowledge dissemination needs to be strengthened. Innovative participatory delivery systems using ICT (information communication technology)-mediated climate early warning systems and weekly agromet advisories would empower farmers of Karnataka to take smart and swift decisions and make them resilient and food secure. There is a need for scaling up the integrated climate smart agricultural practices for bringing in resilience. Social safety nets along with adaptation strategies are needed. Another major activity required under a crop insurance scheme is to identify the weather-based indices for various rainfed crops and varieties.

Acknowledgements

The authors wish to acknowledge the financial support received from the Government of Karnataka and the help and cooperation received from the consortium partners, staff and farmers in conducting the experiments and data collection.

References

Aavudai Anandhi (2010) Assessing impact of climate change on season length in Karnataka for IPCC SRES scenarios. *Journal of Earth System Science* 119(4), 447–460.

Bhatia, V.S., Singh, P., Wani, S.P., Kesava Rao, A.V.R. and Srinivas, K. (2006) *Yield Gap Analysis of Soybean, Groundnut, Pigeonpea and Chickpea in India Using Simulation Modeling.* Global Theme on Agroecosystems Report no. 31. International Crops Research Institute for the Semi-Arid Tropics, Patancheru, Andhra Pradesh, India.

Biradar, V., Ravi, S., Shailendra Kumar, G. and Dayananda, K. (2012) Annual and seasonal rainfall variability of Aurad Taluk, Bidar district, Karnataka. *Asian Journal of Soil Science* 7(1), 39–42.

Cooper, P., Rao, K.P.C., Singh, P., Dimes, J., Traore, P.S., Rao, A.V.R.K., Dixit, P. and Twomlow, S.J. (2009) Farming with current and future climate risk: advancing a 'Hypothesis of Hope' for rainfed agriculture in the semi-arid tropics. *Journal of SAT Agricultural Research* 7. Available at: http://ejournal.icrisat.org/Volume7/Agroecosystems/AES701.pdf (accessed 20 September 2014).

DACNET (2014) *Agricultural Statistics at a Glance 2014.* Directorate of Economics and Statistics, Network of the Department of Agriculture, Cooperation and Farmers' Welfare (DACNET), Ministry of Agriculture, Government of India. Available at: http://eands.dacnet.nic.in/ (accessed 9 October 2014).

Department of Agriculture and Cooperation (DOAC) (2011) *Annual Report 2010-11.* DOAC, Ministry of Agriculture, Government of India. Available at: <http://agricoop.nic.in/Admin_Agricoop/Uploaded_File/Annual Report 2010–11 English OK.pdf> (accessed 9 October 2014).

Devappa, V.M. and Khageshan, P. (2011) Analysis of rainfall in assessing drought using RS and GIS – a case study of Gulbarga district in Karnataka. *Hydrology Journal* 34(1&2), 1–12.

Environmental Management & Policy Research Institute (EMPRI) and The Energy and Resources Institute (TERI) (2012) *Karnataka State Action Plan on Climate Change, 1st Assessment.* EMPRI and TERI, Bengaluru, Karnataka, India.

Government of Karnataka (2013) Results-Framework Document (RFD) for the Department of Agriculture (2013–2014). Available at: <http://raitamitra.kar.nic.in/RFD 2013–14 Targets.pdf> (accessed 24 September 2014).

Government of Karnataka (2014) Agroclimatic Zones of Karnataka. Available at: http://raitamitra.kar.nic.in/stat/kacz.htm (accessed 15 September 2014).

Harris, I., Jones, P.D., Osborn, T.J. and Lister, D.H. (2014) Updated high-resolution grids of monthly climatic observations – the CRU TS3.10 Dataset. *International Journal of Climatology* 34, 623–642.

Indian Institute of Tropical Meteorology (2013) Homogeneous Indian Monthly Rainfall Data Sets (1871–2012). Available at: ftp://www.tropmet.res.in/pub/data/rain/iitm-subdivrf.txt (accessed 3 September 2014).

Kesava Rao, A.V.R., Wani, S.P., Singh, K.K., Irshad Ahmed, M., Srinivas, K., Snehal, D.B. and Ramadevi, O. (2013a) Increased arid and semi-arid areas in India with associated shifts during 1971–2004. *Journal of Agrometeorology* 15(1), 11–18.

Kesava Rao, A.V.R., Wani, S.P., Srinivas, K., Pushparaj Singh, Snehal, D.B. and Ramadevi, O. (2013b) Assessing impacts of projected climate on pigeonpea crop at Gulbarga. *Journal of Agrometeorology* 15 (Special Issue II), 32–37.

Manjappa, K. and Kelaginamani, S.V. (2005) Rainfall characteristics of Sirsi region of Karnataka and its effect on productivity of rice cropping system. *Karnataka Journal of Agricultural Sciences* 18(2), 285–289.

National Bureau of Soil Survey and Land Use Planning (NBSS&LUP) (1985) *Soil Map of India, Scale 1:7 million.* NBSS&LUP, Nagpur, India.

National Institute of Disaster Management (2012) *India Disaster Report 2011*. National Institute of Disaster Management, Ministry of Home Affairs, Government of India, New Delhi.

Pai, D.S., Latha Sridhar, Rajeevan, M., Sreejith, O.P., Satbhai, N.S. and Mukhopadyay, B. (2014) Development of a new high spatial resolution (0.25° × 0.25°) long period (1901–2010) daily gridded rainfall data set over India and its comparison with existing data sets over the region. *MAUSAM* 65(1), 1–18.

Panduranga, B.T., Ravindra Babu, H.L., Guruprasad, N.A., Janardhanagowda and Rajegowda, M.B. (2006) Climate change and agriculture – a case study of Tumkur district in Karnataka state. *Journal of Agrometeorology* 8(2), 274–277.

Rajegowda, M.B. (1992) Effect of variation in forest area on rainfall and temperature. Presented at the National Seminar on Radiation, Environment and Man (REM), Mysore University, Mysore, India, 8–9 October 1992.

Rajegowda, M.B., Muralidhara, K.S., Murali, N.M. and Ashok Kumar, T.N. (2000) Rainfall shift and its influence on crop sowing period. *Journal of Agrometeorology* 2(1), 2002.

Sehgal, J., Mandal, D.K., Mandal, C. and Vadivelu, S. (1992) *Agro-ecological Regions of India*, 2nd edn. Technical Bulletin 24. National Bureau of Soil Survey and Land Use Planning, Nagpur, India.

Shashidhara, K.K. and Reddy, B.S. (2012) Farmers' perceptions and adaptation about changing climate and its variability in UKP area of Karnataka. *Indian Research Journal of Extension Education (Special Issue)*, I, 196–201.

Singh, R.B., Singh, A. and Kumar, A. (2014) Climate change variability in coastal Karnataka, India. Climate change and biodiversity. In: Singh, M., Singh, R.B. and Hassan, M.I. (eds) *Proceedings of IGU Rohtak Conference, Vol. 1, Advances in Geographical and Environmental Sciences*. Available at: http://www.springer.com/cda/content/document/cda_downloaddocument/9784431548379-c1.pdf?SGWID=0-0-45-1454715-p176497885 (accessed 20 September 2014).

Thornthwaite, C.W. and Mather, J.R. (1955) *The Water Balance. Publications in Climatology, Vol. VIII, No. 1*. Drexel Institute of Technology, Laboratory of Climatology, Centerton, New Jersey.

Uppal, K.R., Wani, S.P., Garg, K.K. and Alagarswamy, G. (2015) Balanced nutrition increases yield of pearl millet under drought. *Field Crops Research* 177, 86–97.

Velayutham, M., Mandal, D.K., Mandal, C. and Sehgal, J. (1999) *Agro-ecological Subregions of India for Planning and Development*. National Bureau of Soil Survey and Land Use Planning, Nagpur, India.

Venkatesh, H., Krishna Kumar, K., Aski, S.G. and Kulkarni, S.N. (2008) Impact of climate change on agriculture over northern Karnataka. In: Prasada Rao, G.S.L.H.V., Rao, G.G.S.N., Rao, V.U.M. and Ramakrishna, Y.S. (eds) *Climate Change and Agriculture Over India*. Central Research Institute for Dryland Agriculture, Indian Council of Agricultural Research (ICAR), Hyderabad, India.

Wani, S.P., Rego, T.J., Pathak, P. and Singh, P. (2002) Integrated watershed management for sustaining natural resources in the SAT. In: *Proceedings of International Conference on Hydrology and Watershed Management, 18–20 December 2002, Hyderabad, India*. Jawaharlal Nehru Technological University, Hyderabad, India, pp. 227–236.

Wani, S.P., Singh, H.P., Sreedevi, T.K., Pathak, P., Rego, T.J., Shiferaw, B. and Iyer, S.R. (2003) Farmer-participatory integrated watershed management: Adarsha Watershed, Kothapally, India. An innovative and up-scalable approach. A case study. In: Harwood, R.R. and Kassam, A.H. (eds) *Research Towards Integrated Natural Resources Management: Examples of Research Problems, Approaches and Partnerships in Action in the CGIAR*. Interim Science Council, Consultative Group on International Agricultural Research (CGIAR), Washington, DC, pp. 123–147.

Wani, S.P., Sreedevi, T.K., Rockström, J. and Ramakrishna, Y.S. (2009) Rainfed agriculture – past trends and future prospects. In: Wani, S.P., Rockström, J. and Oweis, T. (eds) *Rainfed Agriculture: Unlocking the Potential*. CAB International, Wallingford, UK, pp. 1–35.

Wani, S.P., Dar, W.D., Guntuku, D.K., Garg, K.K. and Kesava Rao, A.V.R. (2012) Improved livelihoods and building resilience in the semi-arid tropics: science-led, knowledge-based watershed management. In: *Climate Exchange, The Global Framework for Climate Services*. World Meteorological Organization, Geneva, Switzerland, pp. 69–71.

Wani, S.P., Sarvesh, K.V., Sahrawat, K.L., Krishnappa, K., Dharmarajan, B.K., Raju, K.V., Mukherjee, K. and Dar, W.D. (2013) *Bhoochetana: Building Resilience and Livelihoods through Integrated Watershed Management*. Resilient Dryland Systems Report No. 62. International Crops Research Institute for the Semi-Arid Tropics, Patancheru, Andhra Pradesh, India.

8 Crop Yield Estimation Strategy

V. Nageswara Rao* and G. Pardhasaradhi

International Crops Research Institute for the Semi-Arid Tropics (ICRISAT), Patancheru, India

8.1 Introduction

The Indian economy mainly depends on agriculture, which is the major source of rural employment, supporting the livelihoods of 52% of the population and contributing 14% of gross domestic product. Indian agriculture is predominantly rainfed and constitutes 56% of the total cultivated area, exposed to vagaries of weather, including abiotic factors such as droughts, floods and hailstorms, and also biotic factors like pests and diseases, which play their role during the crop growth stages. Accurate estimation of crop yields has never been an easy task in India and other developing countries, and is more challenging in the context of smallholders producing a wide range of diverse crops in rainfed farming systems. Challenges that may occur include among others: (i) absence of cadastral information on land use; (ii) non-uniform plots which cover a wide range of sizes; (iii) occurrence of bimodal rainfall; (iv) rainfed fallowing; (v) intercropping, relay and sequential cropping; and (vi) significant postharvest losses.

8.1.1 Historical background of yield estimations

In India, the earliest mention of agricultural statistics is found in 'Arthashastra' (Wikipedia, no date), the ancient Indian treatise on statecraft,

*v.nageswararao@cgiar.org

economic policies and military strategies as part of the administrative system written in Sanskrit by Kautilya (350–283 BCE). In the Moghul period 'Ain-i-Akbari' contains the details on the manner in which some basic agricultural statistics were collected during the Moghul period to meet the needs of revenue administration (Census of India, 2011). In the history of British India, the great famine of 1860 emphasized the need for information on agricultural statistics. In 1866, the government of British India initiated collection of agricultural statistics mainly by revenue administration, and subsequently used this information for crop forecasts to serve the British trade interests. In India, the wheat (*Triticum aestivum*) yield forecast was first prepared in 1884. Land statistics have been available in the country since then. By 1900, the list of crops for yield estimates extended to rice (*Oryza sativa*), cotton (*Gossypium* sp.), oilseeds, sugarcane (*Saccharum officinarum*), jute (*Corchorus capsularis*) and indigo (*Indigofera tinctoria*) besides wheat.

In the early part of the 19th century, crop yield estimations at the national level were based on a subjective method of visual observation of crops and these evaluations were mostly done by government officials based on the crop condition being factored in, and multiplying it with the normal or standard yield to arrive at an estimated yield. The standard yield was fixed once in 5 years for an average crop under average conditions. Even in advanced countries crop yield estimates during this period were based on periodic reports from crop reporters (selected farmers), voluntarily supplying the statistics based on their personal impression of crop performance.

In British India, crop cutting experiments (CCEs) using a random sampling method were first carried out during 1923 through to 1925 by J.A. Hubback and the work was published by the Government of India, entitled 'Sampling for rice yield in Bihar and Orissa' as Bulletin No. 166 of the Agricultural Research Institute, Pusa (Hubback, 1927). In the late 1940s, Dr V.G. Panse contributed to the development of a scientific method of random sampling techniques for estimating crop yields in India based on crop cutting of small subplots (experiments) within cultivated fields. This led to an all-round impact in the use of statistics in India and abroad for the estimation of crop yields. It subsequently led to a national programme of General Crop Estimation Surveys (GCES) that has been in operation in India since then. This methodology even became a model for adoption in many countries in Asia, Africa and Europe (Panse, 1954; Minhas, 1987). The National Agricultural Statistics Service (NASS) of the USA used these methods for objective yield estimations as a primary tool for yield forecasts and estimates of major feed and food grains in the USA. In the 1950s, the Food and Agriculture Organization of the United Nations (FAO) adopted CCEs as the standard method recommended for estimation of crop production (FAO, 1982; Murphy *et al.*, 1991).

8.1.2 Rationale for crop sampling

Reliable and timely information on crop area and production statistics is of great importance to planners and policy makers for important decision making with respect to procurement, storage, exports, imports, public distribution and food-security-related issues. Yield forecasts can affect the price and sales policies of agricultural commodities, associated storage and handling requirements on farms and market yards locally, ports and terminals nationally and internationally, besides the cost of transporting or shipping to markets.

The most commonly used yield measurement techniques for seasonal crops are discussed here. First, there are farmers' reported crop yields through farmers' surveys. However, total production varies greatly depending on the reported area and yield per acre or hectare. Farmers generally report yields of most crops in measures of containers rather than as weight, resulting in some inaccuracy in the tabulated yields. Farmers may not know the crop yield accurately even after harvest or may not report accurately for various reasons such as: (i) being apprehensive of a levy on a part of their crop produce; (ii) anticipation of the commodity price (cash-crop bias) being affected; (iii) their desire to impress persons with their success in growing the crop; and (iv) anticipation of penalties for reported high production in the event of production controls by the authorities. Despite these possible limitations, growers are probably the most reliable source of data on yields after harvest if data are checked independently (i.e. yield or production) soon after harvest for adjusting biases.

The second yield measurement technique is a survey of marketed or processed quantities of produce from crops, examples being: (i) cotton which is sold at market yards and processed by gins; (ii) sugarcane which is transported to cane-crushing mills; and (iii) palm oil fruits which are processed at oil extraction distilleries. Yields of these crops are generally collected quite accurately on a weight basis from the processing mills rather than from farmers. However, crop yields obtained this way are expressed in terms of marketed volumes, weights or total monetary value after allowances for moisture, foreign material and undercut quantities, rather than that actually harvested by the farmers. Besides, farmers may have used products for domestic consumption from the harvested crop, which include deductions for seed, feed for animals, grain for a labour wage on the threshing floor, seed sold to other farmers as seed and produce wastage due to field drying and storage; thus yield estimates from this method that are reported are lower than the actual yields.

The third yield measurement technique is crop enumeration surveys of farms or crop fields. These provide the only satisfactory direct, accurate and unbiased method of measuring the area for crop yields.

Even though a probability survey of farms growing the crop of interest is the method of data collection which can provide a direct estimate for the cropping area, there are many factors contributing to non-sampling errors. These may result in bias in the estimation of the area and reporting techniques.

In recent years, the thrust of the planning process has shifted from the macro to the micro level, and demands for accurate estimates of various parameters have arisen at the micro level from administrators and policy planners. In view of these demands, the thrust of research efforts has also shifted to development of precise estimators for small areas. An offshoot of this development is that various small area estimation techniques are being proposed by researchers for implementation (Tikkiwal, 1992). The basic rationale of the approach of small area estimation is that certain assumptions/models are conceptualized, which are assumed to hold good for small as well as large areas. Information on some auxiliary variable at the Gram Panchayat level needs to be generated to scale down the crop yield estimates obtained in the present system of GCES at district/taluk (block) level for developing the gram-panchayat-level estimates. In the following section, the methods used for estimation of cropped area, crop yield and production at various administrative scales in the past and at present are examined.

8.2 Estimation of Cropped Area, Crop Yield and Production

8.2.1 Approaches for cropped area estimations

The estimates of crop area and crop mean yield assume prime importance in the entire gamut of agricultural statistics. The accurate estimation of harvested area is as important as the estimation of yield to ensure the accurate determination of total production of a crop. Although the yield estimation assumes significant attention, there are many complications in estimation of area which includes fluctuations during the crop season, distinguishable as early or mid-season and end-of-season. Area estimation is often carried out in three ways: (i) systems using only ground data in the final area estimation; (ii) systems using remote sensing data as a primary input into the final area estimation; and (iii) statistical extrapolation techniques from small areas.

First, the area planted for a given crop may change throughout the growing season. Area estimation may pose difficult challenges for small area sampling, especially in the semi-arid rainfed areas prone to drought or floods or due to very severe events which may take away planted area out of production, since it cannot be harvested or is economically unviable. These unforeseen, one-time conditions may cause a need for a special, one-time measurement of area. Having more than one crop in an area adds another layer of complexity, with widely different seasons between types of crops and planting more than one crop in a field (either sequential crops or intercrops).

In the initial period of agricultural statistics, crop area estimates involved expert opinion of voluntary crop reporters in local areas such as villages, communes or parishes. This is one of the cheapest forms, but is not always accurate. In India, total crop area is by and large compiled on the basis of complete enumeration while the crop yield is estimated on the basis of the crop cutting sample survey approach. The estimates of crop production are obtained by multiplication of area estimates by corresponding yield estimates. Crop cuts and farmer estimates are the two methodologies most commonly used to estimate crop production by scientists and statisticians at the regional to national scale.

Timelines for advance estimates of area and production

The Directorate of Economics and Statistics (DES), Ministry of Agriculture, India releases estimates of area, yield and production of principal crops of food grains, oilseeds, sugarcane, fibre crops, important commercial crops and horticultural crops; these account for nearly 86% of agriculture output. The agricultural crop year starts from July and ends by June in the ensuing year. During this period, various crops are sown and harvested in the agricultural seasons in a crop year. Final Estimates of production are available based on the complete enumeration of area and yield through CCEs, much after the crops are actually harvested. However, advance estimates of production are required by the government planners for taking various policy decisions relating to export/import, distribution within the country, pricing, marketing, etc. Hence, there is a genuine requirement for reliable crop area estimates much before the crops are harvested for various policy purposes. A time schedule of releasing the advance estimates has been evolved and these estimates of crops are prepared by respective state governments; data are consolidated and released by the Government of India four times during a crop year.

A summary of various timelines and approaches followed for advance estimates of area and crop yield projections is given in Table 8.1. Since 2003/04, the Ministry of Agriculture decided that the Final Estimates should be made available in May/June with the Fourth Advance Estimates based on all the data available and the preparation of Final Estimates in December was discontinued.

To ensure accuracy of these estimates, state governments set up High Level Coordination Committees (HLCCs) comprising senior officers from their Departments of Agriculture, Economics and Statistics, Land Records and Field Operations Division (FOD) of the National Sample Survey Organization (NSSO), the Indian Agricultural Statistics Research Institute (IASRI) and the DES from central government,

Table 8.1. Various timelines and methods followed for advance estimates of area and crop yield projections.

Stage of estimate	Procedure	Parameters considered	Timelines (in cropping year)
First Advance Estimates	Rough estimate of area and productivity of rainy season crops	Visual observation of rainy season crops in the fields at maturity	In September
Second Advance Estimates	Revision of advance estimates of rainy season crops; and first advance estimates of post-rainy season crops from state governments	More precise data of rainy season crops from state governments	In January of ensuing year
Third Advance Estimates	Validated with information from State Agricultural Statistical Authority, with remote sensing data from Space Application Centre and Crop Weather Watch Group proceedings	Both rainy and post-rainy season estimates are considered for revision	End of March to beginning of April
Fourth Advance Estimates	On harvest of the post-rainy season crops; more likely availability of summer crop estimates	Estimates are validated against auxiliary data such as rainfall, wet/dry spells and any weather aberrations	In June–July
Final Estimates	Crop estimates from largely varying crop seasons across India	Final yield estimation resulting from compilation and finalization of crop cutting experiments	December–January of next cropping year (discontinued since 2003/04)

for sorting out problems in preparation of these estimates in a timely and orderly manner.

Some of the major issues identified in the case of advance crop production estimates are that the present technique is mostly subjective and is not based on a sound statistical technique. The flow of information from different generating agencies is not time bound and appropriate. The DES is losing the confidence of users' groups due to frequent changes in production figures and mostly the differences in the forecasted estimates are huge. These differences create a lot of confusion and doubt among users. There is a strong need to develop suitable forecasting models which integrate information from different sources on parameters related to crop production such as crop conditions, agro-meteorology and water availability.

Sampling approach for national-level crop area estimation

All the states in the country are divided into four broad categories by the Government of India for the purpose of crop area estimates. The first category covers temporarily settled states and union territories (UTs), which have been surveyed for cadastral details, where area and land-use statistics are built up as a part of the land records maintained by the revenue agencies (referred to as 'Land record states') (this is the first column in Fig. 8.1). This system is followed in 17 major states and four UTs.[1] Together they account for about 86% of the reporting area in the country and they are covered under the Timely Reporting Scheme (TRS). Under this scheme, 20% of villages are selected at random for enumeration of statistics of the complete area.

The second category covers permanently settled states where area statistics are collected on the basis of sample surveys. These states account for about 9% of the reporting area in the country (this is the second column in Fig. 8.1).[2]

In the third category of states (North-Eastern Region states and UTs),[3] the area statistics are based on an impressionistic approach and they account for 5% of the reporting area in the country (this is the third column in Fig. 8.1).

In the fourth category, non-reporting areas (the final column in Fig. 8.1), the areas of Jammu and Kashmir, Pakistan occupied Kashmir (PoK) and adjoining areas of China are covered. In these regions, cropped areas have not been reported due to problems of non-accessability.

Remote sensing approach for cropped area estimation

Remote sensing has evolved from sporadic aerial overflights to frequently repeat high-resolution coverage, and from black and white film coverage to multispectral digital scanners. Computer processing has progressed to digitally segregating points on the ground to distinguish crop types and other land covers. Technology advancements have greatly increased the ease of area frame construction, maintenance and sampling, and this has reduced sampling variation both through better stratification and through regression estimates.

8.2.2 Approaches for crop yield estimations

Crop yield is calculated as total production in a unit area (acre or hectare), i.e. total harvested yield divided by total harvested area of the crop.

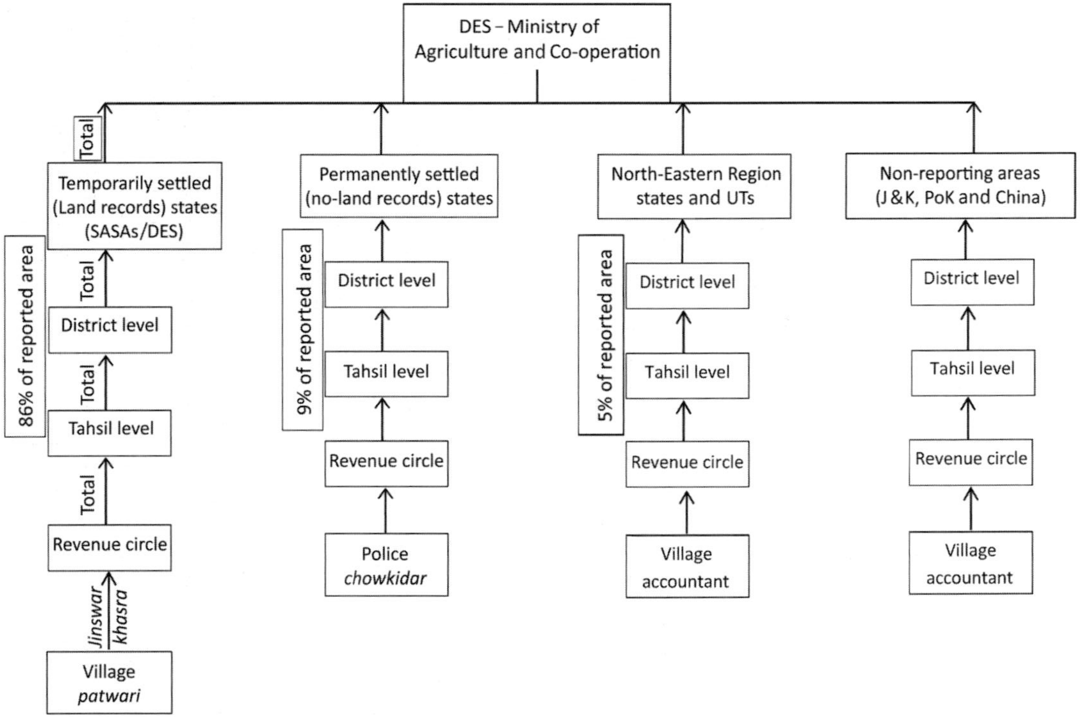

Fig. 8.1. Schematic diagram of the Directorate of Economics and Statistics (DES)' approach for crop area estimation through various ways of sample surveys in India. J&K, Jammu and Kashmir; PoK, Pakistan occupied Kashmir; SASAs, State Agricultural Statistical Authority; UTs, union territories.

Subjective yield surveys: farmers' yield estimations

Before the advent of the crop-cutting-experimental approach, the estimation of yield per acre or hectare was based on a subjective method of judging the crop in a season on visual observation by government officials in terms of a condition factor, the so-called *annewari*, and multiplying it with the normal or standard yield (fixed once in 5 years or so for an average crop under average conditions) to arrive at an estimate of total production in the state or in the country.

Objective yield estimation from crop cutting experiments (CCEs)

The success story of shifting from this subjective approach to an objective approach of CCEs for the estimation of crop yields started during 1940, particularly by the agricultural statisticians. In the past, even in advanced countries, crop yield estimates were based on periodic reports from crop reporters, mostly farmers who voluntarily supplied the statistics based on their personal impression of their crop yield.

8.2.3 CCEs: Government of India approach

Organizational roles and responsibilities

The collection and compilation of crop acreage and yield statistics, consolidation and timely publication of the statistics furnished by the states are primary responsibilities of the DES coordinated by the Ministry of Agriculture, Government of India. In each state, the Ministry of Agriculture and Co-operation coordinates the activities of CCEs assisted by the State Agricultural Statistical Authority (SASA) named in many states as the DES. The activities of CCEs relating to planning the surveys, training of field staff, organization of field work and tabulation of data are done by the DES in each state/UT. At the central level, the NSSO is responsible for the overall planning and organization, including supervision of the field work. The IASRI provides the technical guidance and development of the necessary methodology. The crop yields are estimated on a random sampling basis by conducting CCEs in all the states. Since the introduction of this objective method for estimation of crop statistics, steps have been taken from time to time to improve agricultural statistics in terms of coverage, scope, accuracy, standardization and coordination. In particular, to improve upon the timeliness and quality of statistics, TRS and Improvement of Crop Statistics (ICS) were introduced in the early 1960s and 1970s, respectively. The HLCCs on Agricultural Statistics have been set up in the states by the Government of India; also a Technical Working Group on Agricultural Statistics has been set up to serve as a watchdog authority to look into the progress of crop production in each state and technical problems arising therefrom. Zonal meetings of the SASA, after grouping the states into four zones, are convened from time to time.

Sampling approach for national-level yield estimates

The second most important component of production statistics is yield rate. The yield estimates of major crops are obtained through analysis of CCEs conducted under scientifically designed GCES. At present over 95% of the production of food grains is estimated on the basis of mean yield obtained from the CCEs. The NSSO has been providing technical guidance to the states and UTs for organizing and conducting crop estimation surveys for estimating yield rates of principal crops. In addition, the NSSO in collaboration with states/UTs implements sample check programmes on area enumeration work, area aggregation and conduct of CCEs under the ICS scheme. While executing the programme of sample checks on CCEs, the FOD associates itself with the operational aspects of conducting CCEs, including selection of sample villages, training of field staff and supervision of field work, and in the process, gathers micro-level information relating to the conduct of CCEs for estimation of crop yield. The results of crop estimation surveys are analysed and published annually by the NSSO as reports entitled 'Consolidated Results of Crop Estimation Surveys on Principal Crops'.

The primary objective of GCES is to obtain fairly reliable estimates of average yield of principal food and non-food crops for states and UTs. These estimates of yields are generally adopted for planning, policy formulation and implementation. The main goal of the objective area and yield measurements programme is to estimate or forecast crop yields by direct measurement of plant characteristics; and yield estimation surveys have been on a probability basis from the inception.

Sampling design and scales of estimation

The sampling design of the crop surveys is a stratified two-stage random sampling with Revenue Inspector Circles as strata, villages in the circles as the primary sampling units, fields growing the crop in selected villages as the secondary sampling units and standard plots (usually of size 5 m × 5 m) in selected fields as the ultimate sampling units (as depicted in Fig. 8.2). Two crop fields are selected in each selected village with one crop-cut per field undertaken when the crop is ready for harvest. Generally 80–120 experiments are selected in a major crop-growing district, where the area under the crop in the district exceeds 80,000 ha or when the cropped area ranges between 40,000 ha and 80,000 ha but this area exceeds the average cropped area per district in that state. Otherwise, the district is considered as a minor district for a given crop. On average, about 44 or 46 experiments are planned in a minor district.

When a crop sample is harvested for yield estimation, the produce is weighed and from a subsample of the plot, the produce is stored and reweighed after drying, so that appropriate allowance is made for dryness

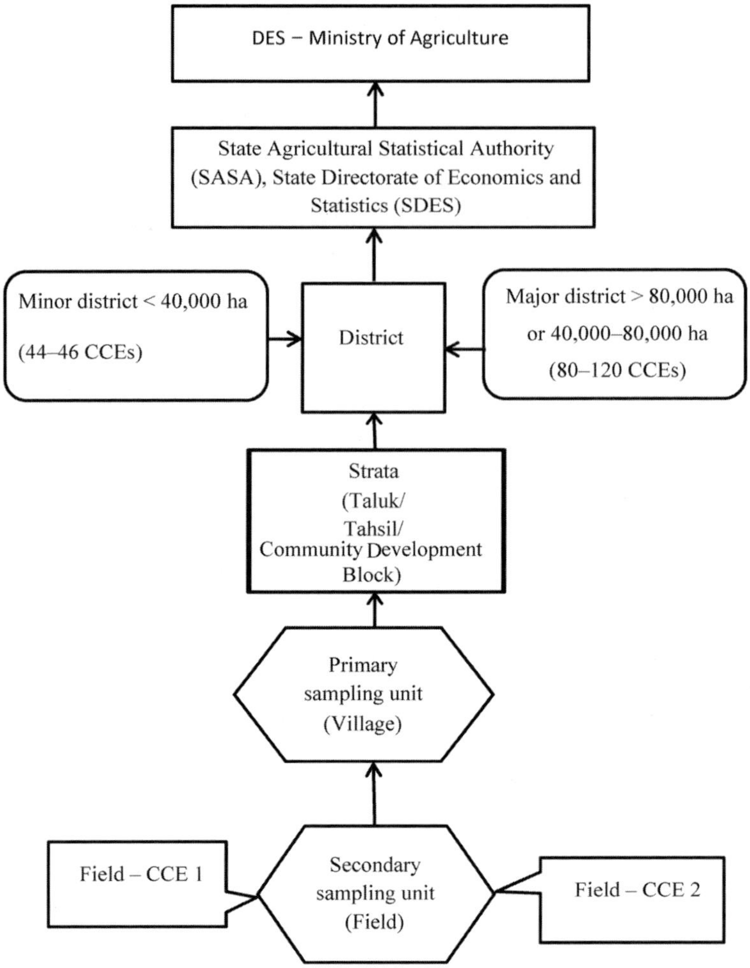

Fig. 8.2. Schematic diagram of the Directorate of Economics and Statistics (DES)' approach to crop cutting experiments (CCEs) for crop yield estimations in India.

in the estimation of the crop yield. In addition, farmers' inquiry-based yield data are collected by an agency which is different from the one used for collecting the CCEs wherever possible. Perhaps the village-based persons who are best suited for the purpose of collecting farmers' yield estimates report the data.

Representative sample size

In India, the crop yield estimates are generated on the basis of scientifically designed CCEs conducted under the scheme of GCES. The GCES covers a total of 68 crops in 25 states and four UTs of India. More than

800,000 CCEs are conducted annually for this purpose. The size of the sample collected through GCES varies with crop type among different states. In the majority of the states in India, cereal and legume crops are generally sampled in an area of 5 m × 5 m per sample except in Bihar, Haryana, Maharashtra and Jharkhand where the sample size has been bigger (10 m × 5 m). In almost all states, samples of wide-spaced row crops such as pigeon pea (*Cajanus cajan*), castor (*Ricinus communis*), cotton and sugarcane are harvested in an area of 10 m × 5 m or 10 m × 10 m. These sample sizes are generally sufficient to provide appropriate estimates of crop yield at district level. However, realizing that collection of such a vast number of samples is time-consuming as the procedure adopted is very tedious for conducting the CCEs (as given in the previous section), the DES introduced the ICS scheme. In order to ensure high quality of data of the GCES and that enumerators follow appropriate techniques for CCEs, NSSO and SASA have been assigned the responsibility of implementing ICS scheme jointly. Under this scheme, a quality check on the field operation of GCES is carried out by supervising around 30,000 CCEs by the NSSO and state government supervisory officers. The findings of the ICS results reveal that the CCEs are generally not carried out properly, resulting in data that lack the desired quality. Due to limitation of infrastructure and constraints of resources, there is a need to drastically reduce the sample size under GCES so that the volume of work of the enumerator is reduced and also better supervision of the operation of CCE becomes possible, leading to improvement in data quality.

Although the CCEs supervised by the ICS are fewer in number, around 30,000 in the entire country, the data collected are of very high quality. The estimates generated using these data are relatively free from various sources of non-sampling errors. Hence, it is recommended that wherever it is not possible to conduct an adequate number of CCEs due to constraints of cost or infrastructure, a small area estimation technique can be gainfully used to generate reliable estimates of crop yield based on a smaller sample to obtain more precise estimates than the ordinary model-based small area estimates.

Sampling error and bias in estimations and level accuracy

The total error in crop yield estimates is a combination of sampling and non-sampling error. Sampling error is the error associated with the selection of a sample population rather than using the whole population. Where a census covers the entire farming population, there is no sampling error. Non-sampling errors are those that arise from data collection, data entry procedures that include incorrect sample listing, response errors and biases, measurement errors and biases, incorrect form filling, and errors made while transferring, entering and processing data (Poate, 1988; Carfagna, 2007). While errors are random deviations from the actual value, biases are consistent in estimation or underestimation of the actual value. The choice of method to estimate crop area, crop production and ultimately crop yield may influence the sampling error, as the choice of method has implications for the required sampling strategy and the non-sampling errors due to method-specific susceptibility to specific response and measurement errors.

The National Statistical Commission (2001) headed by Dr Rangarajan gave a number of suggestions to improve the system of data collection on crop area and crop yield estimations.

8.2.4 A consensus approach

Determining the cropped area of a village/taluk/district

Bhoochetana activities were handled by a consortium of officials from the Department of Agriculture (DoA), Karnataka and the International Crops Research Institute for the Semi-Arid Tropics (ICRISAT) at each level of hierarchy. Emulating the salient features of the approach adopted by DES, Ministry of Agriculture, ICRISAT-DoA consortium adopted a consensus approach in estimating cropped area and crop yields. In the Bhoochetana project, the Government of Karnataka set targets for the crop area to be covered by productivity enhancement options. In each taluk of the district, DoA officials along with ICRISAT staff communicated adoptable technologies by suitable means of dissemination pathways. They initiated enumeration of the area sown to each crop for improved productivity each year during the project period. This enumeration also facilitated estimation of inputs to be placed at the village for farmers' requirement. ICRISAT and DoA field staff visited farmers' fields at different stages of crop growth and photographed the crop status to report on the crop condition. These visits helped to validate the cropped area reported by DoA staff in each village/taluk and the total area covered by the crop in the district. We reported the consensus final estimate of net cropped area against the target area set for the project activities in each district before the harvest season for each crop.

Inspection of fields during the season

Before harvesting the crop for sampling, ICRISAT scientific officers/research technicians in consultation with the concerned Agriculture Officer of the DoA for the area agreed on the number of CCEs to be harvested in different villages of each taluk. They also discussed the timing of crop sample harvests and processing of samples collected for crop yield estimations. A general agreement was reached on what type of drying and shelling process should be used.

District-level CCE Committee for random selection of fields and sampling

A district-level CCE Committee was formed with the responsibility for undertaking the CCEs in Bhoochetana plots. The committee is chaired by the concerned Joint Director of Agriculture (JDA) of the district as the data need to be integrated in the state statistics for agricultural production. The CCE Committee comprises the District JDA (chair) and members representing the DoA, the DES, the Watershed Development Department (WDD), the University of Agricultural Sciences (UAS), and ICRISAT represented by a research technician, farm facilitator and lead farmers serving the committee to ensure ownership for the data. Two major crops were identified for CCEs in each taluk of a district based on the DoA project planning for Bhoochetana at its initiation. The Assistant Director of Agriculture (ADA) and Agricultural Officers (AOs) along with the ICRISAT research technician were responsible for identifying the crops in their districts to ensure the selection of major crops in terms of area coverage under Bhoochetana. Based on the registration, the data with the officials and the technician, ten farmers for each crop in a taluk were selected for the two identified crops. Three to four representative villages were selected, encompassing different zones of soils, seasonal rainfall and area coverage under Bhoochetana. Three to four farmers were selected randomly based on the registrations in the selected village. However, a minimum number of ten farmers were duly selected per crop in each taluk. Each farmer was provided with a unique identification number (UIN) by ICRISAT before the CCEs were initiated in the season. The concerned In-charge Scientist/Scientific Officer at ICRISAT was responsible for ensuring timely supply of harvest bags (muslin cloth bags for stalk and kora cloth bags for pod/head samples) UIN and necessary data sheets for the CCEs in the district.

The improved practice (IP) and farmers' practice (FP) samples were duly collected from the same farmer's field. In each selected farmer's field for IP, a randomly selected representative area of 5 m × 5 m (total area of 25 m^2) at one spot was identified for undertaking CCE. Similarly, in each of the selected Bhoochetana farmer's field for FP, a randomly selected representative area of 5 m × 5 m (total area of 25 m^2) at one spot was identified for undertaking CCE. In the demarcated areas, all crop plants were harvested at the ground level from each plot separately and then pods/heads were separated from the stalk and total fresh weights were recorded separately for pods/heads and stalk (25 m^2) after checking the measuring balance properly each time. A representative minimum subsample of 2 kg was collected from ear heads/pods as well as stalks separately and fresh weights were recorded. These subsamples were placed in the sampling bags and properly labelled with the UIN of the farmer. Care was taken to record the subsample fresh weight of pods/heads and stalk immediately after subsamples were drawn to ensure that it is a minimum of 2 kg for each subsample. Thus two samples were collected (one IP and one FP) from each selected farmer's field, and the weights were properly recorded in the given format and the signatures of all the representatives of the CCE Committee present in the field were obtained. These subsamples were dried in the sun before sending them to ICRISAT for calculating the yield (in kilogrammes per hectare, kg/ha) on a dry weight basis. We ensured that all the identified team members participated in CCEs and the concerned JDAs had delegated the responsibilities to the ADAs and AOs for undertaking CCEs in the respective taluks.

GPS (geographic positioning system)-enabled photographs of CCEs had to be provided to the JDA office. A watershed or a village was considered as a homogeneous unit for selection of fields for crop sampling and yield estimation. Based on the cropped area of the village, 20% of the farmers' fields were randomly selected in a village for crop sampling on different toposequences. However, a minimum of three to five fields were sampled even if the number of farmers' field experiments were fewer. For example, if a village had ten farmers' fields applied with micronutrients or a new variety as an improved management practice, at least three to five of those fields were sampled in a village.

Sample process, drying and yield estimations

The fresh weight of each whole-plant sample was recorded for a unit sample area. A small subsample of whole plants was randomly collected and the pods/heads separated from the stalk for each sample and the fresh weights were recorded. Crop harvest samples were collected with care and accuracy in coordination with staff of the DoA and the sample materials (soybean, cowpea (*Vigna unguiculata*), mung bean, black gram (*Vigna mungo*) or groundnut (*Arachis hypogaea*)) were sun-dried before dispatching to ICRISAT because fresh samples of legume crops are succulent and may get spoiled in transport. At ICRISAT, Patancheru, the samples were oven-dried at 65°C for 48 h to completely dry the plant material until the moisture content was less than 12%, and then dry weights of these samples were recorded. This process facilitated the measurement of dry weights of bone-dry plant parts without excess moisture. In this process, we calculated dry weights of stalk and pod or head per sample area and the final yield estimation was calculated based on sample area (per hectare).

Supervisory responsibility

Crop sampling is generally required in field experiments for estimation of yield attributes such as grain, biomass and fodder per hectare or per acre and also to compare the differences between several crop management options as treatments. In the Bhoochetana project, we followed the guidelines which are statistically valid for small-area crop sampling procedures in farmers' field evaluations with joint participation of the AO of the DoA and farmers for proper (unbiased) estimation of crop yields. Statistically acceptable sampling procedures include collecting randomized samples in a uniformly treated area of crop, to resolve any discrepancy by arbitration and consensus estimation. The supervisory responsibilities at district level also involve cross validation of crop yield estimation at taluk and district levels by comparing with the CCE yield estimation by the DES. A summary of comparisons of the contrasting measures adopted for crop sampling in the Bhoochetana project for crop area and crop yield estimations is provided in Table 8.2.

Statistical approach for the comparison of yield estimates

ICRISAT and DoA staff jointly collected farmers' reported crop yield estimations (subjective yield estimations) soon after these farmers harvested and processed their crop produce but before the cropping year concluded. Studies at IASRI reveal a fairly high degree of correlation between the farmers' estimates and the estimates obtained through the crop cutting approach. The farmer is expected to provide a reasonably accurate estimate of crop production if these data are collected within 2–3 days after harvest or threshing. Studies conducted elsewhere in several African countries (Scott *et al.*, 1988) suggest that such estimates as obtained through farmers' inquiry are in close agreement with the actual production figures. Hence we compiled three data sets: (i) one from DES-CCEs crop yields reported taluk-wise on their website; (ii) the second data set from ICRISAT-DoA estimated crop yields in FP as well as IP treatments from CCEs in Bhoochetana; and (iii) the third data set from farmers' reported crop yields.

We calculated the estimated bias of DES reported crop yields from DES-CCEs and crop yields from FP treatment in Bhoochetana compared with farmers' reported crop yields from farmers in Bhoochetana villages in different taluks of these districts. We also presented crop yields from IP in farmers' fields from the same villages in a taluk. Bias is the difference between the true value/reference value (in this case farmers' reported yields) and the observed/estimated yields from the same village/taluk. Bias is sometimes referred to as the absolute correctness of the measurement system relative to a standard. An estimator or expected value with zero bias is called unbiased, otherwise the estimator is said to be biased. When bias is measured with respect to the mean (expected) value, it is distinguished as the mean-bias:

Bias = Mean of yield observations (estimates) – Mean of reference value (farmers' reported yields)

Table 8.2. Comparison of processes in General Crop Estimation Survey (GCES) crop cutting experiments (CCEs) and Bhoochetana CCEs for crop area and yield estimations.[a]

Process	DES method	ICRISAT method	Reasons for modification
Crop area estimation	Area estimates are prepared by patwari (village assistant)	Research technician (ICRISAT), farm facilitators and lead farmers collect crop area estimates by personal contact with farmers and visits to fields	To have accurate and assured recording of the cropped area
Supervision of work	Multifarious functions of patwari are monitored by revenue officials with minimal supervision by ICS officials	HLCC of ICRISAT and DoA; the joint team of ICRISAT and DoA officials includes agricultural officers and scientists/scientific officers of ICRISAT	To cross-check and improve the quality of field sampling
Sample selection and measurements	Two fields are sampled in a village; measurements are not proper	Samples are randomly collected and a greater number of samples are collected per village, and measurements are counter checked and entered properly	Random sampling improves the accuracy of estimations statistically
Sample processing	Crop-wise dryness is calculated and yields are estimated	Samples are oven-dried at 65°C for 48 h to attain less than 12% moisture in the fodder as well as the grain	To ensure standard weight of marketable produce as farmers' practice

[a]DES, Directorate of Economics and Statistics; DoA, Department of Agriculture; HLCC, High Level Coordination Committee; ICRISAT, International Crops Research Institute for the Semi-Arid Tropics; ICS, Improvement of Crop Statistics.

8.3 Results from Bhoochetana Crop Yield Estimation

We estimated taluk-wise crop yields with crop samples collected from FP and IP treatments in farmers' fields of different villages in a taluk. We focused on estimating crop yields for two to three major crops in a district in both rainy and post-rainy season crops. For a comparison, the number of crop samples taken during the 2009–2013 post-rainy season chickpea crop under the Bhoochetana project and DES-monitored CCE in various districts of Karnataka is shown in Fig. 8.3. The number of crop samples harvested under CCEs was phenomenally high and the extent of staff time and expenditure must have been proportionally high. This exercise might have led to sampling and non-sampling errors due to the volume of work, besides loss of sampled crop material from farmers' fields. District-wise crop yield data of post-rainy season chickpea from farmers' fields during the 2009–2013 crop seasons showed consistently lower grain yield estimates from DES-monitored CCEs (Fig. 8.4). In all other crop yield estimates based on CCEs of the DES, lower grain yield estimates were reported during this period. Bias is very much seen in DES-reported crop yield estimation for post-rainy season chickpea (Fig. 8.5a) and rainy season groundnut (Fig. 8.5b) compared with FP yields of these crops, respectively. Groundnut yield in the rainy season and chickpea yield in the post-rainy season clearly indicated that DES-monitored crop yield estimations were much lower compared with crop yield estimated from FP. When we considered bias between FP crop yield and IP crop yield estimates, the crop yields were higher in IP, which is expected as improved management would generally lead to higher yields. Taluk-wise mean crop yields in IP treatment across the years are normally distributed.

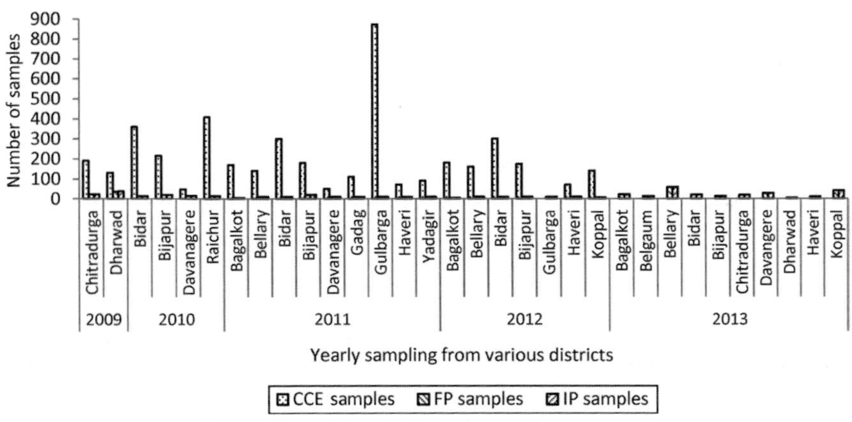

Fig. 8.3. Number of crop samples of post-rainy season chickpea in various districts under the Bhoochetana project and crop cutting experiments (CCEs) of the Directorate of Economics and Statistics (DES) in Karnataka. FP, Farmers' practice; IP, improved practice.

We plotted bias between crop yield estimates as farmers' reported yield, yield from FP and yield estimates from DES-monitored CCEs for maize (Fig. 8.6a), post-rainy season sorghum (Fig. 8.6b), rainy season groundnut (Fig. 8.6c) and chickpea (Fig. 8.6d). In all these plots, it is clearly evident

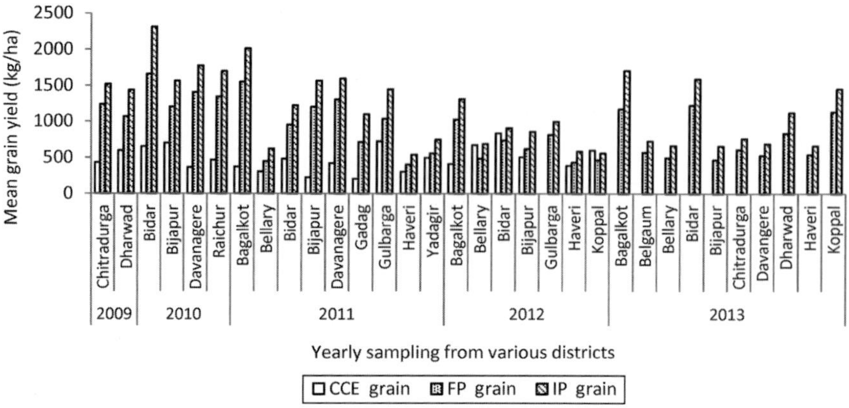

Fig. 8.4. District-wise post-rainy season chickpea yield estimates computed from village-level sampling in farmers' fields during 2009–2013. CCE, Crop cutting experiments; FP, farmers' practice; IP, improved practice.

that crop yields in FP-Bhoochetana treatment are positioned around the 1:1 line, which shows low bias of farmers' reported yields compared with DES-reported crop yield estimates based on CCEs for all crops.

8.4 Discussion

India is a huge country encompassing 1.42 million ha of cultivable land with a crop intensity of around 1.39. In the Indian context, several factors contribute to a complex situation to precisely assess food and fodder security during a crop season, which is the main concern of the planners and policy makers in the government. These factors include availability of irrigation, which varies widely in some cropping years from one crop season to three crop seasons in many regions. Moreover, interannual rainfall variability contributes to production uncertainties in rainfed areas regionally. Fragmented landholdings and several cropping options practised by smallholders to meet domestic food and fodder requirements contribute further to a chaotic situation for policy makers, besides other socio-economic issues. The Ministry of Agriculture, Government of Karnataka required accurate cropping area and yield estimates for major rainfed crops in the state under Bhoochetana for proper planning of input supplies (seed, macro- and micronutrient fertilizers) and policy implementation regarding import of grains in case of deficit production or disposal of a marketable surplus through exports in case of excess production, in order to fetch remunerative prices for farmers in time. These estimates are usually obtained in the state from the analysis of a large number of CCEs and are available much later after the crops are harvested, which leads to chaotic situations at times for policy decisions.

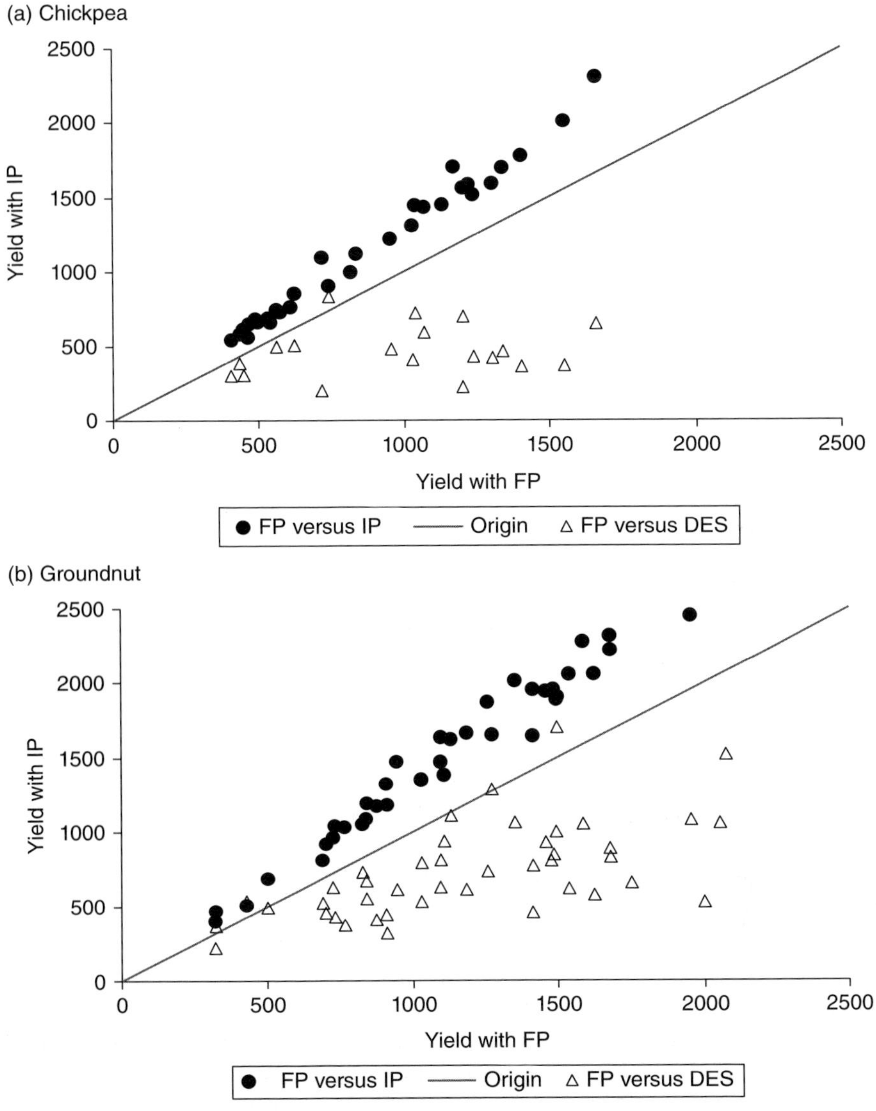

Fig. 8.5. Bias in district-level crop yield estimates (kg/ha) computed from village-level sampling in treatments of farmers' practice (FP), improved practice (IP) and crop cutting experiments (CCEs) from farmers' fields as against farmers' reported crop yield during 2009–2013: (a) chickpea; and (b) groundnut. DES, Directorate of Economics and Statistics.

In the context of scaling up of adoptable technologies under the Bhoochetana project, we reviewed the many approaches that are followed for crop yield estimations since the beginning of organized crop yield estimates started during the 1940s–1950s in India, to the currently proposed advanced approaches in the digital era for the reasons

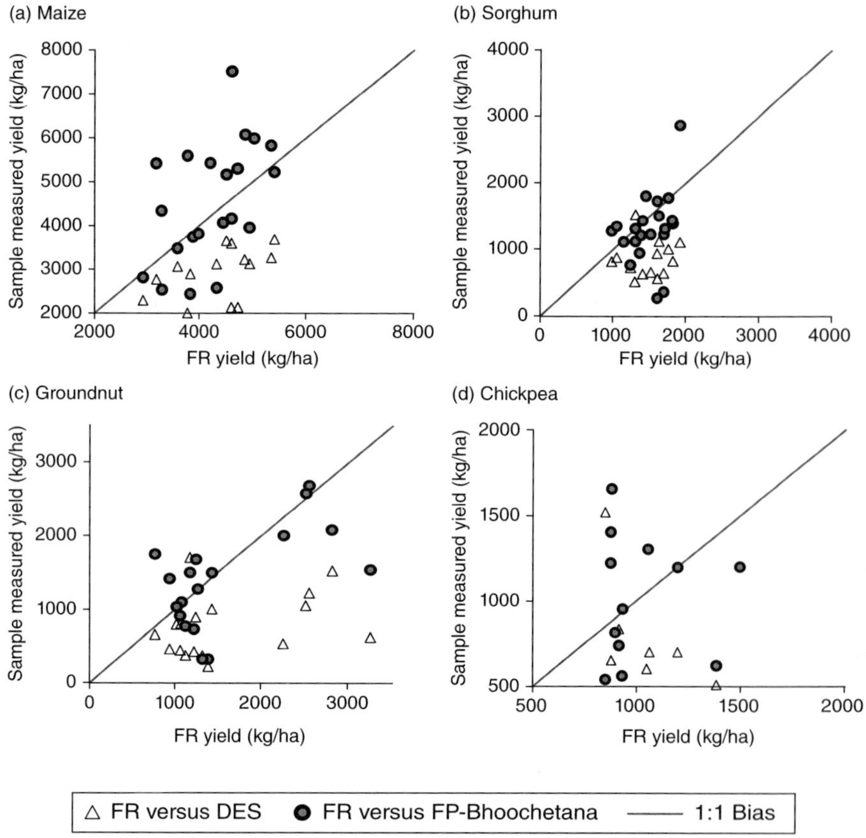

Fig. 8.6. Comparison of mean grain yield of four crops across districts in Karnataka during 2009–2013: (a) maize; (b) sorghum; (c) groundnut; and (d) chickpea. DES, Directorate of Economics and Statistics; FP, farmers' practice; FR, farmers' reported.

of accuracy, appropriateness at scale, practicability and economic feasibility. In the Bhoochetana project, we attempted to address the issue of timely availability of crop yield estimates by considering a statistically acceptable approach of stratified unbiased random sampling of crops based on an objective sampling technique. This method was properly followed to provide reliable estimates of crop yields just before the farmers harvest their crops. We considered the robustness of CCEs in coverage but the ICS scheme aimed to improve the quality of crop statistics and reviewed GCES procedures over a number of years and reported that 20% of the sample estimates were erroneous with several lacunae. The National Commission on Statistics (2001) headed by Dr Rangarajan made several recommendations to improve the accuracy of area statistics and crop yield estimates. Considering these recommendations, we conducted baseline area surveys after thorough training of farm facilitators and experienced ICRISAT field staff before the season commenced. We addressed all recommendations to avoid the lacunae in sampling by employing a sufficient number of farm facilitators and lead farmers along with DoA staff and ICRISAT scientific staff to record and report the exact sown and germinated crop areas, different dates of sowing opportunities, ratio of crop mixtures, sequential crops and management practices from each village in a taluk. Weekly reporting by field staff ensured timely updates of data in the district to avoid lapses.

Some of the precautions we followed in the crop sampling were engaging a sufficient number of field staff by the DoA and ICRISAT. These staff strictly followed the sampling procedures agreed by the joint team of officials to avoid sampling errors and non-sampling errors. Crop sampling at harvest was conducted by ICRISAT field staff who were well versed with random sampling in the experimental fields to avoid sampling errors. ICRISAT field staff collected farmers' reported crop yields immediately after the harvest of crops in order to fetch authentic statistics. The HLCC of senior scientists from ICRISAT and JDAs in the districts participated in supervision to improve the quality of data. We clarified to the participating farmers in villages that rainfall data recorded and crops harvested for yield estimates at taluk level are not disclosed to crop insurance agencies. By this clarification, farmers allowed harvesting of crop samples without any hindrance and farmers revealed observed crop yields without any bias as they recorded. We took maximum care recording fresh sample weights in the field. Samples were dried in the driers at ICRISAT to attain low ($< 12\%$) moisture and recorded dry weights of grain and fodder to attain higher standards in yield estimation. Bias in crop yield estimations between farmers' reported crop yields and crop yields in FP treatment in the Bhoochetana project was much lower compared with DES-reported CCEs recorded crop yields. The CCEs-based crop yield estimates are underreported and distribution of crop yields is skewed, evidence of higher sampling error and non-sampling errors. We did not have access to standard error of means for DES-reported yield estimates used in our comparison. We suspect that farmers' reported crop yields in certain taluks might have been influenced by other local considerations. Furthermore, we did not have information on sowing dates of different crops for crop yield estimates of DES samples to corroborate with corresponding yield estimates from FP.

In the context of smallholders having multiple cropping options and the possibility of two or more cropping seasons in certain regions, CCEs for regional crop yield estimations is considered realistic, but uneconomical and time-consuming. A key financial benchmark is the cost of crop surveys on crop area and CCEs for crop yield estimations. For example, the FAO (2008) reported that average expenditure per country in Africa in 2007 was US$657,000, including estimation of crop area and yield, means of production and socio-economic information. It is a huge investment annually on capital and staff time for harvesting nearly

80,000 CCEs in India besides 30,000 ICS scheme samples for authentication of yield estimates.

New science tools such as remote sensing imagery and crop modelling to estimate agricultural statistics can be sustainable in the long term if their total cost can be budgeted for without endangering the feasibility of surveys that cannot be substituted by satellite technology. Besides, the requirement of yield estimations for rainfed crops in variable rainfall situations to meet the demands of the crop insurance sector has been demanding a greater number of CCEs. A crop growth monitoring system is being applied successfully within the framework of the MARS (Monitoring Agriculture by Remote Sensing) Crop Yield Forecasting System although there are uncertainties related to applying crop growth models over such large areas. However, uncertainties (in the form of: (i) variable sowing opportunity or sowing dates within season; (ii) the effect of drought due to limited sparse weather station density (only at *mandal* or taluk); (iii) highly variable soil parameters like soil texture, soil depth and water holding capacity; and (iv) the lack of information on number/amount of irrigations given and the weightage to individual simulation outputs to an area/strata or administrative region) can all be handled more efficiently with advanced cropping systems simulation models than a very large number of CCEs. However, for simulating crop yield estimations, the immediate focus should be on developing and validating crop models for various rainfed crops for quantifying and reducing uncertainty on the crop model simulations that are used for crop yield estimations and forecasting. The ultimate goal is to improve the accuracy and timeliness of regional crop yield estimations and forecasts. The basis for quantifying uncertainty is to use an ensemble of models where the variability in the outcome of the individual models within the ensemble is an indication of the uncertainty in the final crop yield estimations or forecast. Reducing uncertainty in yield estimates regionally can be attempted by combining crop model simulations with satellite-derived remote sensing information for single cropped large areas such as the wheat belt, rice belt, cotton belt and groundnut belt through an ensemble Kalman filter. However, with the greater number of cropping options practised by smallholders in India, use of remote sensing for crop area estimation instead of physical crop area enumeration is a distant possibility, as not much evidence is available in the literature, but to minimize the cost of CCEs for crop yield estimations, validated crop simulations would be the way forward in larger economies like India.

8.5 Conclusions

Various approaches practised for cropped area estimation and crop yield estimation in various countries, and especially in India, have been summarized in this chapter. A comparison of crop yield estimation prepared by DES, ICRISAT-DoA in Bhoochetana and farmers' reported crop yields in the project for different crops during the project period have been presented. The small area enumeration technique can be satisfactorily applied to produce reliable district-level estimates of crop yield using a CCE approach with good supervision and commitment from staff under the Bhoochetana project. Although the sample size is very small for a district, in the ICRISAT-DoA, Karnataka-supervised CCEs, the procedure adopted (e.g. for field selection, random sampling and measurement of yield parameters, post-processing of samples for moisture estimation and data recording) is of very high quality, eliminating sampling and non-sampling errors. The estimates generated using this data are expected to be relatively free from various sources of non-sampling errors. It is, therefore, suggested that wherever crop productivity enhancement activities are scaled up over larger areas, it is possible to conduct a small number of planned CCEs qualitatively, by gainfully employing small-area estimation techniques for crop area estimations supported by cropping systems modelling for more reliable crop yield estimation from smaller regions scaled-up to generate reliable yield forecasts in the season and crop yield estimates regionally.

Acknowledgements

We gratefully acknowledge the editors of this book, Dr Suhas P. Wani, Director, ICRISAT Development Center (IDC) and Dr K.V. Raju, Assistant Director and Principal Scientist, IDC, for giving us an opportunity to contribute this chapter. We sincerely thank all the JDAs in the districts of Karnataka for their commitment and support as Chairs of CCEs committees. We also acknowledge the staff of the DoA and farm facilitators in all districts for their active participation and meticulous execution of crop sampling.

We sincerely appreciate the efforts of coordinators, resident scientific officers and research technicians of ICRISAT for careful handling of sampling material and providing accurate data on crop sampling. We acknowledge the efforts of scientific officers and research technicians at ICRISAT for pre- and post-processing of crop samples for crop yield estimation from the Bhoochetana project in Karnataka.

Notes

[1] States: Andhra Pradesh (undivided), Assam (excluding hilly districts), Bihar, Chhattisgarh, Gujarat, Haryana, Himachal Pradesh, Jammu and Kashmir, Jharkhand, Karnataka, Madhya Pradesh, Maharashtra, Punjab, Rajasthan, Tamil Nadu, Uttar

Pradesh and Uttarakhand; and UTs of Chandigarh, Delhi, Dadra and Nagar Haveli, and Pondicherry.

[2] States: Kerala, Odisha, West Bengal, Arunachal Pradesh, Nagaland, Sikkim and Tripura.

[3] The hilly districts of Assam and the rest of the states in the North-Eastern Region (other than Arunachal Pradesh, Nagaland, Tripura and Sikkim), Goa, and UTs of Andaman and Nicobar Islands, Daman and Diu and Lakshadweep.

References

Carfagna, E. (2007) Crop area estimates with area frames in the presence of measurement errors. Paper presented at the Fourth International Conference on Agricultural Statistics, 22–24 October 2007, Beijing, China.

Census of India (2011) *Provisional Population Totals, Paper 1 of 201*. Dadra and Nager Haveli (Series 27). Directorate of Census Operations, Gujarat, India.

Food and Agriculture Organization of the United Nations (FAO) (1982) *The Estimation of Crop Areas and Yields in Agricultural Statistics*. Economic and Social Development Paper No. 22. FAO, Rome.

Food and Agriculture Organization of the United Nations (FAO) (2008) *The State of Food and Agricultural Statistical Systems in Africa – 2007*. RAF publication 2008/E. FAO Regional office for Africa, Accra, Ghana.

Hubback, J.A. (1927) Sampling for rice yield in Bihar and Orissa. Bulletin No. 166 of the Agricultural Research Institute, Pusa.

Minhas, B.S. (1987) Estimating food crop acreage and yields: Indian experience and its relevance in Africa. In: *Eurostat News*, Special Edition. Statistical Office of the European Communities, Brussels, pp. 184–192.

Murphy, J., Casley, D.J. and Curry, J.J. (1991) *Farmers' Estimations as a Source of Production Data*. World Bank Technical Paper 132. World Bank, Washington, DC.

National Statistical Commission (2001) *Dr Rangarajan Commission Report*. The National Statistical Commission, 5 September 2001, Vol. II, Agricultural Statistics. Available at: http://statsinfoindia.weebly.com/dr-rangarajan-commission-report.html (accessed 15 June 2015).

Panse, V.G. (1954) *Manual on the Estimation of Crop Yields*. Food and Agricultural Organization of the United Nations, Rome (reprinted in 1964).

Poate, C.D. (1988) A review of methods for measuring crop production from smallholder producers. *Experimental Agriculture* 24, 1–14.

Scott, C., Merchant, T. and Verma, V. (1988) Evaluation of crop-cut methods and farmer reports for estimating crop production: results of a methodological study in five African countries. Report issued by Longacre Agricultural Development Centre Limited, London.

Tikkiwal, B.D. (1992) Modelling through survey data for small domains. Paper presented at the Proceedings of International Scientific Conference on Small Area Statistics and Survey Design held on 30 September–3 October 1992 at Warsaw, Poland.

Wikipedia (no date) Arthashastra. Available at: http://en.wikipedia.org/wiki/Arthashastra#Economic_ideas (accessed 15 June 2015).

9

An Integrated Approach for Productivity Enhancement

Gajanan Sawargaonkar,* Sudi Raghavendra Rao and Suhas P. Wani

International Crops Research Institute for the Semi-Arid Tropics (ICRISAT), Patancheru, India

9.1 Introduction

During the past six decades, state agricultural universities (SAUs) in India, the Indian Council of Agricultural Research (ICAR) institutes and international centres have invested efforts to develop improved agricultural technologies for poor tropical and subtropical countries targeting the innovations that could yield quick benefits. However, agricultural research emphasizes mainly component- and commodity-based research involving improvement of the genetic base of potential crops in order to enhance productivity and improve resistance to pests and diseases, development of animal breeding, farm implements and machinery, fertilizer use, and other production and protection technologies. These technologies are mostly conducted in isolation and at the institute level, which enables the farmers to increase productivity, but this has led to overexploitation of natural resources, resulting in decreased factor productivity and resource use efficiency, and ultimately decreased farm productivity and profitability (Chadha *et al.*, 2004). Similarly, the products of these improvements were largely confined to areas of high agricultural potential, and they often benefited the more prosperous farmers, and not the poorest of the poor. In fact in many cases, these interventions have produced short-term gains, but resulted in long-term degradation of natural resources, namely, soil, water, biodiversity and non-cultivated land (Conway, 1997).

*g.sawargaonkar@cgiar.org

The Green Revolution also underestimated the complexity of the systems for crop production. For example, agriculture is usually only a small part of a broad livelihood portfolio that may encompass a wide variety of off-farm activities, including the raising of livestock and poultry, fishery and goat rearing among others. In this context, productivity enhancement is important, but risk reduction, improved food security and the maintenance of social capital have greater importance. Farming by poor people in the tropics and subtropics is subject to a multitude of exogenous influences, namely, rainfall variability, fluctuating input costs and market prices, dynamic land use changes and various other episodic events.

Therefore, it has been well understood with the past experiences that a different kind of approach is needed for sustainable agricultural growth to have a positive impact on the well-being of poor farmers in developing countries. Even though cutting-edge agricultural technology is the requisite, it has to be set in local contexts and be applied in ways that recognize the special conditions of poor farmers. This implies that researchers can no longer remain exclusively external actors, but need to engage themselves in action research to develop appropriate solutions together with resource users (Hagmann *et al.*, 2002). More emphasis should be given to judicious use of natural resources, management of risks (Korten, 1995), reduction of dependence on agricultural inputs, avoidance of long-term depletion of productive potential, and careful control of environmental externalities (Conway, 1997).

Apart from this, ever increasing population and decline in per capita availability of land in the country indicates practically no scope for horizontal expansion of land for agriculture and therefore vertical expansion is the only possible solution that can be achieved through integrating all farming components (Kuruvilla and Thomas, 2009). Thus the farm as a unit is to be considered and planned for effective integration of crop production and other allied enterprises; however, its complexity makes an interdisciplinary approach necessary to build the ability of the farmers to cope with uncertainty and complexity (Sanjeev *et al.*, 2010).

9.2 Yield Gap Analysis

Globally, rainfed areas are hot spots of poverty, malnutrition and degradation of natural resources. In India, out of 142 million ha of arable lands, 60% (85.2 million ha) is rainfed. The state of Karnataka in India has the second largest area (5 million ha) under rainfed agriculture after Rajasthan. Crop yields in dryland areas are quite low (1–1.5 t/ha) and are lower by two- to threefold of the achievable potential yield (Fig. 9.1) largely due to low rainwater use efficiency (35–45%). The potential of dryland agriculture could be unlocked using available scientific technologies for improving rural livelihoods through sustainable intensification of rainfed areas.

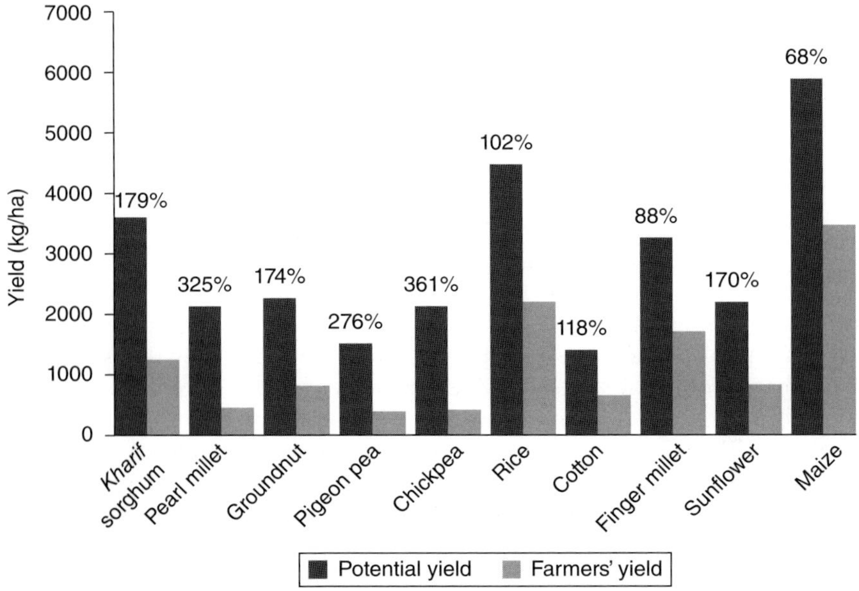

Fig. 9.1. Current farmers' yields are lower by two- to threefold than achievable potential yields. Values above the bars for potential yield are the percentage increase of the potential over the farmers' yield.

9.3 The Concept of an Integrated Approach

The integrated approach aims at an interrelation of a complex matrix of soil, water, plant, animal and environment and their interaction with each other, enabling the system to be more viable and profitable than the arable farming system and this leads to production of quality food. Thus a holistic approach (Fig. 9.2) is essential to consider the concepts of minimizing risk, increasing production and profits while improving the utilization of organic wastes and crop residues (Edwards, 1997; Jayanthi *et al.*, 2000; Radhammani *et al.*, 2003). Similarly, there is a need to have effective linkages and complementarities of various components to develop a holistic system (Gill *et al.*, 2009; Kuruvilla and Thomas, 2009).

Thus, to achieve rapid progress in the agriculture sector besides maintaining sustainability, our strategy must focus on conserving natural resources, enhancing efficient use of resources, increasing productivity and profitability, and improving quality and competitiveness through reducing the unit cost of production (Singh *et al.*, 2011).

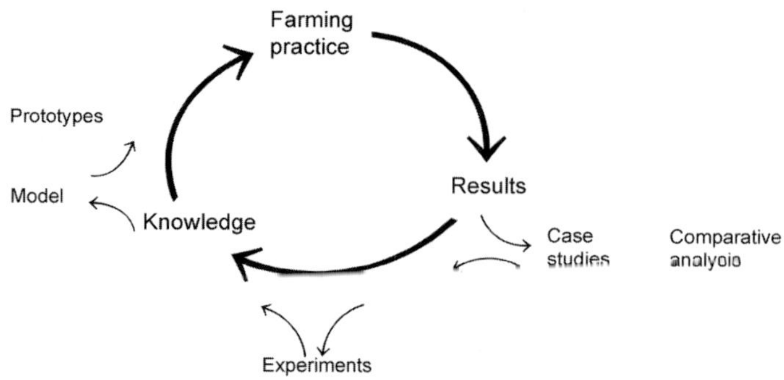

Fig. 9.2. The integrated farming system approach (Mogensen and Kristensen, 2000).

A major constraint in the agriculture of developing countries is the lack of strong linkages. Therefore, linkages between education, research and extension need to be strengthened to develop a holistic system approach to provide a cost-effective and eco-friendly agro-technological package for improving productivity, profitability and employment on a long-term basis in rural areas. In order to sustain a positive growth rate in agriculture, agricultural professionals including farmers, will need different types of reorientation and training that emphasize holistic concepts (Kuruvilla and Thomas, 2009).

Integration of technologies at the production system level results in several advantages, including:

- It gives a better understanding of farmers' livelihoods influencing their decision making and information processing capacity.
- It identifies and introduces multi-scale approaches connecting the farm to the landscape and farmers to the markets.
- It derives a set of resource development and utilization practices, which will lead to substantial and sustained increase in agricultural production.
- It improves long-term agricultural productivity, profitability and sustainability through adoption of a holistic approach by integrating all cost-effective and eco-friendly agrotechnologies.
- It minimizes pest infestations, namely pests such as insects, pathogens and weeds through management of a natural cropping system.
- It reduces the usage of chemical fertilizers, including harmful agrochemicals and pesticides, thereby protecting against environmental degradation.

In the mainstream productivity enhancement approach in Bhoochetana, the prime objective was to promote the standardized area-based technologies developed by the International Crops Research Institute for the Semi-Arid Tropics (ICRISAT) and SAUs as a package to the farmer in order to improve yields of the dominant crops. Here, we considered the farmers, extension officers, researchers and policy makers as stakeholders, involving them from the initiation of the development work. Lovell *et al.* (2002) concluded that scaling up to benefit many people is largely a function of planning and investment (Harrington *et al.*, 2001; Jones and Thornton, 2002; Lovell *et al.*, 2002) at the outset to create the enabling environment that will meet various preconditions for scaling up. Scaling up is most likely to happen if top-down and bottom-up approaches to development are properly reconciled as both the approaches are likely to be needed for an effective delivery of benefits from integrated genetic and natural resource management (IGNRM) (Douthwaite *et al.*, 2001; Lovell *et al.*, 2002).

The innovative Bhoochetana programme has adopted the integrated science-based approach to converge all the productivity enhancement interventions at production system level, as an area-based approach, funnelled into a holistic complete package to unlock the potential of rainfed agriculture. The adopted holistic approach provides a cost-effective and eco-friendly agrotechnological package to improve long-term rural productivity, profitability and employment in rural areas. Therefore, in this chapter, the strategic interventions, integrated and introduced at watershed scale, in Karnataka are addressed.

9.4 Integrated Watershed Management for Sustainable Development of Dryland Agriculture

Based on the evidence gathered from 3700 farmers' fields through the scaling-up phase of the Sujala–ICRISAT initiative in six districts of Karnataka, vast potential was revealed to exist in rainfed agriculture in Karnataka which was waiting to be harnessed. Hence, a strategy was designed and developed by a consortium for enhancing the productivity of rainfed crops in Karnataka through the innovative Bhoochetana initiative taking into consideration the watershed approach. Thus, Bhoochetana has adopted an integrated watershed management approach to bring together all the improved technologies to scale up the results of demonstrations on station, as well as on farm, that were carried out in different states of the country including Karnataka (Wani *et al.*, 2009). Thus, the results from the watershed case studies from the Sujala watershed programme formed the basis for the Bhoochetana initiative.

Integrated watershed development is the strategy adopted in the country for sustainable development of dryland areas and a recent comprehensive assessment of watershed programmes in India undertaken by an ICRISAT-led consortium revealed that the integrated watershed can become the growth engine for sustainable development of dryland areas by improving the performance of two-thirds of watersheds in the country

(Wani *et al.*, 2008). In most of the developed watersheds with concerted efforts to manage rainwater, the groundwater availability is improved in the watershed and also the downstream areas benefit from the increased groundwater recharge (Wani *et al.*, 2003; Sreedevi *et al.*, 2006; Pathak *et al.*, 2007). Increased water availability also had a positive impact in improving the welfare of women, reducing drudgery and protecting the environment. In a few well-managed watersheds, the productivity per unit of land and water increased substantially (Wani *et al.*, 2003). However, in most of the watersheds the agricultural production increased but the productivity per unit of land and water did not increase (Sreedevi *et al.*, 2006). Thus, more efforts are needed to harness the untapped potential of rainfed agriculture by adopting an integrated watershed management approach (Rockström *et al.*, 2007; Wani *et al.*, 2009).

Generally, water scarcity for food production refers to water for irrigation. Globally, green water (soil moisture) used for food production is threefold more (5000 km^3/year) than blue (irrigated) water (1800 km^3/year) for food production. The current rainwater use efficiency for food production is 35–45% and thus vast scope exists to improve the rainwater use efficiency (Wani *et al.*, 2003, 2009). There is a need to adopt more water use efficiency measures along with integrated management of water resources in watersheds for sustaining the development measures. In view of the dynamism and complexity of the production system, one of the prime objectives of the Bhoochetana programme was to improve the adaptive capacity of the system (i.e. its ability to sustain a flow of the diverse products and services that poor people depend upon) and thus increase its flexibility and ability to respond to exogenous influences (Lal *et al.*, 2001; Hagmann *et al.*, 2002; Lynam *et al.*, 2002).

Thus, an integrated approach at the production system level is the optimal solution for the future and includes: (i) pooling and sharing of resources/inputs at farm level; (ii) efficient use of family labour; and (iii) conservation, preservation and utilization of farm biomass, including non-conventional feed and fodder resources, effective use of manure/animal wastes and regulation of soil fertility. It involves a holistic IGNRM approach to increase sustainable economic returns and focus the activities on a few selected, interdependent, interrelated and often interlinked production systems management aspects.

9.5 Bhoochetana: An Integrated Approach at the Production System Level

Looking at the changing complex scenario, the Bhoochetana initiative undertaken by the Department of Agriculture (DoA), Government of Karnataka is considered as the pathbreaking approach for the development and inclusive agricultural growth through enhanced productivity in dryland agriculture. Before the start of the Bhoochetana programme in Karnataka, a huge yield gap existed in farmer's fields due to lack of outreach of the technologies developed by researchers in farmers' fields. Also, natural resource management is like jazz as the stakeholders change their aspirations in the context of exogenous factors which have unpredicted influences on farming. In view of all these challenges, the approach to promote IGNRM interventions must be based upon continuous dialogue (Zadeh, 1973; Holling and Meffe, 1996) and deliberation among stakeholders (Daniels and Walker, 1999; Hagmann, 1999; Maarleveld and Dangbégnon, 1999) and needs to engage all the stakeholders in positive action to develop appropriate solutions together with resource users (Hagmann *et al.*, 2002).

In this context, the Bhoochetana programme was planned to make a difference in the lives of the farmers of Karnataka through use of a science-based technological approach along with sustainable natural resource management. It is well proven that for enhancing the overall agricultural productivity and crop quality of the rainfed systems, balanced plant nutrition, the choice of crops and adapted cultivars along with soil, water and nutrient management practices need to be integrated at the farm level (Wani *et al.*, 2009; Sahrawat *et al.*, 2010a). Indeed, ICRISAT and its research partners most appropriately advocate the integration of genetics (crops and cultivars, social aspects) and natural resource management for targeting technology and greater impact of agricultural research in the semi-arid tropics (SAT). Thus, the Bhoochetana programme has adopted a holistic IGNRM approach and has also given equal importance to an integrated approach in soil, water and crop management starting from soil-test-based fertilizer management as an entry point, promoting good quality seeds of high-yielding as well as stress-tolerant cultivars of important agricultural crops, *in situ* soil and water conservation through appropriate landform treatments and low-cost integrated pest management modules in the entire state besides building the capacity of the stakeholders (farmers and consortium partners).

The strategy for using improved crop varieties was to introduce varieties that are adapted to the harsh conditions of the semi-arid tropical regions, especially water stress and nutrient deficiencies, and thereafter soil, water and nutrient management practices were developed and promoted around the adapted cultivars to realize the potential of the cultivars in diverse production systems (Ae *et al.*, 1990; Condon *et al.*, 2004; Passioura, 2006; Hiradate *et al.*, 2007; Bationo *et al.*, 2008; Sahrawat, 2009; Passioura and Angus, 2010). To achieve this, a convergence approach was adopted to integrate research and extension support and for backstopping along with capacity building of all the stakeholders (Wani, 2008; Sahrawat *et al.*, 2010a). The concept of farm facilitators and lead farmers as trainers to train large number of farmers was used for large-scale adoption of the developed integrated watershed management approach.

In order to bring the new science to the doorstep of the farmers using Bhoochetana as a platform, ICRISAT planned to use the experiences of the field demonstrations adopting an integrated watershed management approach in a more precise manner. The tested, improved agricultural technologies identified and introduced through the Bhoochetana programme in Karnataka are now discussed.

9.5.1 Integrated nutrient management

Soil and water are vital natural resources for human survival. The growing world population and increasing standard of living are placing tremendous pressure on these resources. Because the soil and water resources are finite, their optimal management without adverse environmental consequences is necessary, if human survival is to be assured and development is to be sustained. There is growing realization throughout the world that we can no longer afford to misuse these resources. Furthermore, these resources have to be managed using an integrated approach. In India, the problem of soil and water resource degradation has been in existence in the past; however, the pace of degradation has greatly increased in recent times due to the burgeoning population and the enhanced means of exploitation of natural resources. Apart from these anomalous situations, the problem of severe soil infertility exists in rainfed systems (Rego *et al.*, 2007; Bekunda *et al.*, 2010; Sahrawat *et al.*, 2010a) and managing water stress alone cannot sustainably enhance the productivity of rainfed systems. Hence for achieving sustainable gains in rainfed productivity both water shortage and soil fertility problems need to be simultaneously addressed through effective natural resource management practices (Wani *et al.*, 2009; Sahrawat *et al.*, 2010b).

Indian soils show deficiency of not only nitrogen (N), phosphorus (P) and potassium (K) but also secondary nutrients (sulfur (S), calcium (Ca), magnesium (Mg), sodium (Na)) and micronutrients (boron (B), zinc (Zn), copper (Cu), iron (Fe), manganese (Mn)) in most parts of the country. Field studies carried out by different researchers clearly indicate that integrated nutrient management (INM) comprising a soil-test-based nutrient management approach can be an important entry point activity and also a mechanism to diagnose and manage soil fertility in practical agriculture (Wani, 2008). Integrated nutrient management is considered as the timely application of all essential plant nutrients (which include primary, secondary and micronutrients) in readily available form, in optimum quantities and in the right proportion, through the correct method, suitable for specific soil/crop conditions. Components of INM include judicious use of chemical fertilizers based on deficient soil nutrients as established by soil testing in conjunction with other sources of plant nutrients such as organic manures and biofertilizers. Use of soil amendments for acidic/alkaline soils also need to be promoted to improve soil health and its fertility, thereby ensuring adequate availability of nutrients in soils to meet the requirement of plants at critical stages of growth and thus ensuring adequate soil humus to improve physicochemical and biological properties of the soil. Therefore soil sampling and its chemical analysis assumes greater importance for developing soil-test-based fertilizer recommendations more precisely at micro level.

Taluk-wise soil-test-based fertilizer recommendations

Soil and plant tests have long been used as tools to diagnose and manage soil fertility problems in the intensified irrigated systems and commercial crops, including fruit and vegetable crops, to maximize productivity (Dahnke and Olson, 1990; Black, 1993; Mills and Jones, 1996; Reuter and Robinson, 1997). However, soil testing has not been used to diagnose and manage nutrient problems in farmers' fields in the semi-arid tropical regions at a larger scale. Soils below the critical limits of the nutrients evaluated responded to the application of nutrients; although the overall crop response was regulated by the rainfall received during the cropping season (Rego *et al.*, 2007; Sahrawat *et al.*, 2007, 2010a). Soil-test-based nutrient application also allows judicious and efficient use of nutrient inputs at the local and regional levels (Black, 1993; Sahrawat *et al.*, 2010a). Our research at ICRISAT and several on-farm benchmark watersheds demonstrated that soil testing is a useful tool for diagnosing the nutrient deficiencies in farmers' fields (Sahrawat *et al.*, 2010a). To characterize the soil fertility status of soils, 92,904 soil samples were collected from farmers' fields in watersheds spread over 30 districts of Karnataka. Using stratified random sampling, we collected eight to ten cores of surface (0–15 cm depth) soils to make one composite sample and analysed for pH, electrical conductivity (EC), organic carbon (C), total N and extractable P, K, Ca, Mg, Na, S, Zn, Mn, Fe, Cu and B. Based on the results of soil analysis, soil-test-based fertilizer recommendations have been made for different crops at taluk level. A soil fertility atlas (Wani *et al.*, 2011) has been brought out, which is available in the public domain, making precious knowledge available to stakeholders such as policy makers, development agencies, extension workers, researchers and farmers of Karnataka.

Based on the use of critical limits for various plant nutrients, farmer fields were categorized as deficient or sufficient. Fertilizer recommendations were provided to farmers accordingly. Individual farmers were provided with soil health cards based on the nutrient status of the soils of the farmers sampled in the village following a stratified soil sampling methodology. The soils analysed were representative of the village soils and mean data were used for wall writing and booklet preparation. Soil nutrient status maps were developed and provided for each district using interpolated soil nutrient data at taluk level and geographical information system (GIS) techniques for the benefit of policy makers. Balanced fertilizer rates differ from area to area and also from crop to crop. Hence we used soil analyses results (e.g. in Raichur district, see Table 9.1) and seasonal rainfall as the basis to recommend fertilizer doses. Availability of organic manures, crop residues and biofertilizers was also considered to provide crop-specific taluk-level fertilizer recommendations in all 30 districts of Karnataka. One example of Raichur district is presented in Table 9.2 to show the fertilizer recommendations considering all the above-mentioned factors.

Table 9.1. Soil health status of Raichur district, Karnataka.

	Percentage of farmers' fields deficient in nutrients					
Taluk	Organic carbon (%)	Available phosphorus	Available potassium	Available sulfur	Available zinc	Available boron
Raichur	77	45	14	73	78	63
Manvi	79	49	0	70	85	18
Sindhanur	40	31	0	26	53	12
Lingasur	64	54	4	84	86	52
Devadurga	85	54	6	70	84	46
Raichur district	69	47	4	64	78	37

Promoting organic matter usage

The most complex and least understood component of soils is soil organic matter (SOM). SOM comprises all living soil organisms and the remains of previous living organisms in their various degrees of decomposition. The Soil Science Society of America defines SOM as the organic fraction of soil after the removal of undecomposed plant and animal residues. However, Magdoff (1992) proposed to consider SOM to be diverse organic materials, such as living organisms, slightly altered plant and animal organic residues, and well-decomposed plant and animal tissues that vary considerably in their stability and susceptibility to further degradation. Put simply, SOM is any soil material that comes from the tissues of organisms (plants, animals or microorganisms) that are currently or were once living. SOM is rich in nutrients such as N, P, S and micronutrients, and is composed of approximately 50% C. SOM plays a major role in maintaining soil quality as it positively influences a wide range of soil properties such as the provision of nutrients, water retention and release as well as reducing the risks of soil compaction, surface crusting and soil erosion.

Table 9.2. Taluk-level crop-specific fertilizer recommendation (kg/ha) for Raichur district, Karnataka.

		Fertilizer recommendation (kg/ha)					
Crop	Taluk	Urea	DAP[a]	Potash	Gypsum	Zinc sulfate	Borax
Pearl millet	Raichur	98	27	0	200	25	2.5
	Manvi	98	27	0	200	25	2.5
	Sindhanur	44	27	0	200	25	2.5
	Lingasur	88	54	0	200	25	2.5
	Devadurga	88	54	0	200	25	2.5
Pigeon pea	Raichur	22	54	21	200	25	2.5
	Manvi	22	54	21	200	25	2.5
	Sindhanur	0	54	21	200	25	2.5
	Lingasur	0	109	21	200	25	2.5
	Devadurga	0	109	21	200	25	2.5
Cotton	Raichur	157	43	33	200	25	2.5
	Manvi	157	43	33	200	25	2.5
	Sindhanur	70	43	33	200	25	2.5
	Lingasur	14	87	33	200	25	2.5
	Devadurga	140	87	33	200	25	2.5
Groundnut	Raichur	22	54	21	200	25	2.5
	Manvi	22	54	21	200	25	2.5
	Sindhanur	0	54	21	200	25	2.5
	Devadurga	0	109	21	200	25	2.5
Sunflower	Raichur	55	54	33	200	25	2.5
	Manvi	55	54	33	200	25	2.5
	Sindhanur	17	54	33	200	25	2.5
	Lingasur	33	109	33	200	25	2.5
	Devadurga	33	109	33	200	25	2.5
Paddy	Raichur	185	82	75	200	25	2.5
	Manvi	185	82	75	200	25	2.5
	Sindhanur	77	82	75	200	25	2.5
	Lingasur	154	163	75	200	25	2.5
	Devadurga	154	163	75	200	25	2.5

[a]DAP, Diammonium phosphate.

The Green Revolution followed by the era of modernization of agriculture overlooked organic matter as a component of fertilizer recommendations. In fact, the practices of continuous tillage, lack of replenishing SOM and imprudent crop rotation are the missing links in today's agricultural management. Therefore, in the Bhoochetana programme, the focus was to promote the use of organic sources, namely, farmyard manure, compost, vermicompost and green manuring crops (e.g. *Gliricidia)*.

In order to convert farm residues and organic wastes available in villages, vermicomposting technology was introduced involving earthworms which efficiently convert these residues into valuable manure.

Several compost pits were constructed in the targeted villages during the project period by converging different schemes of the DoA. Technology components mainly include selection and use of a non-burrowing type of earthworms (*Eisenia* sp. and *Eudrilus* sp.), use of weeds and crop residues and sericulture residues, animal and poultry manures and rock phosphate as materials. Training was imparted to rural women, self-help group members, farmers and farm facilitators to promote vermicomposting as an alternative rural livelihood option.

Green manuring

Green manuring was promoted through the Bhoochetana programme by distribution of *Gliricidia* seeds and seedlings to the farmers to plant on field bunds and boundaries (Fig. 9.3). Farmers were encouraged to plant 3–4-month-old plants from the nursery or cuttings of tender branches of *Gliricidia* at 50 cm spacing on field bunds. *Gliricidia* plants produce green leaves and succulent green branches abundantly, which are rich in N. The *Gliricidia* plants grown on bunds not only strengthen the bunds preventing soil erosion but also provide N-rich green biomass, fodder and fuel. The cross section of the earthen bund can also be reduced. A study conducted at the ICRISAT Development Center indicated that by adding N-rich green biomass from *Gliricidia* plants planted on a bund at 50 cm spacing for a length of 700 m could provide about 30–45 kg N/ha/year (Wani and Kumar, 2002). From the first year, farmers harvested the green leaves and loppings, leaving plants that were 1 m in height, and applied the harvested material to the topsoil for enriching organic C and nutrients in the soil. *Gliricidia* can be harvested two to three times a year and applied before sowing of the rainy season crop, post-rainy season crop and summer crop.

Fig. 9.3. *Gliricidia*: (a) planting on a bund in Joga village, Bellary District, Karnataka; and (b) used as a green manuring crop.

Biofertlizers

Besides the use of compost and green manure crops, the Bhoochetana programme also promoted the use of biofertilizers which are very important, low-cost, eco-friendly, organic agro-inputs, supplementary to chemical fertilizers. *Rhizobium*, *Azospirillum* and *Azotobacter* add N to the soil and phosphate-solubilizing bacteria make citrate-soluble P available to crops and also secrete certain growth-promoting substances. Biofertilizers are considered harmless and help to improve the soil structure (porosity) and water holding capacity, increase fertilizer use efficiency and ultimately help in increasing crop yield by 15–20%. Therefore, farmers were educated on biofertilizers and seed treatment.

Thus, the Bhoochetana programme ensured the adoption of good agricultural management practices and comprising organic materials as an integral part of fertilizer recommendations along with seed treatment. This approach therefore resulted in a 'win–win' situation for all the stakeholders with respect to increasing crop productivity besides positively improving soil fertility and at the same time maintaining the physical, chemical and biological conditions of the soil.

9.5.2 Soil and water conservation methods

Water-shortage-related plant stress is the primary constraint to crop production and productivity in the rainfed systems in the SAT and consequently the importance of water shortage has been rightly emphasized globally (Wani *et al.*, 2003; Molden, 2007; Pathak *et al.*, 2009). In rainfed agriculture, demand for rainwater can be met through efficient rainwater conservation and management. For this, both *in situ* and *ex situ* rainwater management play crucial roles in increasing and sustaining crop productivity. The comprehensive assessment of water management in agriculture (Molden, 2007) describes a large untapped potential for upgrading rainfed agriculture and calls for increased water investments in the sector. Based on experiences from the various watershed programmes and research station works in India, soil and water conservation practices for different agroclimatic zones of Karnataka were identified (Table 9.3). It clearly shows that for different regions the problems of soil and water conservation are quite different. This information was used to determine the appropriate *in situ* and *ex situ* soil and water conservation practices for various regions.

In situ *soil and water conservation*

In the Bhoochetana programme, efforts were focused on different low-cost *in situ* soil and water conservation measures, which can be practically implemented and adopted by the farmers. Field-based soil and water conservation measures are essential for *in situ* conservation of soil and water. The main aim of these practices is to reduce or prevent either water erosion or wind erosion, while achieving the desired moisture for sustainable production. The suitability of any *in situ* soil and water management practices depends greatly upon

Table 9.3. Proposed field-based soil and water conservation measures for various rainfall zones in Karnataka.[a] (From Pathak *et al.*, 2009.)

Seasonal rainfall			
< 500 mm	500–700 mm	750–1000 mm	> 1000 mm
Contour cultivation with: Conservation furrows Ridging Sowing across slopes Mulching Scoops Tied ridges Off-season tillage Small basins Contour bunds Field bunds *Khadin*[b]	Contour cultivation with: Conservation furrows Ridging Sowing across slopes Scoops Tide ridges Mulching Zingg terrace Off-season tillage BBF[a] Small basins Field bunds *Khadin*[b]	BBF (Vertisols)[a] Conservation furrows Sowing across slopes Tillage Small basins Field bunds Vegetative bunds Graded bunds *Nadi*[c] Terraces	BBF (Vertisols)[a] Field bunds Vegetative bunds Graded bunds Terraces

[a]BBF, Broad-bed and furrow.
[b]*Khadin*, Cultivation in a tank bed after receding the water in the tank.
[c]*Nadi*, Cultivation in a bigger tank bed after receding the water in the tank.

soil, topography, climate, cropping system and the farmer's resources. Based on past experience, several field-based soil and water conservation measures have been proposed and found promising for the various rainfall zones in India (see Table 9.3). Some of the promising *in situ* soil and water conservation practices promoted for the rainfed regions of Karnataka are contour cultivation, conservation furrows, broad-bed and furrow (BBF) and related systems, vegetable barriers, off-season tillage, etc.

Conservation furrow system

The conservation furrow is a simple and low-cost *in situ* soil and water conservation practice for rainfed areas with a moderate slope (Fig. 9.4). This practice is suitable for soils with severe problems of crusting, sealing and hard setting. Due to these problems, early runoff is quite common on these soils. In this system, a series of furrows are opened on the contour or across the slope at 3–5 m apart (Fig. 9.4). The spacing between the furrows and the size of the furrow can be chosen based on the rainfall, soils, crops and topography. The furrows can be made either during planting time or during the interculture operation using a country plough (Pathak *et al.*, 2007). Two to three passes in the same furrow may be needed to obtain the required furrow size. These furrows harvest the local runoff water and improve the soil moisture in the adjoining crop rows, particularly during the period of water stress. The practice has been found to increase the crop yields by 10–25% and it costs around ₹250–350/ha. To further improve its effectiveness, this system should be used along with contour cultivation (Tables 9.4 and 9.5).

Fig. 9.4. Conservation furrow system at Hedigonda watershed, Haveri, Karnataka.

Table 9.4. Mean crop yields in improved land and water management system in Karnataka.

Crop	Mean crop yield (t/ha)		Increase (%) in yield over farmers' practice
	Farmers' practice	Contour cultivation with conservation furrows	
Groundnut	1.19	1.49	25
Finger millet	1.63	2.29	24
Maize	3.33	3.91	19
Sunflower	1.25	1.42	14

Table 9.5. Comparison of benefit–cost ratio of improved versus farmers' land management systems in Karnataka.

District	Watershed	Crop	Benefit–cost ratio		Increase in yield (%)
			Farmers' practice	Improved practice	
Chitradurga	Maradihalli	Sunflower	1.78	2.05	15
	Toparmalige	Maize	1.83	2.05	12
Haveri	Aremallapur	Maize	1.74	2.06	18
Dharwad	Anchatageri	Soybean	1.77	2.18	23
Tumkur	Kanakapura	Groundnut	1.36	1.62	19
Kolar	Belaganhalli	Groundnut	2.14	2.34	10
	Belaganhalli	Finger millet	1.26	1.40	11
	Machanhalli	Finger millet	1.22	1.44	18

Broad-bed and furrow (BBF) and related systems
Soils with a high clay content, namely Vertisols, are often prone to waterlogging and thereby result in crop failure. Out of 30 districts of Karnataka, around 12–13 districts have Vertisols as the predominant soil type and often face the waterlogging problem. Therefore, *in situ* soil and water conservation technology called the 'broad-bed and furrow system' that can protect the soil from erosion throughout the season, helps in proper drainage and controls the water at the place where the rain falls. This raised land configuration system has been found to satisfactorily attain these goals. The BBF system consists of a relatively raised flatbed or ridge approximately 95 cm wide and a shallow furrow about 55 cm wide and 15 cm deep (Fig. 9.5). The system is laid out on a grade of 0.4–0.8% for optimum performance. This BBF system is most effectively implemented in several operations or passes (Kampen, 1982).

Fig. 9.5. The broad-bed and furrow (BBF) system at Raichur, Karnataka.

The technology was found efficient in targeted districts of Karnataka and the results clearly revealed that the BBF system proved effective in conserving the rainwater and increasing the soil water in the profile and thus increased the winter sorghum (*Sorghum bicolor*) grain yield by 28–41% and safflower (*Carthamus tinctorius*) yield by 18–22% as compared with flat sowing. Apart from these soil and water conservation measures, compartmental bunding was also practised in several rainfed areas of Karnataka, which involves dividing fields, depending upon the land slopes, into small land parcels of square or rectangular shapes, by providing small bunds (Mishra *et al.*, 2002). The results from demonstrations at Bijapur revealed that compartmental bunding and ridges and furrows increased the sorghum grain yield by 23% and 26%, respectively, over these crops grown on a flatbed.

Ex situ *water conservation measures*
Along with *in situ* methods, *ex situ* soil moisture conservation measures were also included in the Bhoochetana programme, basically to concentrate on different low-cost water harvesting structures (WHS), which can be practically implemented and adopted by the farmers.

In Karnataka, community and farmer-based soil and water conservation interventions such as check dams, farm ponds, field bunding and percolation pits were undertaken to improve the surface and groundwater availability (Fig. 9.6). Findings at most of the sites reveal that open wells located near WHS have significantly higher water levels compared with those away from WHS (Fig. 9.7). Improved water availability in the sites resulted in not only increased crop productivity but also a significant shift in the area under cultivation towards high-value cereals, cash crops, vegetables, flowers and fruits.

Fig. 9.6. Different soil and water conservation structures constructed in Bellary, Karnataka.

The above results clearly indicate the excellent opportunities of implementing water harvesting, groundwater recharging and supplemental irrigation at the watershed scale. The key advantage of this approach is that these interventions can be implemented both at the farmer's field level as well as the community level. Also, the watershed-based community organizations and institutions assist in sustainable management of WHS.

9.5.3 Cropping system management

The greatest challenges for agricultural science and technology today occur in rapidly developing countries such as China, India, Brazil,

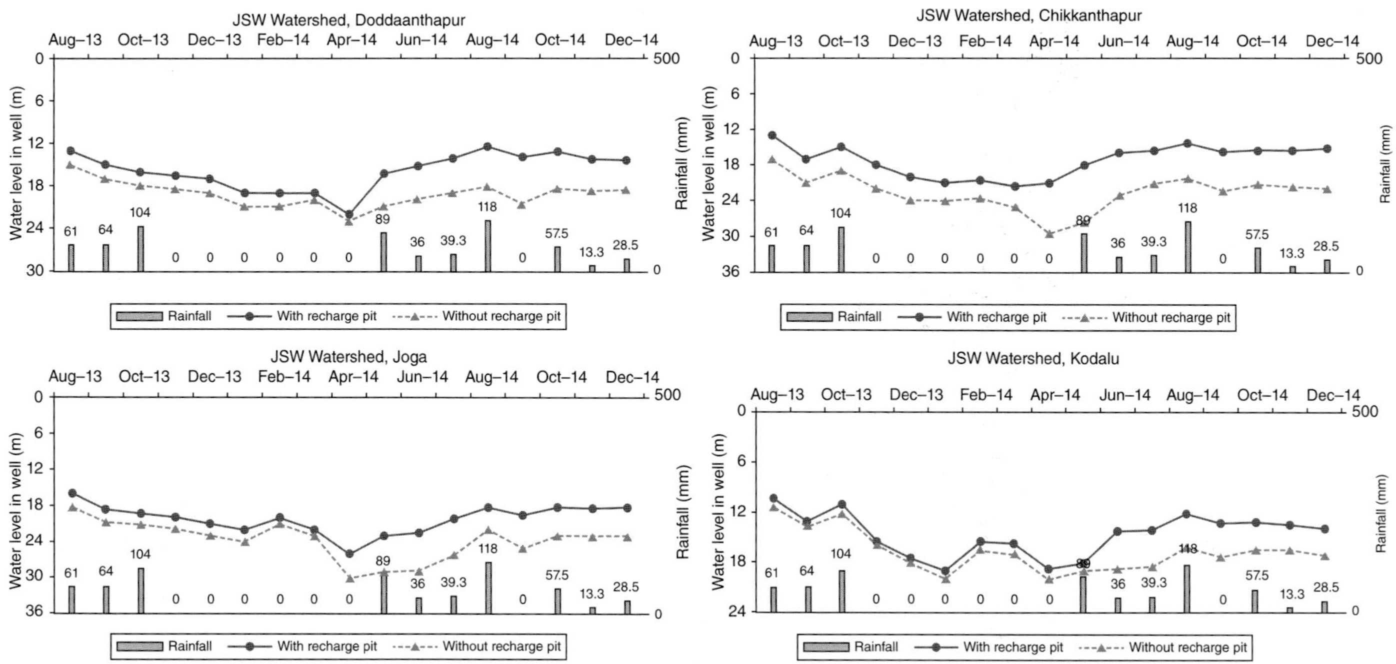

Fig. 9.7. The impact of watershed interventions on groundwater levels at two benchmark sites in Bellary, Karnataka, during 2013–2014.

Mexico, Indonesia, Vietnam, Pakistan and Sri Lanka. Although fertility rates have dropped substantially in these countries, populations are continuing to grow rapidly in most of them as a consequence of demographic momentum. Moreover, all are experiencing increasing per-capita demands for food, as some seek to overcome substantial regional malnutrition, and as all experience increasing demand for meat and other animal products. Therefore, balancing productivity, profitability and environmental health is a key challenge for today's agriculture for ensuring long-term sustainability (Robertson and Swinton, 2005; Foley *et al.*, 2011). However, most crop production systems in the world are characterized by low species and management diversity, high use of fossil energy and agrichemicals, and large negative impacts on the environment. Therefore there is an urgent need to focus our attention towards the development of crop production systems with improved resource use efficiencies and more benign effects on the environment (Tilman *et al.*, 2002; Foley *et al.*, 2011). Cropping system design provides an excellent framework for developing and applying integrated approaches to management because it allows new and creative ways of meeting the challenge of sustaining the agricultural productivity.

Crop diversification

Crop diversification in India is generally viewed as a shift from traditionally grown less remunerative crops to more remunerative crops. Here we can reconnoitre the potential benefits of diversifying cropping systems through efficient crop rotations as a means of controlling decline in crop productivity while simultaneously enhancing other desirable agroecosystem processes. Crop diversification involves incorporating a potential substitute crop in the existing cropping system in order to maintain soil health, bring stability into the production system and reduce the risk of insect pest incidence. Crop rotation is intended to give a wider choice in the production of a variety of remunerative crops in a given area so as to expand production-related activities on various crops and thereby ensure more profitability. Similarly, the aim in diversifying the cropping system, particularly in rainfed ecology, is to reduce the risk factor of crop failures due to drought or less rain or a declining market.

In the Bhoochetana initiative, efforts were made to promote diverse potential crops depending upon the existing agroecological conditions. There was introduction of some new crops and crop cultivars which are completely new to their cropping system and therefore efforts were also made to educate the farmer about improved agronomic management technologies for these new introductions. The new crop introductions were castor (*Ricinus communis*) to Bengaluru Rural and Bengaluru Urban, cluster bean (*Cyamopsis tetragonoloba*) to most of the districts, fodder maize (*Zea mays*), sorghum and maize to non-maize growing areas, namely, Tumkur, Koppal, etc. (Tables 9.6, 9.7 and 9.8).

Farmers' participatory varietal evaluation

The participatory varietal evaluation programme works towards increasing farm productivity by facilitating the delivery of high-yielding, profitable varieties that are well adapted to a wide range of soil types, environments and farming systems. This is achieved by providing accredited, unbiased information to farmers on better adapted crop varieties, or new and better cultivars, at the earliest opportunity.

In all 30 districts of Karnataka, farmers were given the option to choose improved varieties of preferred dryland crops from the list of varieties provided to farmers' groups. ICRISAT and SAUs released improved varieties and proprietary hybrids of crops were evaluated in the Bhoochetana Mission Program with the objective to select varieties with suitable traits for better adaptation to biotic and abiotic stresses to enhance or sustain productivity and further scale up the spread of these varieties to satellite taluks. Each demonstration was laid out on approximately 0.5–1 acre of the farmer's field. In this trial, there were two treatments: (i) local/traditional cultivar + farmers' inputs; and (ii) improved variety + best-bet inputs. The layout of the varietal trial was designed to assess the performance of the local variety with the traditional way of input management as compared to the improved variety with best-bet management. Best-bet management was done with the application of 70 kg/ha diammonium phosphate (DAP), 100 kg/ha urea fertilizers, 5 kg/ha borax, 50 kg/ha zinc sulfate and 200 kg/ha gypsum for cereal crops and a reduction in urea application from 100 kg/ha to 40 kg/ha for legumes.

With these trials, farmers were exposed to several improved varieties of each crop grown in their watershed and had the option of evaluating the performance of each variety more or less in the same climatic and soil conditions with different levels of input management. Participatory varietal selection trials were confined to two or three main rainfed cropping systems of the district/region during the crop season (Table 9.9). During the rainy and post-rainy season, crops evaluated included:

- cereals and millets (sorghum, pearl millet (*Pennisetum glaucum*), barley (*Hordeum vulgare*), finger millet (*Eleusine coracana*));
- pulses (pigeon pea (*Cajanus cajan*), chickpea (*Cicer arietinum*), cluster bean); and
- oilseed crops (groundnut (*Arachis hypogaea*), soybean (*Glycine max*), sunflower (*Helianthus annuus*), castor).

The activity is promoted through Bhoochetana programme in all the 30 districts of Karnataka with active involvement of the DoA and ICRISAT. The programme collects and delivers the data, which not only assists farmers with their choice of suitable varieties but also facilitates the registration and commercialization of new varieties by plant breeders. The experimental protocol has been established to evaluate the performance of improved varieties under balanced nutrition against a common set of traditional varieties to characterize their yield, quality, disease resistance/tolerance and agronomic characteristics. The information on yield performance of the improved cultivars was collected through crop cutting experiments by ICRISAT staff and farm facilitators in the presence of staff/officials of DoA.

Field demonstrations on cereals, namely, finger millet, pearl millet, sorghum, etc., revealed that mere introduction of improved varieties recorded significantly higher (15–30%) grain yield compared with the traditional local variety. The new introduced varieties are high-yielding, early maturing, and tolerant to mid-season and end-of-season drought. Due to their early and uniform maturity, attractive ear head and seed shape and high harvest index, the varieties are becoming popular among the farmers in many districts of Karnataka.

In participatory varietal evaluation trials of oilseeds and pulses, there was a distinct yield advantage with the introduction of improved varieties as compared with the farmers' traditional preferred cultivar. There was 24–46% increase in crop yield with the same amount of resources and inputs. Therefore farmers started procuring the new improved varieties from *Raithu Samparka Kendra* (RSK) located in each block.

Improving seed replacement ratio

The important issue of low crop productivity is poor seed replacement ratio. Farmers generally tend to use the same varietal seed for 5–10 years and thus there is a sharp reduction in purity of the seed and thereby decline in crop productivity. Therefore, the farmers in targeted districts of Karnataka were educated frequently on replacing the varietal seed every 3 years, whereas hybrid seed should be purchased every year. Similarly, in order to make the farmer-preferred varieties available to a large number of farmers, we encouraged the farmers to form farmer groups and also introduced the seed bank concept so that every farmer can deposit his/her share and when needed, he/she can use the seed for cultivation.

9.5.4 Innovative knowledge sharing approach

Dissemination of agricultural technology information plays a major role in transmitting the latest farm technology to the farmers through various means (www.agripb.gov.in). On the same line, in order to bring in the science-led innovative technologies to the doorstep of the farmers, the Bhoochetana programme adopted innovative modules of dissemination. The DoA in Karnataka, with technical backstopping

Table 9.6. Crop cultivars demonstrated in farmers' fields in different districts of Karnataka.

District	Pigeon pea	Castor	Cluster bean	Green gram	Groundnut	Sorghum	Pearl millet
Belgaum	Asha (ICPL 87119), Puskal (ICPH 2671)	–	HG 563	SML 668	ICGV 91114	CSV 15, CSV 23	–
Davangere	Lakshmi (ICPL 85063), Asha (ICPL 87119), ICPH 2740, Puskal (ICPH 2671)	DCH 177, Jyothi	–	–	ICGV 91114	CSV 15, CSV 23	–
Haveri	Asha (ICPL 87119), Lakshmi (ICPL 85063), Puskal (ICPH 2671), ICPH 2740	–	–	–	–	CSV 15, CSV 23	–
Bijapur	Asha (ICPL 87119), Lakshmi (ICPL 85063)	–	HG 563	SML 662	–	–	HHB 67, ICTP 8203
Chikkamagalur	Lakshmi (ICPL 85063), Asha (ICPL 87119), Puskal (ICPH 2671), ICPH 2740	DCH 177, Jyothi	HG 563	SML 668	–	–	–
Hassan	Lakshmi (ICPL 85063), Puskal (ICPH 2671)	DCH 177, Jyothi	HG 563	SML 668	–	–	–
Ramanagara	Lakshmi (ICPL 85063), Asha (ICPL 87119), Puskal (ICPH 2671)	DCH 177, Jyothi	–	–	–	–	–
Chamarajanagar	Lakshmi (ICPL 85063), Puskal (ICPH 2671)	DCH 177, Jyothi	HG 563	SML 668	–	–	–
Bagalkot	Asha (ICPL 87119), Lakshmi (ICPL 85063), Puskal (ICPH 2671)	DCH 177, Jyothi	HG 563	SML 668	ICGV 91114	CSV 23, CSV 15	HHB 67, ICTP 8203
Gadag	Lakshmi (ICPL 85063), Puskal (ICPH 2671), Asha (ICPL 87119)	–	–	–	ICGV 91114	–	–

Dharwad	Lakshmi (ICPL 85063), Puskal (ICPH 2671), Asha (ICPL 87119)	–	HG 563	SML 668	ICGV 91114	–	–
Udupi	Asha (ICPL 87119), Lakshmi (ICPL 85063), Puskal (ICPH 2671)	–	–	–	ICGV 91114	–	–
Uttara Kannada	Puskal (ICPH 2671)	DCH 177, Jyothi	–	–	–	–	–
Shimoga	Asha (ICPL 87119), Puskal (ICPH 2671)	–	–	–	–	–	–
Mangalore	Asha (ICPL 87119), Puskal (ICPH 2671)	–	–	–	ICGV 91114	–	–
Bangalore 1	Lakshmi (ICPL 85063), Asha (ICPL 87119), Puskal (ICPH 2671)	Jyothi, DCH 519, DCH 177	HG 563	SML 668	–	–	–
Bangalore 2	Lakshmi (ICPL 85063), Asha (ICPL 87119), Puskal (ICPH 2671)	Jyothi, DCH 177	–	–	–	–	–
Kodagu	Lakshmi (ICPL 85063), Asha (ICPL 87119), Puskal (ICPH 2671)	–	–	–	–	–	–
Chikkaballapur	Lakshmi (ICPL 85063)	Jyothi, DCH 519, DCH 177	HG 563	–	ICGV 91114	–	–
Kolar	Lakshmi (ICPL 85063)	Jyothi, DCH 519, DCH 177	HG 563	–	ICGV 91114	–	–

Continued

Table 9.6. Continued.

District	Pigeon pea	Castor	Cluster bean	Green gram	Groundnut	Sorghum	Pearl millet
Tumkur	Lakshmi (ICPL 85063)	DCH 519, DCH 177	HG 563	–	ICGV 91114, ICGV 02266, ICGV 00308, ICGV 00351	–	–
Chitradurga	Lakshmi (ICPL 85063), Asha (ICPL 87119), Puskal (ICPH 2671), ICPH 2740	Jyothi, DCH 519, DCH 177	HG 563	–	ICGV 91114	–	–
Koppal	Asha (ICPL 87119), Lakshmi (ICPL 85063)	DCH 177	–	–	–	CSV 15, CSV 23	HHB 67, ICTP 8203
Yadgiri	Puskal (ICPH 2671), Asha (ICPL 87119), Lakshmi (ICPL 85063)	DCH 177	–	–	–	CSV 15, CSV 23	HHB 67, ICTP 8203
Gulbarga	Puskal (ICPH 2671), Asha (ICPL 87119), Lakshmi (ICPL 85063), ICPH 2740	DCH 177	HG 563	SML 668	–	CSV 15, CSV 23	HHB 67, ICTP 8203
Bidar	Puskal (ICPH 2671), Asha (ICPL 87119), Lakshmi (ICPL 85063), ICPH 2740	–	HG 563, N87, RGE-986	SML 668	–	CSV 15, CSV 23	HHB 67
Bellary	Puskal (ICPH 2671), Asha (ICPL 87119), Lakshmi (ICPL 85063), ICPH 2740	DCH 177, Jyothi, DCH 519	HG 563	SML 668	ICGV 91114	CSV 15	HHB 67, ICTP 8203
Raichur	Puskal (ICPH 2671), Asha (ICPL 87119), Lakshmi (ICPL 85063), ICPH 2740	DCH 177, Jyothi, DCH 519	HG 563	SML 668	ICGV 91114, ICGV 02266, ICGV 00308, ICGV 00351	–	HHB 67, ICTP 8203

Table 9.7. Contingent new crop cultivars demonstrated in farmers' fields in different districts of Karnataka.

District	Finger millet	Soybean	Sunflower	Maize
Belgaum	–	JS 9560	–	–
Chikkamagalur	MR 1	–	–	–
Hassan	MR 1	–	–	–
Ramanagara	MR 1	–	–	–
Chamarajanagar	MR 1	–	DRSH 1	–
Bagalkot	–	–	DRSH 1	–
Dharwad	–	JS 9560	–	–
Uttara Kannada	MR 1	–	–	–
Chikkaballapur	MR 1	–	–	–
Kolar	MR 1	–	–	–
Tumkur	MR 1	–	–	Pusa 3
Chitradurga	–	–	DRSH 1	–
Bidar	–	JS 9560	DRSH 1	–
Bellary	–	–	DRSH 1	Bioseed 9220
Raichur	–	–	DRSH 1	–

Table 9.8. Improved crop cultivars demonstrated in farmers' fields in different districts of Karnataka.

District	Sorghum	Chickpea	Barley	Groundnut
Gulbarga	CSV 22	JG 11, JAKI 9218	–	ICGV 91114
Bidar	CSV 22	JG 11, JAKI 9218	–	–
Bellary	CSV 22, SPV 1411	JG 11, JAKI 9218, KAK 2	–	–
Raichur	CSV 22	JG 11, JAKI 9218, KAK 2	–	–
Belgaum	CSV 22	JG 11, JAKI 9218	DWB73, HB 902	–
Davangere	CSV 22	JG 11, JAKI 9218, ICCC 37	–	–
Haveri	CSV 22	ICCC 37, JG 11, JAKI 9218	DWB73	–
Bijapur	CSV 22, SPV 1411	KAK 2, ICCC 37, JG 11, JAKI 9218	–	–
Chikkamagalur	–	ICCC 37, JG 11, JAKI 9218	–	–
Chamarajanagar	–	KAK 2, ICCC 37, JG 11, JAKI 9218	–	–
Yadgir	CSV 22	JG 11, JAKI 9218	–	ICGV 91114
Gadag	CSV 22	JG 11, JAKI 9218	–	–
Udupi	–	–	–	ICGV 91114
Tumkur	–	JG 11, JAKI 9218	–	ICGV 91114
Chitradurga	CSV 22	JG 11, JAKI 9218	–	–

Table 9.9. Yield improvement under landform management.

District	Crop	Variety	Grain yield (kg/ha) BBF landform[a]	Farmers' practice	Increase in yield (%)
Dharwad	Maize	Kavri	2866	2550	12
		Prabhat	4704	4250	11
	Soybean	JS 9305	3715	3400	9
		JS 335	4353	3262	33
	Chickpea	JG 11	1389	1051	32
	Maize	Prabhat	6577	5936	11
Vijayapura	Pigeon pea	TS3R	800	720	11
(Bijapur)	Pearl millet/ pigeon	HHB 67	800	700	14
	pea (2:1)	(I)/TS3R	500	430	16
Raichur	Pigeon pea	Maruti	1050	900	17
	Pigeon pea	ICPH 2740	1400	1250	12

[a]BBF, Broad-bed and furrow.

from ICRISAT, started with a simple but effective approach of soil health assessment and sharing with the farmers.

Soil atlas and sharing soil health cards

In order to know the soil health status and site-specific nutrient recommendations, ICRISAT developed a soil atlas encompassing analysed soil test results and soil-test-based fertilizer recommendations and put it in the public domain to benefit all the stakeholders, policy makers, development agencies, extension workers, researchers and farmers of Karnataka (Wani *et al.*, 2011).

The soil health card is a field-specific detailed report of the soil fertility status and other important soil parameters that affect crop productivity. Besides soil health, it also provides advice on soil-test-based use of fertilizers and amendments. Soil health cards were provided to individual farmers in their local language (Kannada) with details of individual nutrient status and critical limits along with a comment on the nutrient status of the field. The second side of the card contains the recommended dose of nutrients for each crop as well as the quantity of nutrients available in commercially marketed fertilizers, so that the farmers can understand this. The reference nutrient recommendations for each crop were extracted from recommendations of the University of Agricultural Sciences (Bangalore, Dharwad, Raichur) for their respective operational regions.

This novel initiative of Karnataka has attracted and inspired several policy makers and state governments and even central government to conduct extensive soil analysis and issue soil health cards to every farmer and prepare the soil fertility atlas of the state.

Digital extension technologies

Innovative agricultural extension systems are necessary to ensure deep penetration of available agricultural technologies into farming community. In the Bhoochetana project, we have introduced innovative information communication technology models to rejuvenate agricultural extension systems for knowledge sharing and dissemination. These tools include a tablet-based extension system (*Krishi Gyan Sagar*) and farmer-to-farmer video documentation. More information about digital technologies used for agricultural extension in the Bhoochetana project are discussed in Chapter 5 of this book.

Publicity

Under Bhoochetana in Karnataka, while distributing farm inputs such as seeds, micronutrients and biofertilizers as a kit to farmers, pamphlets containing the improved production practice of major crops undertaken by Bhoochetana were also provided to farmers for effective use of the inputs and implementation of improved technology, rather like a prescription (Fig. 9.8).

Wall writing is one of the effective ways of publicizing the project activities to farmers in rural areas. This has great impact and attracts the attention of a large number of farmers, creating awareness about the novel initiative of Bhoochetana. The objectives and technology components of Bhoochetana were displayed through two wall writings written in Kannada in each village at prominent locations where groups of farmers assemble or meet such as the RSK office, the Gram Panchayat office, bus shelters or the market/central locations in the village (Fig. 9.9).

Entertainment-education or 'edutainment' is a particularly useful strategy that entails the 'process of purposely designing and implementing a media message to both entertain and educate, in order to increase audience members' knowledge about an educational issue, create favourable attitudes, shift social norms, and change overt behaviour' (Singhal *et al.*, 2004). Dissemination of Bhoochetana technologies to farmers, particularly illiterate folks who cannot make use of wall writings or various print media, the street plays, folk songs/dances and village fairs, are found to be the most effective and powerful tool to reach mass farmers. This is a process of communication to the masses instead of mass communication. In this method, the farm message is conveyed through entertainment, namely plays and songs, which are more attractive to the farmers.

Jathas (vehicles) and cycle rallies were organized to bring in awareness and sensitize farmers about the project (Fig. 9.10). This has good impact as social gathering of youths and other supporters promote and publicize the message of the Bhoochetana project in Karnataka.

Farm facilitator

The new concept of engaging a local farmer from the village as a farm facilitator (para-worker) was developed for effective knowledge

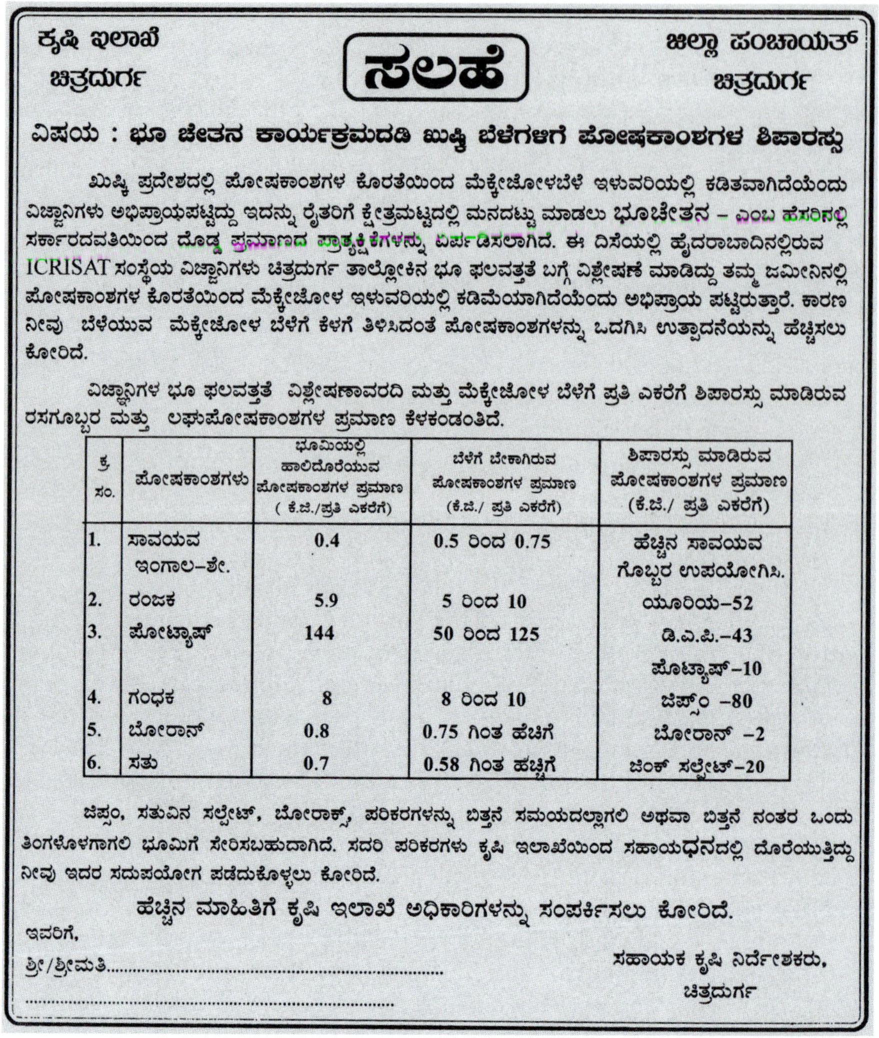

ಕೃಷಿ ಇಲಾಖೆ
ಚಿತ್ರದುರ್ಗ

ಸಲಹೆ

ಜಿಲ್ಲಾ ಪಂಚಾಯತ್
ಚಿತ್ರದುರ್ಗ

ವಿಷಯ : ಭೂ ಚೇತನ ಕಾರ್ಯಕ್ರಮದಡಿ ಖುಷ್ಕಿ ಬೆಳೆಗಳಿಗೆ ಪೋಷಕಾಂಶಗಳ ಶಿಪಾರಸ್ಸು

ಖುಷ್ಕಿ ಪ್ರದೇಶದಲ್ಲಿ ಪೋಷಕಾಂಶಗಳ ಕೊರತೆಯಿಂದ ಮೆಕ್ಕೇಜೋಳಬೆಳೆ ಇಳುವರಿಯಲ್ಲಿ ಕಡಿತವಾಗಿದೆಯೆಂದು ವಿಜ್ಞಾನಿಗಳು ಅಭಿಪ್ರಾಯಪಟ್ಟಿದ್ದು ಇದನ್ನು ರೈತರಿಗೆ ಕ್ಷೇತ್ರಮಟ್ಟದಲ್ಲಿ ಮನದಟ್ಟು ಮಾಡಲು ಭೂಚೇತನ – ಎಂಬ ಹೆಸರಿನಲ್ಲಿ ಸರ್ಕಾರದವತಿಯಿಂದ ದೊಡ್ಡ ಪ್ರಮಾಣದ ಪ್ರಾತ್ಯಕ್ಷಿಕೆಗಳನ್ನು ಏರ್ಪಡಿಸಲಾಗಿದೆ. ಈ ದಿಸೆಯಲ್ಲಿ ಹೈದರಾಬಾದಿನಲ್ಲಿರುವ ICRISAT ಸಂಸ್ಥೆಯ ವಿಜ್ಞಾನಿಗಳು ಚಿತ್ರದುರ್ಗ ತಾಲ್ಲೋಕಿನ ಭೂ ಫಲವತ್ತತೆ ಬಗ್ಗೆ ವಿಶ್ಲೇಷಣೆ ಮಾಡಿದ್ದು ತಮ್ಮ ಜಮೀನಿನಲ್ಲಿ ಪೋಷಕಾಂಶಗಳ ಕೊರತೆಯಿಂದ ಮೆಕ್ಕೇಜೋಳ ಇಳುವರಿಯಲ್ಲಿ ಕಡಿಮೆಯಾಗಿದೆಯೆಂದು ಅಭಿಪ್ರಾಯ ಪಟ್ಟಿರುತ್ತಾರೆ. ಕಾರಣ ನೀವು ಬೆಳೆಯುವ ಮೆಕ್ಕೇಜೋಳ ಬೆಳೆಗೆ ಕೆಳಗೆ ತಿಳಿಸಿದಂತೆ ಪೋಷಕಾಂಶಗಳನ್ನು ಒದಗಿಸಿ ಉತ್ಪಾದನೆಯನ್ನು ಹೆಚ್ಚಿಸಲು ಕೋರಿದೆ.

ವಿಜ್ಞಾನಿಗಳ ಭೂ ಫಲವತ್ತತೆ ವಿಶ್ಲೇಷಣಾವರದಿ ಮತ್ತು ಮೆಕ್ಕೇಜೋಳ ಬೆಳೆಗೆ ಪ್ರತಿ ಎಕರೆಗೆ ಶಿಪಾರಸ್ಸು ಮಾಡಿರುವ ರಸಗೊಬ್ಬರ ಮತ್ತು ಲಘುಪೋಷಕಾಂಶಗಳ ಪ್ರಮಾಣ ಕೆಳಕಂಡಂತಿದೆ.

ಕ್ರ ಸಂ.	ಪೋಷಕಾಂಶಗಳು	ಭೂಮಿಯಲ್ಲಿ ಹಾಲಿದೊರೆಯುವ ಪೋಷಕಾಂಶಗಳ ಪ್ರಮಾಣ (ಕೆ.ಜಿ./ಪ್ರತಿ ಎಕರೆಗೆ)	ಬೆಳೆಗೆ ಬೇಕಾಗಿರುವ ಪೋಷಕಾಂಶಗಳ ಪ್ರಮಾಣ (ಕೆ.ಜಿ./ ಪ್ರತಿ ಎಕರೆಗೆ)	ಶಿಪಾರಸ್ಸು ಮಾಡಿರುವ ಪೋಷಕಾಂಶಗಳ ಪ್ರಮಾಣ (ಕೆ.ಜಿ./ ಪ್ರತಿ ಎಕರೆಗೆ)
1.	ಸಾವಯವ ಇಂಗಾಲ–ಶೇ.	0.4	0.5 ರಿಂದ 0.75	ಹೆಚ್ಚಿನ ಸಾವಯವ ಗೊಬ್ಬರ ಉಪಯೋಗಿಸಿ.
2.	ರಂಜಕ	5.9	5 ರಿಂದ 10	ಯೂರಿಯ–52
3.	ಪೋಟ್ಯಾಷ್	144	50 ರಿಂದ 125	ಡಿ.ಎ.ಪಿ.–43 ಪೊಟ್ಯಾಷ್–10
4.	ಗಂಧಕ	8	8 ರಿಂದ 10	ಜಿಪ್ಸಂ –80
5.	ಬೋರಾನ್	0.8	0.75 ಗಿಂತ ಹೆಚಿಗೆ	ಬೋರಾನ್ –2
6.	ಸತು	0.7	0.58 ಗಿಂತ ಹಚ್ಚಿಗೆ	ಜಿಂಕ್ ಸಲ್ಫೇಟ್–20

ಜಿಪ್ಸಂ, ಸತುವಿನ ಸಲ್ಫೇಟ್, ಬೋರಾಕ್ಸ್, ಪರಿಕರಗಳನ್ನು ಬಿತ್ತನೆ ಸಮಯದಲ್ಲಾಗಲಿ ಅಥವಾ ಬಿತ್ತನೆ ನಂತರ ಒಂದು ತಿಂಗಳೊಳಗಾಗಲಿ ಭೂಮಿಗೆ ಸೇರಿಸಬಹುದಾಗಿದೆ. ಸದರಿ ಪರಿಕರಗಳು ಕೃಷಿ ಇಲಾಖೆಯಿಂದ ಸಹಾಯಧನದಲ್ಲಿ ದೊರೆಯುತ್ತಿದ್ದು ನೀವು ಇದರ ಸದುಪಯೋಗ ಪಡೆದುಕೊಳ್ಳಲು ಕೋರಿದೆ.

ಹೆಚ್ಚಿನ ಮಾಹಿತಿಗೆ ಕೃಷಿ ಇಲಾಖೆ ಅಧಿಕಾರಿಗಳನ್ನು ಸಂಪರ್ಕಿಸಲು ಕೋರಿದೆ.

ಇವರಿಗೆ,
ಶ್ರೀ/ಶ್ರೀಮತಿ..
....

ಸಹಾಯಕ ಕೃಷಿ ನಿರ್ದೇಶಕರು,
ಚಿತ್ರದುರ್ಗ

Fig. 9.8. A pamphlet describing Bhoochetana technologies that was distributed to farmers.

dissemination of Bhoochetana technologies to farmers in villages on a seasonal basis. Under the Bhoochetana programme, the Government of Karnataka hired 10,000 farm facilitators. They were selected from local communities (on a seasonal basis) considering that villagers are more likely to trust a local person rather than a stranger. Farm facilitators are the link between the DoA and farmers. Logistically, the government provided seeds, fertilizers and micronutrients to distribute to farmers through cluster points/RSKs with the help of farm facilitators. Under this new system, farmers were also trained on gender-related

Fig. 9.9. Wall writings in Kannada on the goals of Bhoochetana and appropriate technologies for enhancing crop productivity in farmers' fields in Tumkur and Chamarajanagar districts in Karnataka.

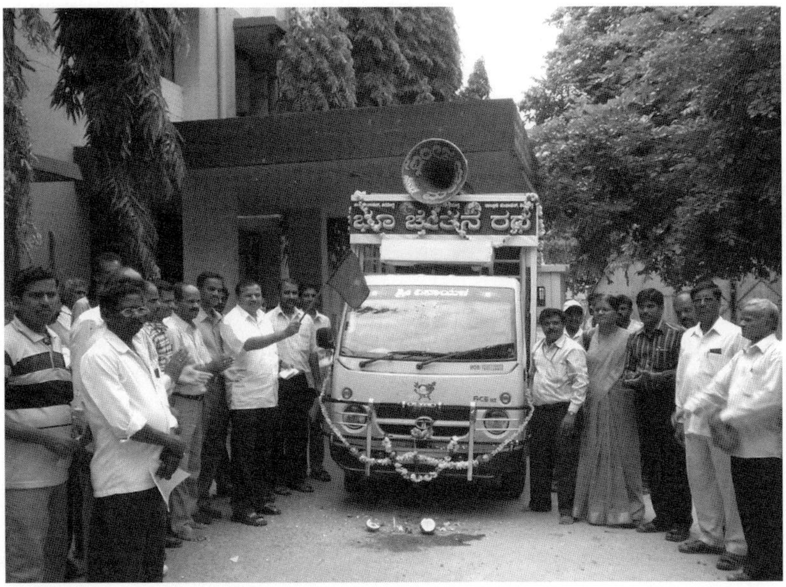

Fig. 9.10. Bhoochetana publicity vehicle in Davangere, Karnataka.

issues to be used in the field. Women farm facilitators were hired and they comprised 10% of the total number. Trained farm facilitators raised awareness about Bhoochetana among other farmers, which is one of a kind in the country.

Farmer field school (FFS)

Farmer field school (FFS) is a unique way to educate farmers and an effective platform for sharing of experiences and collectively solving agriculture-related problems. The FFS is facilitated by farm facilitators and consists of 25–30 farmers who meet every week for an entire crop growing season. Empowering non-formal education methods, the field is used as the primary resource for discovery-based learning. The process is facilitative and respects the experience that farmers bring with them. The focus of FFS is on crop productivity enhancement with various improved technologies recommended in the Bhoochetana programme. Farmers work in small groups to ensure that everyone's ideas are shared to focus on enhancing crop productivity. In FFS, there is acceptance of the uniqueness of each participant. The activities are designed to respond to the immediate needs of farmers and are geared towards encouraging creativity and independence. The FFS trainers play a crucial role in ensuring that the environment and all resources contribute to the farmers' learning experiences.

Our experience in Karnataka has shown that it is possible to adapt the FFS approach effectively to the dryland context. In the dryland system, the facilitators have to skilfully use this principle in its broader farming system perspective. In this context, the farm facilitator plays a greater role in ensuring farmers' participation and approaching the problems in a skilful manner. The Government of Karnataka is promoting FFSs with a focus on productivity enhancement using indigenous technologies to protect crops and to share local wisdom in farm management. To enhance learning and to ensure relevance of the field school contents, farmers design the studies. The FFS includes a process of: (i) problem identification; (ii) prioritization of topics to address; and (iii) designing the field studies. The approach encourages farmers to learn science by developing their skills in designing experiments that address their field problems even beyond the initial FFS experience. Therefore, farmers themselves can become expert in upscaling the FFS approach in and around the locality.

9.6 Summary

The agriculture sector in India is particularly vulnerable to present-day climate variability, including multiple years of low and erratic rainfall. Declining size of landholdings without any alternative income-augmenting opportunity is also resulting in a fall of total farm income, causing agrarian distress, and a large number of smallholders have to move to non-farm activities to augment their incomes. Equally, modern agricultural technologies emphasized mainly the component and commodities-based innovations and are mostly conducted in isolation and have led to overexploitation of natural resources, resulting in decreasing resource use efficiency and ultimately decreased farm productivity and profitability. Thus, these interventions have produced short-term gains, but resulted in long-term degradation of natural resources, namely, soil, water, biodiversity and non-cultivated land. There is a need to introduce a change in farming approach which addresses maximum production of a cropping pattern with optimal utilization of resources.

Therefore, in order to bridge the gap between the 'desired' and 'achieved' and to bring quantitative as well as qualitative improvement in fulfilling the national food needs to sustain the agricultural resource base, and to provide livelihood security to millions of rural masses, the DoA, Government of Karnataka has implemented the innovative programme Bhoochetana, which has benefited millions of farmers in Karnataka. This programme is aimed at taking science to the doorstep of farmers by integrating agricultural research far beyond cropping systems and principles of the production system. This approach has been widely recognized and advocated as the important way for harmonious use of inputs and their compounded response to make the production system more viable and sustainable. Bhoochetana has created an active ground force of para-agronomists (farm facilitators) who are directly involved in information dissemination and help the resource-poor farmers to use agricultural technologies. Thus, the importance of Bhoochetana is more precise as an important approach for scaling up the technologies to achieve agricultural sustainability. This innovative flagship programme targets integration of management interventions, including integrated nutrient management, soil and water conservation, cropping system management as well as a different outreach approach, as a complete technological tool.

9.7 Conclusion

Based on the above discussion and elaboration, the following conclusions can be drawn:

- An integrated approach is better than the traditional system to enhance productivity, profitability and economic viability of agribusiness.
- This helps the family labourers of rainfed farmers to work for the farm round the year.
- The approach enables maintenance of soil health and the environment through reduction in purchase of fertilizers for crops by recycling the manures by vermicomposting, bio-composting, etc.
- The approach is positive and helps in conserving the resource base through efficient recycling of residues within the system and will help in sustaining food security.
- The integrated approach helps in substantial improvement of income and thereby livelihood of resource-poor farmers in terms of standard of living, capabilities, assets, economic security, health security and educational security.

References

Ae, N., Arihara, J., Okada, K., Yoshihara, T. and Johansen, C. (1990) Phosphorus uptake by pigeon pea and its role in cropping systems of the Indian subcontinent. *Science* 248, 477–480.

Bationo, A., Kihara, J., Vanlauwe, B., Kimetu, J., Waswa, B.S. and Sahrawat, K.L. (2008) Integrated nutrient management: concepts and experience from sub-Saharan Africa. In: Aulakh, M.S. and Grant, C.A. (eds) *Integrated Nutrient Management for Sustainable Crop Production*. The Haworth Press, Taylor and Francis Group, New York, pp. 467–521.

Bekunda, M., Sanginga, N. and Woomer, P.L. (2010) Restoring soil fertility in sub-Saharan Africa. *Advances in Agronomy* 108, 183–286.

Black, C.A. (1993) *Soil Fertility Evaluation and Control*. Lewis Publishers, Boca Raton, Florida.

Chadha, G.K., Sen, S. and Sharma, H.R. (2004) *State of the Indian Farmer: A Millennium Study, Volume 2: Land Resources*. Department of Agriculture and Cooperation, Ministry of Agriculture, Government of India. Academic Foundation, New Delhi, India.

Condon, A.G., Richards, R.A., Rebetzke, G.J. and Farquhar, G.D. (2004) Breeding for high water-use efficiency. *Journal of Experimental Botany* 55, 2447–2460.

Conway, G. (1997) *The Doubly Green Revolution, Food for All in the 21st Century*. Penguin Books, London.

Dahnke, W.C. and Olson, R.A. (1990) Soil test calibration, and the recommendation. In: Westerman, R.L. (ed.) *Soil Testing and Plant Analysis*, 3rd edn. Soil Science Society of America Series 3. Soil Science Society of America, Madison, Wisconsin, pp. 45–71.

Daniels, S. and Walker, G. (1999) Rethinking public participation in natural resources management: concepts from pluralism and five emerging approaches in FAO. In: *Pluralism and Sustainable Forestry and Rural Development, Proceedings of an International Workshop*. Food and Agriculture Organization of the United Nations (FAO), Rome, pp. 29–48.

Douthwaite, B., de Haan, N., Manyong, V.M. and Keatinge, J.D.H. (2001) Blending 'hard' and 'soft' science: the 'Follow the Technology' approach to catalysing and evaluating technology change. *Conservation Ecology* 5(2). Available at: http://www.ecologyandsociety.org/vol5/iss2/art13/ (accessed 4 April 2016).

Edwards, P. (1997) Sustainable food production through aquaculture. In: *Aquaculture Asia*, Volume 2. School of Environment, Resources and Development, Asian Institute of Technology, Pathumthani, Thailand, pp. 4–6.

Foley, J.A., Ramankutty, N., Brauman, K.A., Cassidy, E.S., Gerber, J.S., Johnston, M., Müller, N.D., O'Connell, C., Ray, D.K., West, P.C., Balzer, C., Bennett, E.M., Carpenter, S.R., Hill, J., Monfreda, C., Polasky, S., Rockström, J., Sheehan, J., Siebert, S., Tilman, D. and Zaks, D.P.M. (2011) Solutions for a cultivated planet. *Nature* 478, 337–342.

Gill, M.S., Singh, J.P. and Gangwar, K.S. (2009) Integrated farming system and agriculture sustainability. *Indian Journal of Agronomy* 54, 128–139.

Hagmann, J. (1999) Learning together for change. Facilitating innovation in natural resource management through learning process approaches in rural livelihoods in Zimbabwe. In: *Kommunikation und Beratung*, Number 29. Margraf Verlag, Weikersheim, Germany.

Hagmann, J., Chuma, E., Muriwira, K., Connolly, M. and Ficarelli, P. (2002) Success factors in integrated natural resource management R&D: lessons from practice. *Conservation Ecology* 5(2), 29. Available at: http://www.consecol.org/vol5/iss2/art29 (accessed 4 April 2016).

Harrington, L., Jeffrey, W., Grace, P., Hodson, D., Agnes Dewi Hartkamp, Vaughan, C. and Meisner, C. (2001) Delivering the goods: scaling out results of natural resource management research. *Conservation Ecology* 5(2), 19. Available at: http://www.consecol.org/vol5/iss2/art19 (accessed 4 April 2016).

Hiradate, S., Ma, J.F. and Matsumoto, H. (2007) Strategies of plants to adapt to mineral stresses in problem soils. *Advances in Agronomy* 96, 65–132.

Holling, C.S. and Meffe, G.K. (1996) Command and control and the pathology of natural resource management. *Conservation Biology* 10, 328–337.

Jayanthi, C., Rangasamy, A. and Chinnusamy, C. (2000) Water budgeting for components in lowland integrated farming systems. *Agricultural Journal* 87, 411–414.

Jones, P.G. and Thornton, P.K. (2002) Spatial modeling of risk in natural resource management. *Conservation Ecology* 5(2), 27. Available at: http://www.consecol.org/vol5/iss2/art27 (accessed 4 April 2016).

Kampen, J. (1982) *An Approach to Improved Productivity on Deep Vertisols*. Information Bulletin no. 11. International Crops Research Institute for the Semi-Arid Tropics, Patancheru, Andhra Pradesh, India.

Korten, D.C. (1995) *When Corporations Rule the World*. Earthscan, London.

Kuruvilla, V. and Thomas, M. (2009) Integrated farming systems for sustainability in coastal ecosystem. *Indian Journal of Agronomy* 54, 120–127.

Lal, P., Lim-Applegate, H. and Scoccimarro, M. (2001) The adaptive decision-making process as a tool for integrated natural resource management: focus, attitudes, and approach. *Conservation Ecology* 5(2), 11. Available at: http://www.consecol.org/vol5/iss2/art11 (accessed 4 April 2016).

Lovell, C., Mandondo, A. and Moriarty, P. (2002) The question of scale in integrated natural resource management. *Conservation Ecology* 5(2), 25. Available at: http://www.consecol.org/vol5/iss2/art25 (accessed 4 April 2016).

Lynam, T., Bousquet, F., Le Page, C., d'Aquino, P., Barreteau, O., Chinembiri, F. and Mombeshora, B. (2002) Adapting science to adaptive managers: spidergrams, belief models, and multiagent systems modeling. *Conservation Ecology* 5(2), 24. Available at: http://www.consecol.org/vol5/iss2/art24 (accessed 4 April 2016).

Maarleveld, M. and Dangbégnon, C. (1999) Managing natural resources: a social learning perspective. *Agriculture and Human Values* 16, 267–280.

Magdoff, F. (1992) *Building Soils for Better Crops: Organic Matter Management*. University of Nebraska Press, Lincoln, Nebraska.

Mills, H.A. and Jones, J.B., Jr (1996) *Plant Analysis Handbook II. A Practical Sampling Preparation, and Interpretation Guide*. MicroMacro Publishing, Athens, Georgia.

Mishra, P.K., Sastry, G. and Osman, M. (2002) *Dividends from Soil and Water Conservation Practices: A Brief Review of Work Done in Rainfed Eco-system*. Central Research Institute for Dryland Agriculture, Hyderabad, India.

Mogensen, L. and Kristensen, T. (2000) Organic milk production in Denmark by using private farms for research. *Livestock Farming Systems: Integrating Animal Science Advances into the Search for Sustainability. Proceedings of the Fifth International Symposium on Livestock Farming Systems, Posieux, Fribourg, Switzerland, 19–20 August, 1999*, pp. 96–101.

Molden, D. (ed.) (2007) *Water for Food, Water for Life: A Comprehensive Assessment of Water Management in Agriculture*. Earthscan, London and International Water Management Institute, Colombo, Sri Lanka.

Passioura, J.B. (2006) Increasing crop productivity when water is scarce – from breeding to field management. *Agricultural Water Management* 80, 176–196.

Passioura, J.B. and Angus, J.F. (2010) Improving productivity of crops in water-limited environments. *Advances in Agronomy* 106, 37–74.

Pathak, P., Wani, S.P., Sudi, R., Chourasia, A.K., Singh, S.N. and Kesava Rao, A.V.R. (2007) *Rural Prosperity through Integrated Watershed Management: A Case Study of Gokulpura-Goverdhanapura in Eastern Rajasthan*. Global Theme on Agroecosystems Report No. 36. International Crops Research Institute for the Semi-Arid Tropics, Patancheru, Andhra Pradesh, India.

Pathak, P., Sahrawat, K.L., Wani, S.P., Sachan, R.C. and Sudi, R. (2009) Opportunities for water harvesting and supplemental irrigation for improving rainfed agriculture in semi-arid areas. In: Wani, S.P., Rockström, J. and Oweis, T. (eds) *Rainfed Agriculture: Unlocking the Potential*. Comprehensive Assessment of Water Management in Agriculture Series. CAB International, Wallingford, UK, pp. 197–221.

Radhammani, S., Balasubramanian, A., Ramamoorthy, K. and Geethalakshmi, V. (2003) Sustainable integrated farming systems for dry lands: a review. *Agricultural Reviews* 24, 204–210.

Rego, T.J., Sahrawat, K.L., Wani, S.P. and Pardhasaradhi, G. (2007) Widespread deficiencies of sulfur, boron, and zinc in Indian semi-arid tropical soils: on-farm crop responses. *Journal of Plant Nutrition* 30, 1569–1583.

Reuter, D.J. and Robinson, J.B. (eds) (1997) *Plant Analysis: An Interpretation Manual*, 2nd edn. Commonwealth Scientific and Industrial Research Organisation (CSIRO), Clayton, Victoria, Australia.

Robertson, G.P. and Swinton, S.M. (2005) Reconciling agricultural productivity and environmental integrity: a grand challenge for agriculture. *Frontiers in Ecology and the Environment* 3, 38–46. Available at: http://dx.doi.org/10.1890/1540-9295(2005)003[0038:RAPAEI]2.0.CO;2 (accessed 4 April 2016).

Rockström, J., Nuhu, H., Oweis, T. and Wani, S.P. (2007) Managing water in rainfed agriculture. In: Molden, D. (ed.) *Water for Food, Water for Life: A Comprehensive Assessment of Water Management in Agriculture*. Earthscan, London and International Water Management Institute, Colombo, Sri Lanka, pp. 315–348.

Sahrawat, K.L. (2009) The role of tolerant genotypes and plant nutrients in reducing acid-soil infertility in upland rice ecosystem: an appraisal. *Archives of Agronomy and Soil Science* 55, 597–607.

Sahrawat, K.L., Wani, S.P., Rego, T.J., Pardhasaradhi, G. and Murthy, K.V.S. (2007) Widespread deficiencies of sulphur, boron and zinc in dryland soils of the Indian semi-arid tropics. *Current Science* 93, 1428–1432.

Sahrawat, K.L., Wani, S.P., Pardhasaradhi, G. and Murthy, K.V.S. (2010a) Diagnosis of secondary and micronutrient deficiencies and their management in rainfed agroecosystems: case study from Indian semi-arid tropics. *Communications in Soil Science and Plant Analysis* 41, 346–360.

Sahrawat, K.L., Wani, S.P., Pathak, P. and Rego, T.J. (2010b) Managing natural resources of watersheds in the semi-arid tropics for improved soil and water quality: a review. *Agricultural Water Management* 97, 375–381.

Sanjeev, M.V., Venkatasubramanian, V. and Singha, A.K. (2010) *Farming Systems of North East India Research and Development Strategies for KVKs*. Indian Council of Agricultural Research, New Delhi, India.

Singh, S., Singh, P., Verma, N. and Kumar, D. (2011) Integrated fish farming – rationale and scope. *Ascot International* 14(2), 13.

Singhal, A., Sharma, D., Papa, M.J. and Witte, K. (2004) Air cover and ground mobilization: integrating entertainment-education broadcasts with community listening and service delivery in India. In: Singhal, A., Cody, M., Rogers, E.M. and Sabido, M. (eds) *Entertainment-Education and Social Change: History, Research, and Practice*. Lawrence Erlbaum, Mahwah, New Jersey, pp. 351–374.

Sreedevi, T.K., Wani, S.P., Sudi, R., Patel, M.S., Jayesh, T., Singh, S.N. and Tushaar Shah (2006) *On-site and Off-site Impact of Watershed Development: A Case Study of Rajsamadhiyala, Gujarat, India*. Global Theme on Agroecosystems Report no. 20. International Crops Research Institute for the Semi-Arid Tropics, Patancheru, Andhra Pradesh, India.

Tilman, D., Cassman, K.G., Matson, P.A., Naylor, R. and Polasky, S. (2002) Agricultural sustainability and intensive production practices. *Nature* 418, 671–677.

Wani, S.P. (2008) Taking soil science to farmers' doorsteps through community watershed management. *Journal of the Indian Society of Soil Science* 56, 367–377.

Wani, S.P. and Kumar, M.S. (2002) On-farm generation of N-rich organic material. In: Wani, S.P., Pathak, P. and Rego, T.J. (eds) *A Training Manual on Integrated Management of Watersheds, ICRISAT, Patancheru, India*. International Crops Research Institute for the Semi-Arid Tropics, Patancheru, Andhra Pradesh, India, pp. 30–31.

Wani, S.P., Singh, H.P., Sreedevi, T.K., Pathak, P., Rego, T.J., Shiferaw, B. and Iyer, S.R. (2003) Farmer-participatory integrated watershed management: Adarsha Watershed, Kothapally, India. An innovative and up-scalable approach. A case study. In: Harwood, R.R. and Kassam, A.H. (eds) *Research Towards Integrated Natural Resources Management: Examples of Research Problems, Approaches and Partnerships in Action in the CGIAR*. Interim Science Council, Consultative Group on International Agricultural Research, Washington, DC, pp. 123–147.

Wani, S.P., Joshi, P.K., Raju, K.V., Sreedevi, T.K., Wilson, J.M., Shah Amita, Diwakar, P.G., Palanisami, K., Marimuthu, S., Jha, A.K., Ramakrishna, Y.S., Meenakshi Sundaram, S.S. and D'Souza, M. (2008) *Community Watershed as a Growth Engine for Development of Dryland Areas. A Comprehensive Assessment of Watershed Programs in India*. Global Theme on Agroecosystems Report No. 47. International Crops Research Institute for the Semi-Arid Tropics, Patancheru, Andhra Pradesh, India.

Wani, S.P., Sreedevi, T.K., Rockström, J. and Ramakrishna, Y.S. (2009) Rainfed agriculture – past trends and future prospects. In: Wani, S.P., Rockström, J. and Oweis, T. (eds) *Rainfed Agriculture: Unlocking the Potential*. Comprehensive Assessment of Water Management in Agriculture Series. CAB International, Wallingford, UK, pp. 1–35.

Wani, S.P., Sahrawat, K.L., Sarvesh, K.V., Baburao, M. and Krishnappa, K. (2011) *Soil Fertility Atlas for Karnataka,* India. International Crops Research Institute for the Semi-Arid Tropics, Patancheru, Andhra Pradesh, India.

Zadeh, L.A. (1973) Outline of a new approach to the analysis of complex systems and decision processes. *IEEE (Institute of Electrical and Electronics Engineers) Transactions on Systems, Man, Cybernetics* 3(1), 28–44.

10 Water Productivity and Income

Kaushal K. Garg,* Suhas P. Wani, Girish Chander, K.H. Anantha and G. Pardhasaradhi

International Crops Research Institute for the Semi-Arid Tropics (ICRISAT), Patancheru, India

10.1 Introduction

Food security and reducing poverty for the ever-growing population in India is a challenging task. India's agricultural land is 142 million ha with 135% cropping intensity (NAAS, 2009) and 60% is rainfed, which is characterized by water scarcity, land degradation, low use of inputs and low productivity. Agricultural productivity of these areas oscillates between 0.5 t/ha and 2 t/ha with an average of 1 t/ha (Rockström *et al.*, 2010; Wani *et al.*, 2011a, b). Of the total agricultural area, the 40% that is irrigated land contributes 55% of total food production in the country (GoI, 2012) but on the other hand it consumes almost 70% of freshwater resources and has left limited scope for further expansion of the irrigated area (Central Water Commission, 2005; CGWB, 2012). Thus, achieving food security of the country at present and in the future is largely dependent on rainfed agriculture (Wani *et al.*, 2009, 2012a). Despite several constraints and limitations of rainfed areas, huge untapped potential exists for enhancing crop yield through improved land, water, nutrient and other natural resource management options (Rockström *et al.*, 2007; Garg *et al.*, 2012a, 2013; Wani *et al.*, 2012a; Singh *et al.*, 2014).

Karnataka, in southern India, covers nearly 70% of the total cultivable area under rainfed conditions and is the second largest rainfed

*k.garg@cgiar.org

state after Rajasthan (NRAA, 2012). Agriculture and allied sectors' contribution to the gross domestic product of the state of Karnataka was about 17% in 2009/10 (Wani *et al.*, 2013). Poor investment capacity, financial structures and extension support are the major reasons for keeping rainfed farming at the subsistence level in the state (Rockström *et al.*, 2010). Further, land fragmentation with the burgeoning population is adding to the problem, consequently land share and livelihood opportunities are reducing (Wani *et al.*, 2011a, b). In addition, poor land and water management practices along with exploitation of available natural resources over the years, coupled with the rainfall pattern and climate variability, has contributed to poor agricultural growth in the state.

A large section of the rural population in the state is dependent on agriculture and allied activities. The rural population in the state is largely suffering with various internal and external stresses/shocks such as: (i) weather/climatic variability (drought, flood); (ii) pest and disease infestation; (iii) market failures; and (iv) health-related stress (GoI, 2010). Upcoming challenges such as global warming and climate change bring further uncertainty on available resources and increased risk in the agricultural sector (Boomiraj *et al.*, 2010). It is estimated that a geographical area of approximately 3 million ha in Karnataka is being shifted from the subhumid tropics to the semi-arid tropics, which shows the increasing water stress situation in the region (Rao *et al.*, 2013). Therefore, science-led interventions need to be scaled up in millions of farmers' fields to address current and future food security challenges, rural livelihood, employment and sustainability of the system.

The Government of Karnataka started the innovative and large-scale mission-mode project called 'Bhoochetana' in 2009 with the help of the International Crops Research Institute for the Semi-Arid Tropics (ICRISAT) and its partners. This programme not only focused on soil, nutrient, land and crop management practices but also strengthened institutional capacity through regular capacity- and awareness-building programmes for various stakeholders. Bhoochetana was initially aimed at increasing the productivity of rainfed agriculture using an integrated approach. After realizing the potential of this approach, the programme expanded to irrigated agriculture within a short period of 2 years. Soil fertility assessment has been undertaken as an entry point activity and crop-specific soil-test-based fertilizer application was recommended at the taluk level as against the blanket recommendation followed earlier at state level as there was indiscriminate use of fertilizer both in rainfed and irrigated areas. This approach was adopted in a phased manner and covered all the 30 districts of Karnataka within a span of 4 years, covering 3.73 million ha with major dryland and irrigated crops. With this background, the impact of the Bhoochetana programme using data collected from a large number of farmers' fields during the 4-year period from 2009 to 2012 is discussed in this chapter. These data capture large variability in meteorological, biophysical, socio-economic and agronomic factors in the state.

10.2 Soil Fertility Mapping

The Government of Karnataka along with consortium partners like ICRISAT started the Bhoochetana programme in 2009. To define the soil-test-based fertilizer application at taluk (administrative boundary comprising several villages) and village levels, soil nutrient mapping was considered as the first and foremost entry point activity. A statistically proven stratified soil sampling technique was adopted to collect representative samples from rainfed agricultural land covering the entire state of Karnataka. A large number of samples (92,904) were collected, covering huge spatial variability in terms of rainfall and topography, cropping system, field size and its management. Soil samples were analysed in a state-of-the-art laboratory and a range of soil health parameters (i.e. organic carbon, available phosphorus (P), potassium (K), zinc (Zn), boron (B), sulfur (S), pH and electrical conductivity (EC)) were analysed. Data were used to map the soil nutrient status using a geographical information system (GIS) interface and nutrient-deficient hot spots were identified. Crop-specific nutrient recommendations were prepared and results were shared among consortium partners and stakeholders.

Soil analysis results from the entire state have clearly shown that the majority of the farmers' fields (52%) were low in organic carbon (Wani *et al.*, 2012b). In Karnataka as a whole 41% of farms were deficient in P, indicating the majority of farms had sufficient P and so had the opportunity through site-specific nutrient management to cut costs on current recommendations of P application. K as such was not a problem in the state. Across the state only 23% of sampled fields tested low and a science-led approach calls for a reduction in recommended K. Interestingly, the diagnosis revealed widespread deficiencies of secondary and micronutrients on most farms in Karnataka such that 52% of farms were deficient in S, 55% in Zn and 62% in B. Figure 10.1 shows spatial variability of different soil nutrients for different taluks in the state. It was found that soils are deficient largely in micro and secondary nutrients (S, Zn and B). Deficiency of P was largely found in north-western districts of Karnataka. Test results also showed that organic matter is poor, which largely varied from 0.25% to 0.50%. Western Ghats were found to be relatively good in soil organic carbon content, which could be due to the large area under forest and plantation crops. Moreover, soils in Western Ghats were acidic in nature due to heavy rainfall. Soil pH was found to increase from west to east and from a south to north direction as per the changing rainfall pattern.

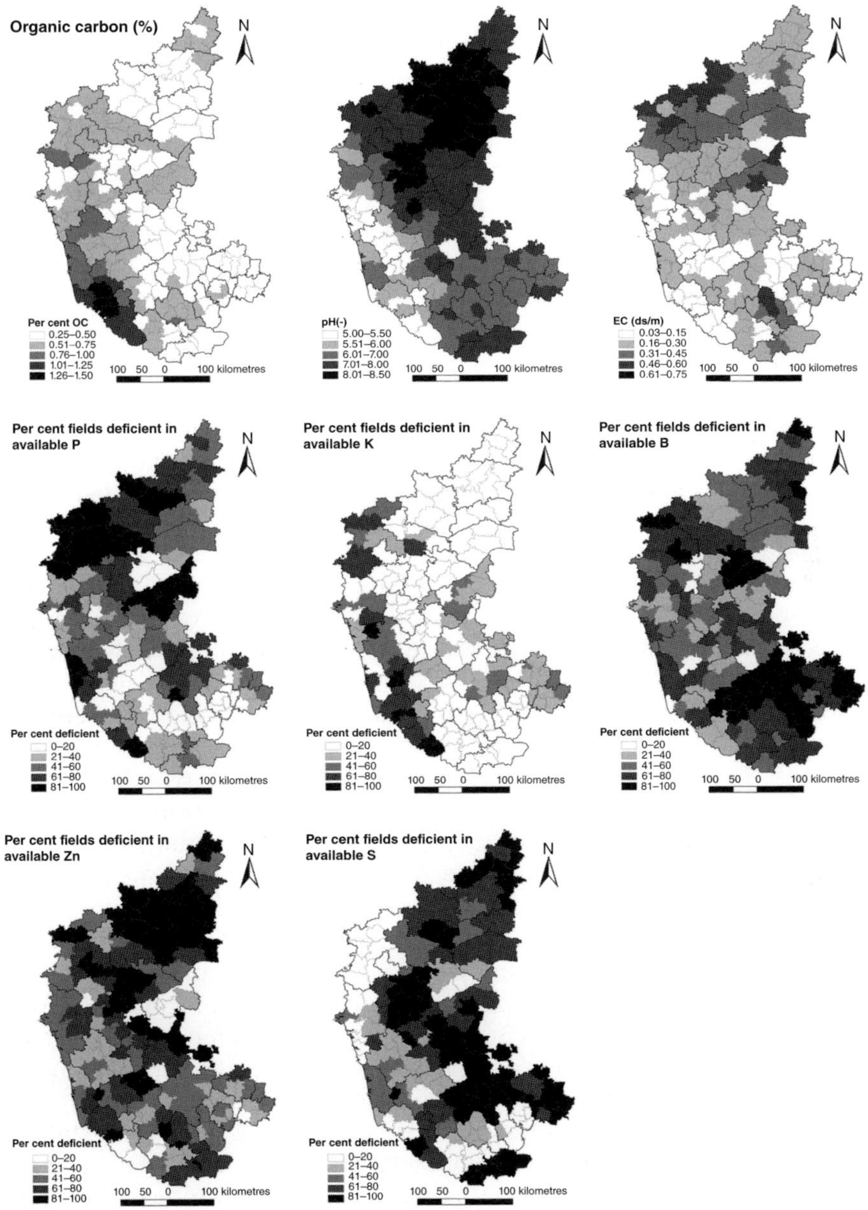

Fig. 10.1. Organic carbon (OC) content, pH and electrical conductivity (EC) of soil and percentage of farmers' fields deficient in available P, K, B, Zn and S in different taluks of Karnataka.

Soil health mapping was the first important output/milestone of the Bhoochetana project which convinced multi-stakeholders to apply crop- and site-specific nutrient application rather than following a common recommendation. Most of the farmers and stakeholders were unaware of widespread deficiencies of secondary and micronutrients and were not including them in their fertilizer management strategies. Considering how essential nutrients are for crop growth, deficiencies were holding back the realization of higher yields. But in a quest to get higher yields, farmers in many parts of the state were adding more than the required amounts of tested macronutrients such as nitrogen (N), P and K although these nutrients were not deficient. Soil health mapping indicated individual nutrient deficiencies scattered differently and thus provided a basis to design new fertilizer recommendations aimed at the level of a cluster of villages (i.e. the block level) to meet varying soil fertility needs as opposed to the current state-level blanket recommendations. Secondary and micronutrients were included in the recommendations, while the amounts of macronutrients were also optimized according to soil test values.

10.3 Spatial and Temporal Variability of Water Resources

ICRISAT developed a one-dimensional water balance model called the 'Water Impact Calculator' (WIC) used for analysing water balance components. The WIC is a generic decision-making tool, which could be applied for any land use and cropping system by providing minimum sets of biophysical and management inputs for partitioning rainfall into different hydrological components. WIC requires the following details as inputs:

- soil (water retention, soil depth);
- weather (evapotranspiration (ET), rainfall);
- crop growth (biomass (kc), root growth function);
- topography (land slope, landform conditions); and
- crop management (date of crop sowing and harvesting, irrigation method).

The model analyses the water balance components on a daily timescale as shown in Eqn 10.1:

$$\text{Rainfall} = \text{Surplus water (Runoff + Groundwater recharge)} + \text{ET} + \text{Change in soil moisture storage} \quad (10.1)$$

The model was run for all the 176 taluks in 30 districts of Karnataka for selected major crops. Water balance components were derived for all 4 years between 2009 and 2012. Agriculture in Karnataka is largely dependent on rainfall and its distribution as 70% of the total agricultural area is under rainfed conditions. The amount of rainfall received during the monsoon period (June–October) in 2009–2012 is presented in Plate 1. Rainfall varied from less than 500 mm in Central and Northern Karnataka to 5000 mm in Western Ghats. Large variability was also found in rainfall distribution in different months (data not shown). There was surplus rainfall in 2009 as most of the taluks received rainfall above normal. In 2010, rainfall was normal but 2011 and 2012 experienced deficit rainfall. The total rainfall received in 2009 was less than 500 mm in seven taluks out of the total 176 taluks of the entire state, whereas 101 and 127 taluks experienced rainfall less than 500 mm during 2011 and 2012, respectively, and thus there was severe drought in the state (Fig. 10.2).

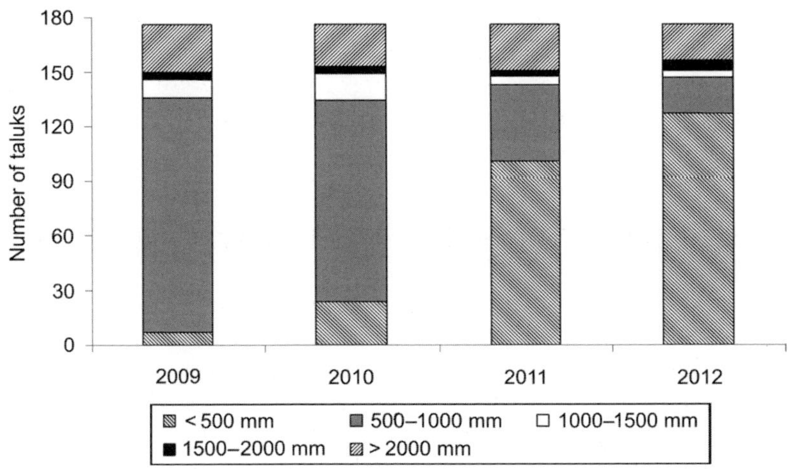

Fig. 10.2. Rainfall variability across Karnataka during 2009–2012.

Water balance components for each of the taluks were derived by hydrological modelling. Total rainfall received during the monsoon period was partitioned into three components: (i) surplus water (surface runoff and groundwater recharge); (ii) ET; and (iii) change in soil moisture content. Surplus water or blue water is the amount of water in which a portion of water flows on the soil surface as surface runoff and joins the riverine ecosystem; and the other portion of blue water contributes to recharge which is available in groundwater aquifers. ET is the amount of water which was initially harvested into soil layers (also known as green water) and subsequently utilized by crops and evaporated from the soil surface. The change in soil moisture indicates the amount of moisture that is left out in soil layers at the end of the monsoon which further could be utilized in the following season.

The water balance showed that a significant portion (nearly 70%) of the total rainfall received in dry regions has been utilized by crops as ET and little is partitioned as blue water. For example, out of 176 taluks in Karnataka state, 101 taluks received an average rainfall of 363 mm during 2011. Out of this, 241 mm (i.e. 66%) of rainfall received partitioned into ET, 32 mm (8%) into blue water and 90 mm (25%) remained as soil moisture at the end of the monsoon (Table 10.1). As rainfall distribution was poor, significant rainfall in October enhanced the soil moisture, which in fact was not useful for a monsoonal crop. Moreover, ET between June and October ranged from 100 mm to 900 mm depending on the rainfall amount and its distribution across the state. Due to poor rainfall in 2011 and 2012, ET in more than 80% of taluks was found to be less than 300 mm during the monsoon period, which shows poor soil moisture availability (Plate 2). This was further translated in terms of poor crop yield. Analysis of crop yield and income with rainfall variability will

Table 10.1. Water resource availability and its distribution in Karnataka during the monsoon (June–October) in 2009–2012.

Rainfall classes (mm)	No. of taluks	Monsoonal rainfall (mm)	Surplus amount (mm)	ET (mm)[a]	Change in soil moisture content (mm)
2009					
< 500	7	442	70	303	69
500–1000	129	691	179	423	89
1000–1500	10	1117	396	626	95
1500–2000	4	1693	812	766	116
> 2000	26	3203	2269	814	120
2010					
< 500	24	452	55	303	94
500–1000	110	644	131	419	94
1000–1500	15	1144	398	644	102
1500–2000	4	1750	830	809	111
> 2000	23	3242	2307	814	121
2011					
< 500	101	363	32	241	90
500–1000	42	634	102	433	98
1000–1500	5	1253	416	735	101
1500–2000	3	1801	882	810	109
> 2000	25	3250	2315	814	121
2012					
< 500	127	292	24	187	81
500–1000	20	643	107	453	83
1000–1500	4	1245	403	752	90
1500–2000	5	1827	921	809	97
> 2000	20	2729	1806	813	110

[a]ET, Evapotranspiration.

be discussed in the next section of this chapter. On the other hand, ET at Western Ghats was found to be at a maximum as paddy (*Oryza sativa*) is the dominating crop under the surplus water conditions; most of the crop water demand was met through rainfall and a significant amount of water was also found to be evaporated from the soil surface.

Taluks having rainfall more than 1500 mm were the main source of blue water for the riverine ecosystem. Out of 176 taluks, three received rainfall of 1500–2000 mm in 2011 and 25 taluks received more than 2000 mm rainfall (Table 10.1); runoff coefficients on average for these taluks were 48% and 71%, respectively.

10.4 Impact of Bhoochetana on Crop Yield

10.4.1 Crop yield with balanced fertilizer application

Along with large-scale implementation of the Bhoochetana programme, farmers' participatory demonstration trials were conducted in selected fields covering the major cropping systems in each taluk. Fields were divided into two parts: (i) improved practice (IP); and (ii) farmers' practice (FP). A balanced fertilizer dose was applied under IP and another part of the plot was cultivated as per the normal FP. Crop yield was estimated by conducting crop cutting experiments. Entire biomasses from a 3 m × 3 m sample area were harvested in both the plots and fresh biomass was measured in the presence of an ICRISAT representative, the Department of Agriculture (DoA) (Karnataka) staff and the farmer. Subsamples were taken and dried in the oven at 65°C and actual yields were calculated. Data were used to monitor the impact of balanced fertilizer application on crop yield and income from 2009 onwards.

The Bhoochetana programme, which initially started addressing soil nutrient deficiency from six districts in the first year, was scaled up to all 30 districts of the state in the third and fourth years, which has made a huge impact on crop productivity. Cropping systems in Karnataka are very diverse and a large number of crops were grown; however, to understand the impact of Bhoochetana, in this chapter we have selected four important crops: (i) maize (*Zea mays*); (ii) finger millet (*Eleusine coracana*); (iii) chickpea (*Cicer arietinum*); and (iv) groundnut (*Arachis hypogaea*). Crop productivity measured by crop cutting experiments under IP was compared with FP over 4 years from 2009 to 2012. The IP enhanced crop yields by 20–66% compared with FP. Crop productivity decreased with decreasing rainfall from 2009 to 2012 but yields under IP were consistently higher compared with FP even during deficit rainfall years in 2011 and 2012.

Average maize yields were 5500 kg/ha and 7600 kg/ha in 2009, and 3900 kg/ha and 5100 kg/ha in 2012 under FP and IP, respectively.

Large variation in crop productivity (maximum to minimum range) is recorded during the dry years compared with wet and normal years. With supplemental irrigation, yield levels obtained by farmers in dry years were even higher than the wet years but under rainfed conditions the crops suffered water stress resulting in poor yield. Finger millet is a drought-tolerant crop and largely grown in the southern part of Karnataka. Average finger millet yields were 1750 kg/ha and 2700 kg/ha in 2009, and 1250 kg/ha and 1680 kg/ha in 2012 under FP and IP, respectively. However, productivity decreased from 2009 to 2012 but IP helped farmers to harness better yields despite high water stress conditions. Chickpea is a post-rainy season crop, which is generally grown with residual soil moisture. It also showed better productivity under IP compared with FP but this difference decreased with increasing soil moisture stress especially in 2012. Average chickpea yields were 1050 kg/ha and 1400 kg/ha in 2009, and 600 kg/ha and 780 kg/ha in 2012 under FP and IP, respectively. Average groundnut yields in the first 3 years (2009–2011) were 1300 kg/ha and 1800 kg/ha but dropped significantly to 600 kg/ha and 780 kg/ha in 2012 under FP and IP, respectively.

10.4.2 Crop yield with rainfall variability

Crop production functions (rainfall versus crop yields) describing the crop sensitivity with available water and its further interaction with nutrient management were derived for:

- major cereals (maize, rice, finger millet, pearl millet (*Pennisetum glaucum*), sorghum (*Sorghum bicolor*));
- oilseeds (groundnut, soybean (*Glycine max*)); and
- pulses (pigeon pea (*Cajanus cajan*), chickpea, green gram (*Vigna radiata*)).

The Food and Agriculture Organization of the United Nations (FAO) described the linear relationship between crop yield and water use, where relative yield reduction is related to the corresponding relative reduction with available water for crop use (Stewart *et al.*, 1977; Doorenbos and Kassam, 1979; Lovelli *et al.*, 2007) as shown in Eqn 10.2.

$$\left[\frac{Y_x - Y_a}{Y_x}\right] = K_y\left[\frac{\text{Available water}}{\text{Actual water need}}\right] \tag{10.2}$$

where Y_x is maximum yield and Y_a is actual yield, and K_y is the correlation or proportionality factor between the related productivity loss (Lovelli *et al.*, 2007). In the current study, the production function for selected crops is established from large-scale farmers' field data from 2009 to 2012.

To understand yield sensitivity with monsoonal rainfall, the average yield measured in different taluks was plotted against rainfall for important cereals (paddy, maize, pearl millet, finger millet, sorghum), pulses (chickpea, green gram, pigeon pea) and oilseeds (soybean, groundnut) both under FP and IP (Figs 10.3 and 10.4). In general, crop yield increased with increasing rainfall amount but huge variability was recorded even in the same rainfall class. This variability was due to variation in rainfall distribution, soil types (nutrient status, moisture holding capacity, etc.) and a number of management factors (fertilizer input, time of sowing, etc.) among taluks.

Paddy is largely grown under irrigated conditions in drylands. It is also grown under rainfed conditions where rainfall is high. As we have not acquired total water inputs (rainfall plus irrigation) for paddy, it is at least important to analyse crop yield with rainfall. Most importantly, paddy yield under IP is higher (Student *t*-test, $P < 0.001$, nearly an additional 600–1000 kg as shown by the trend line) than FP, which indicates the importance of micro and secondary nutrients even under the higher yielding scenarios. Data plotted from a large number of crop cutting samples showed that maize yield in general increased with increasing rainfall. Farmers apply supplemental irrigation as maize is a water-demanding crop. Achieving relatively moderate to higher grain yield in poor rainfall scenarios (200–300 mm) indicates application of supplemental irrigation. The maize yield irrespective of irrigation inputs with IP was 800–1500 kg higher ($P < 0.001$) compared with FP (Fig. 10.3). Farmers generally cultivate pearl millet, finger millet and sorghum in low rainfall regions (less than 800 mm). Yields of these crops also increased with better water availability and a 500–800 kg difference in crop yield was found between FP and IP ($P < 0.001$). Similar observations were recorded for pulses and oilseeds (Fig. 10.4).

Further, yields for every 100 mm rainfall range were grouped together and averaged; productivity functions for selected cereals, pulses and oilseeds were developed (Figs 10.5 and 10.6). A linear trend was found in maize productivity with increasing rainfall, whereas productivity of pearl millet and finger millet increased linearly up to 600–700 mm but started declining with increasing rainfall, indicating that these crops are resilient in dry climatic conditions. The production function developed for pulses (chickpea, green gram and pigeon pea) and oilseeds also showed a strong linear response with increasing rainfall (Fig. 10.6).

To compare the economic benefit from different crops, we translated crop yields into net income after subtracting the cost of cultivation from gross income both for FP and IP scenarios and also compared it with rainfall variability (Figs 10.7 and 10.8). Paddy and maize were more remunerative crops compared with millets and sorghum. Net income from maize and pearl millet cultivation was ₹50,000–60,000/ha and ₹20,000–25,000/ha, respectively, under moderate to good rainfall

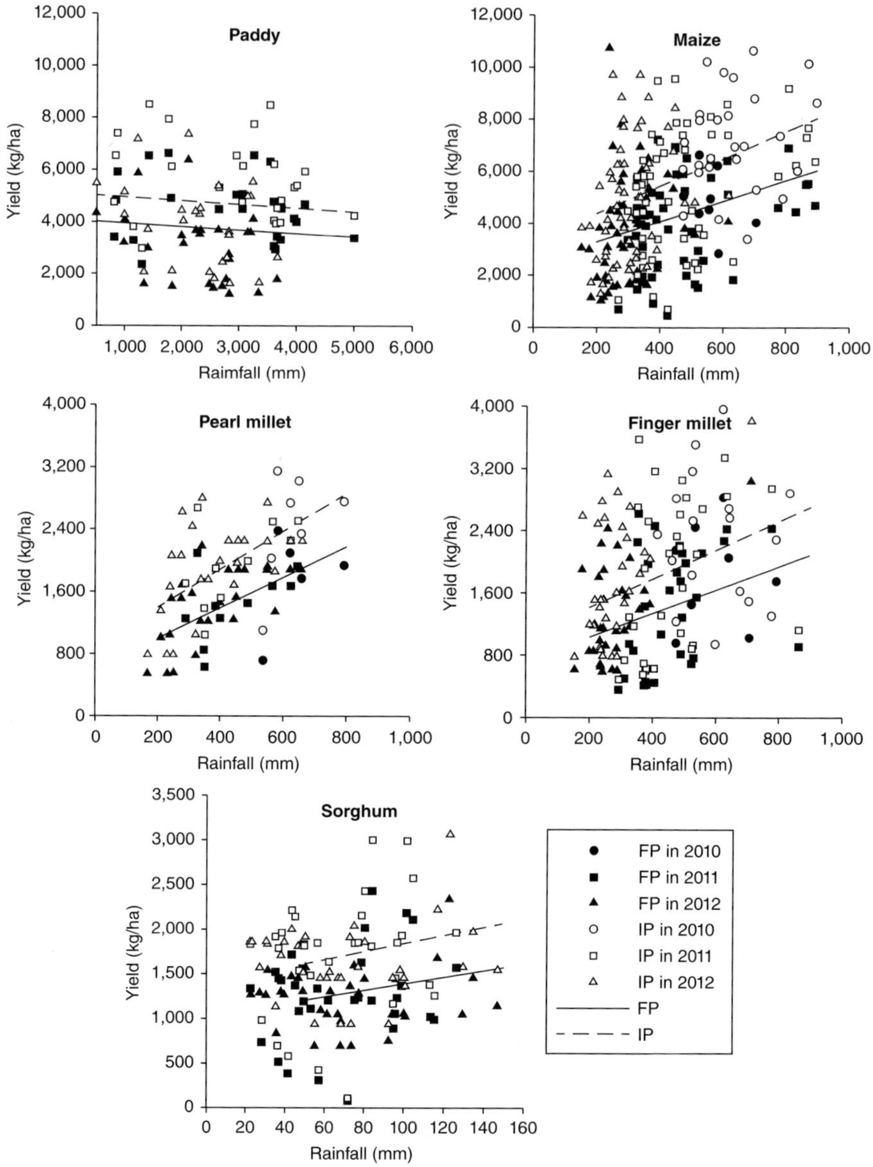

Fig. 10.3. Yield of selected cereals with rainfall and balanced fertilizer application in Karnataka during 2010–2012. Filled and open symbols represent crop yield under farmers' practice (FP) and improved practice (IP), respectively; solid and broken lines further show the yield trend with increasing rainfall amount under FP and IP, respectively.

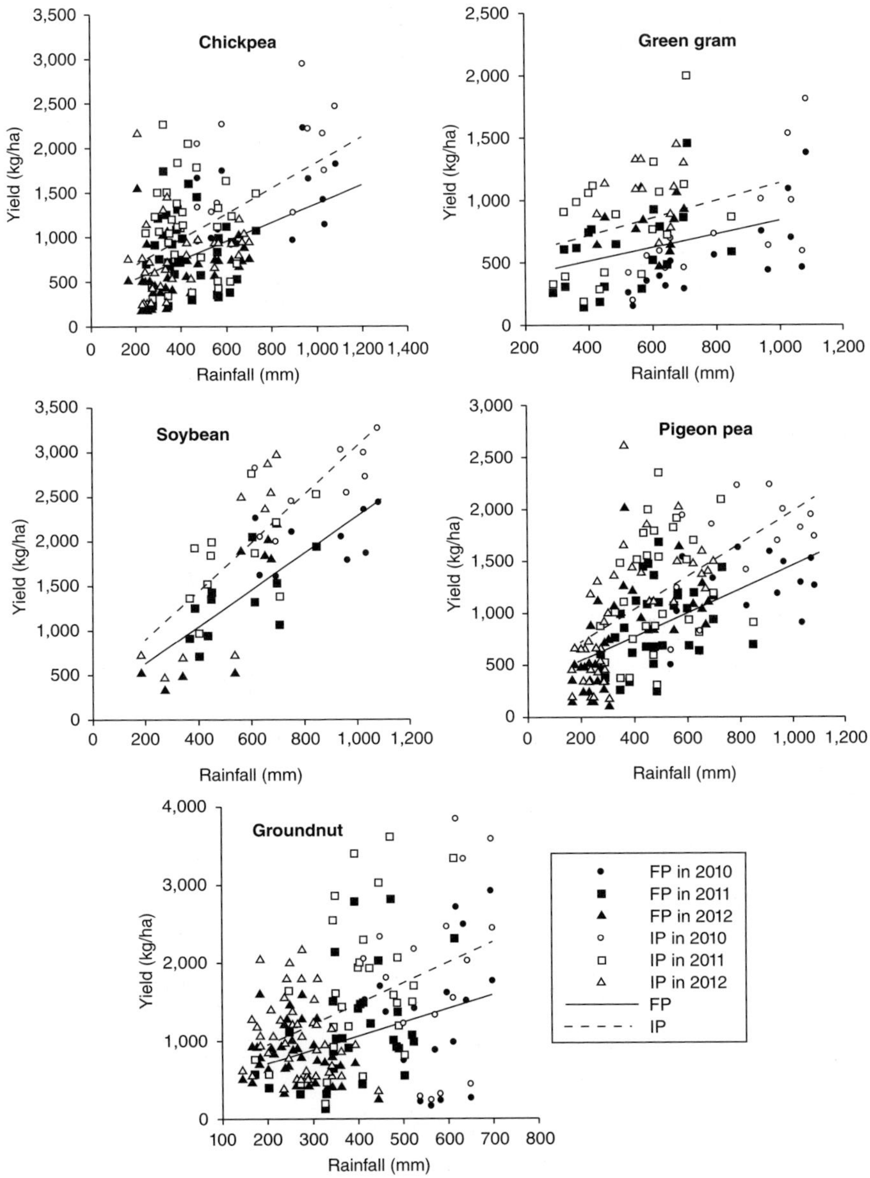

Fig. 10.4. Yield of selected pulses and oilseeds with rainfall and balanced fertilizer application in Karnataka during 2010–2012. Filled and open symbols represent crop yield under farmers' practice (FP) and improved practice (IP), respectively; solid and broken lines further show the yield trend with increasing rainfall amount under FP and IP, respectively.

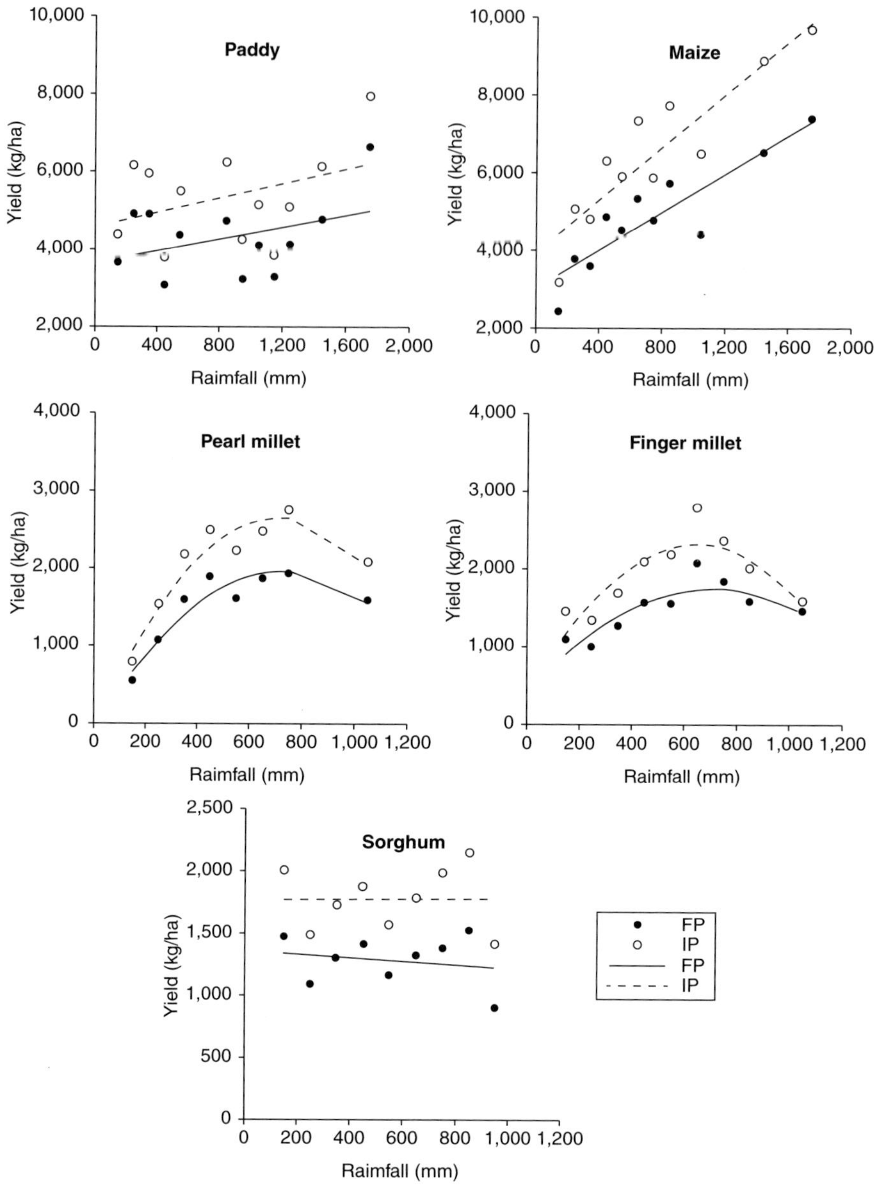

Fig. 10.5. Crop production function of major selected cereals obtained by averaging-up large-scale crop cutting experiments data under different rainfall classes with every 100 mm increment. Filled and open circles represent crop yield under farmers' practice (FP) and improved practice (IP), respectively; solid and broken lines further show the crop yield trend with increasing rainfall amount under FP and IP, respectively.

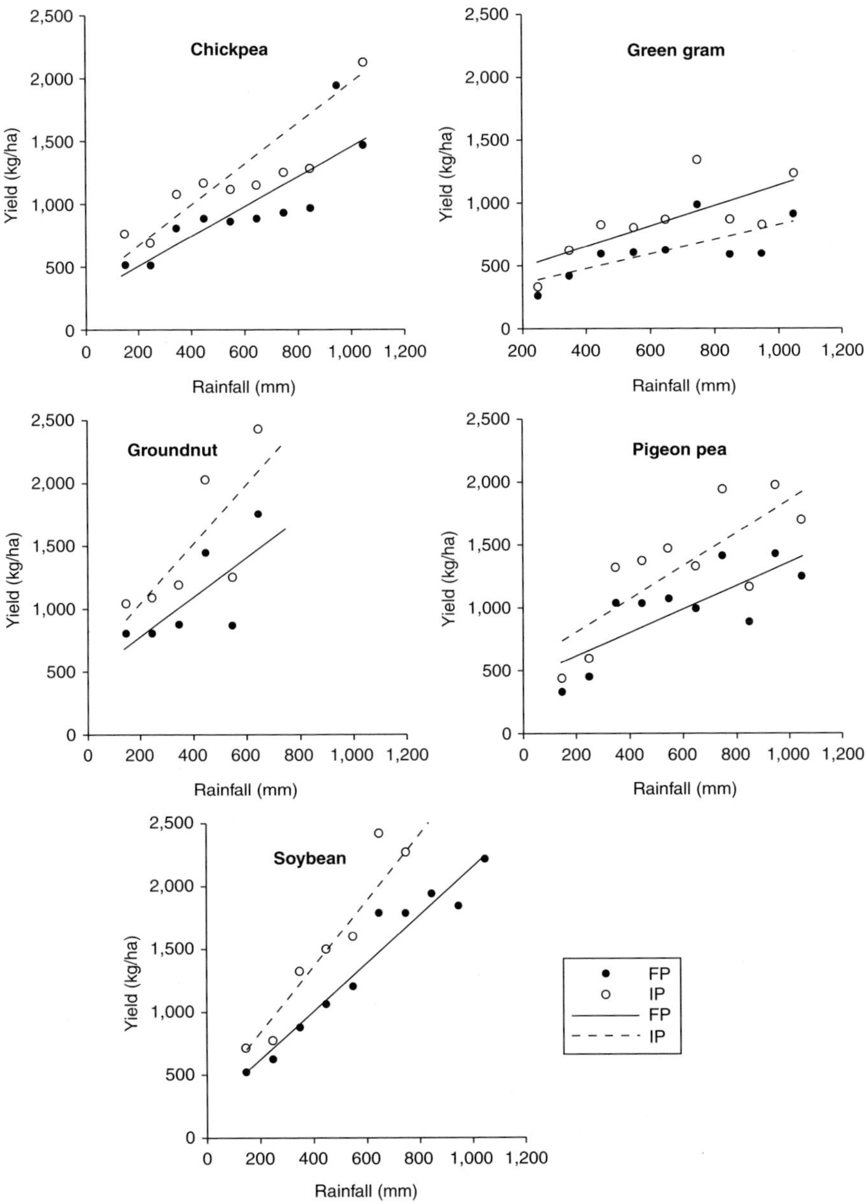

Fig. 10.6. Crop production function of major selected pulses and oilseeds obtained by averaging-up large-scale crop cutting data under different rainfall classes with every 100 mm increment. Filled and open circles represent crop yield under farmers' practice (FP) and improved practice (IP), respectively; solid and broken lines further show the yield trend with increasing rainfall amount under FP and IP, respectively.

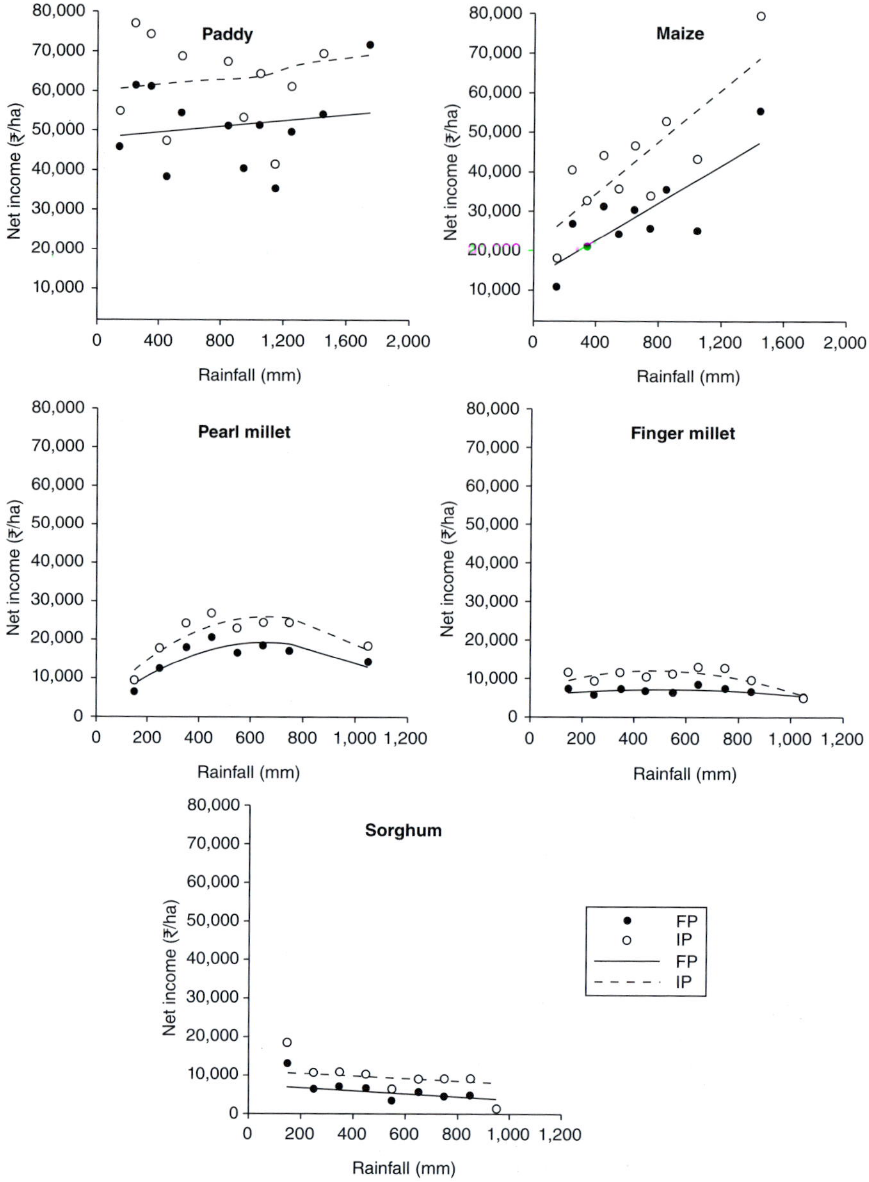

Fig. 10.7. Economic gain (net income) and its sensitivity with increasing rainfall for selected cereals. Filled and open circles represent estimated net income under farmers' practice (FP) and improved practice (IP), respectively; solid and broken lines further show the net income trend with increasing rainfall amount under FP and IP, respectively.

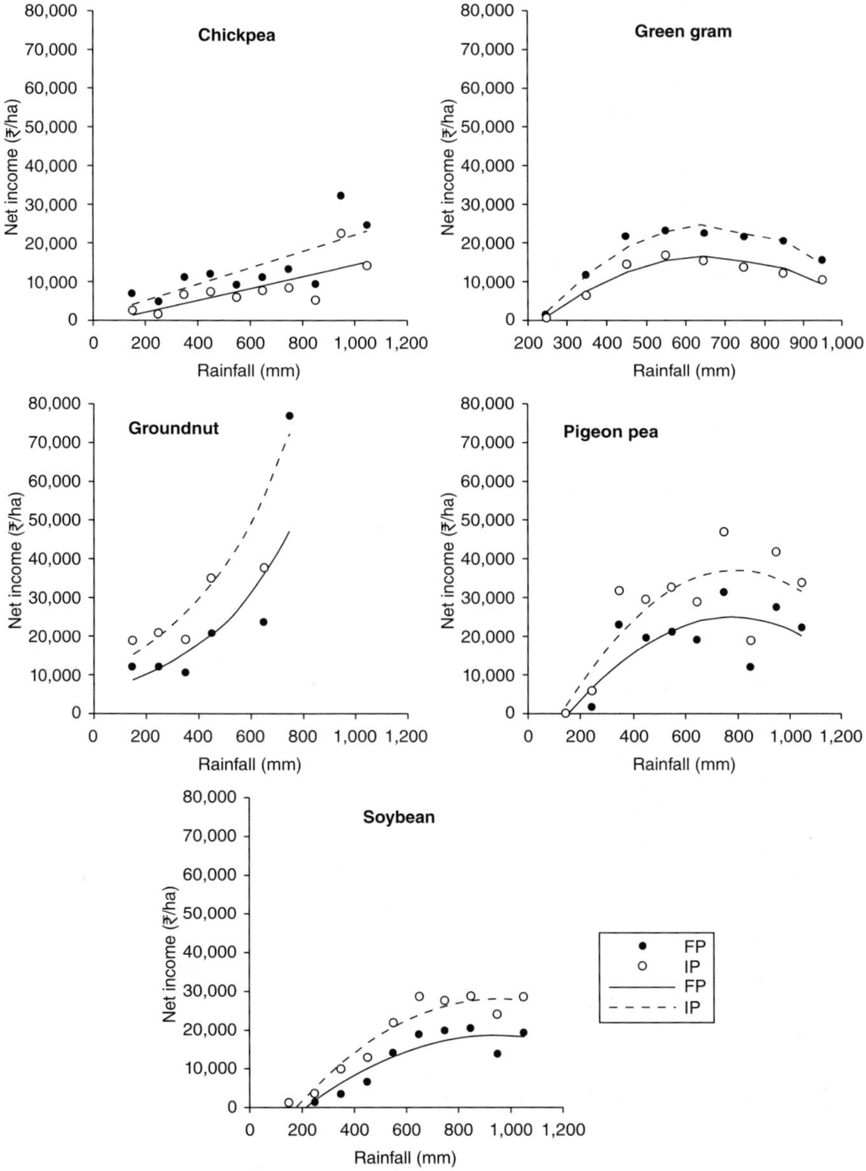

Fig. 10.8. Economic gain (net income) and its sensitivity with increasing rainfall for selected pulses and oilseeds. Filled and open circles represent the estimated net income under farmers' practice (FP) and improved practice (IP), respectively; solid and broken lines further show the net income trend with increasing rainfall amount under FP and IP, respectively.

conditions (600–800 mm) under IP. IP enhanced the net income by ₹8000–10,000/ha under maize and nearly ₹3000–5000/ha under millet and sorghum production. Among pulses, pigeon pea was more remunerative as net income obtained from this crop varied from ₹20,000 to 25,000/ha at moderate rainfall of 800 mm. IP enhanced the net income further to ₹8000–10,000/ha. On the other hand, groundnut was very sensitive with application of micro and secondary nutrients. Net income for groundnut cultivation increased by ₹5000–15,000/ha with IP as compared with FP.

10.4.3 Spatial distribution of crop yields

Spatial variability in cereal productivity among different taluks was analysed. Yield and income obtained from cereal crops under FP and IP in 2012 are shown in Plate 3. Several crop layers overlaid each other; therefore, some of the plotted data are hidden, as they are covered by other layers, however, a comprehensive overview of the cropping system is depicted. Paddy is found largely in the high rainfall zone such as the western part of Karnataka and hilly regions (Western Ghats) where annual rainfall is higher than 1500–2000 mm. Maize is cultivated in the central part of Karnataka covering the state from north to south. Pearl millet and sorghum are cultivated in northern Karnataka and finger millet in southern Karnataka. Results showed that IP enhanced crop productivity in most of the taluks compared with FP. Thus, farmers benefited with additional income of about ₹2500/ha (minimum) to ₹30,000–350,000/ha (maximum). Net additional income was maximum under maize cultivation followed by paddy and pearl millet. Furthermore, the additional benefit–cost ratio ranged from 2 to 20, indicating impressive returns with IP. Similarly, results are also depicted for pulse and oilseed crops in Plate 4. Yield and income levels increased with IP compared with FP.

10.4.4 Crop yields with improved varieties

The Bhoochetana programme has provided opportunities to harness the potential of rainfed areas and showcased it to several stakeholders for large-scale adoption and helped in enhancing their learning and capacity. Land, water, nutrient and crop management interventions were demonstrated under on-farm conditions. Farmers in various parts of Karnataka are still using indigenous and low-yielding crop varieties. Farmers' participatory field trials were conducted for different crops in different districts of Karnataka. Improved crop varieties were compared with the local variety by growing both varieties side by side under the same management conditions. Data obtained from a number of trials indicated 50–150% higher crop yields with improved varieties compared with the local variety (Fig. 10.9). Plate 5 shows the variability in yield among cereals (finger millet, pearl millet), pulses (chickpea, pigeon pea) and oilseed crop (groundnut) between districts. Results indicate that crop productivity could easily be doubled by introducing new and improved varieties along with application of balanced fertilization application.

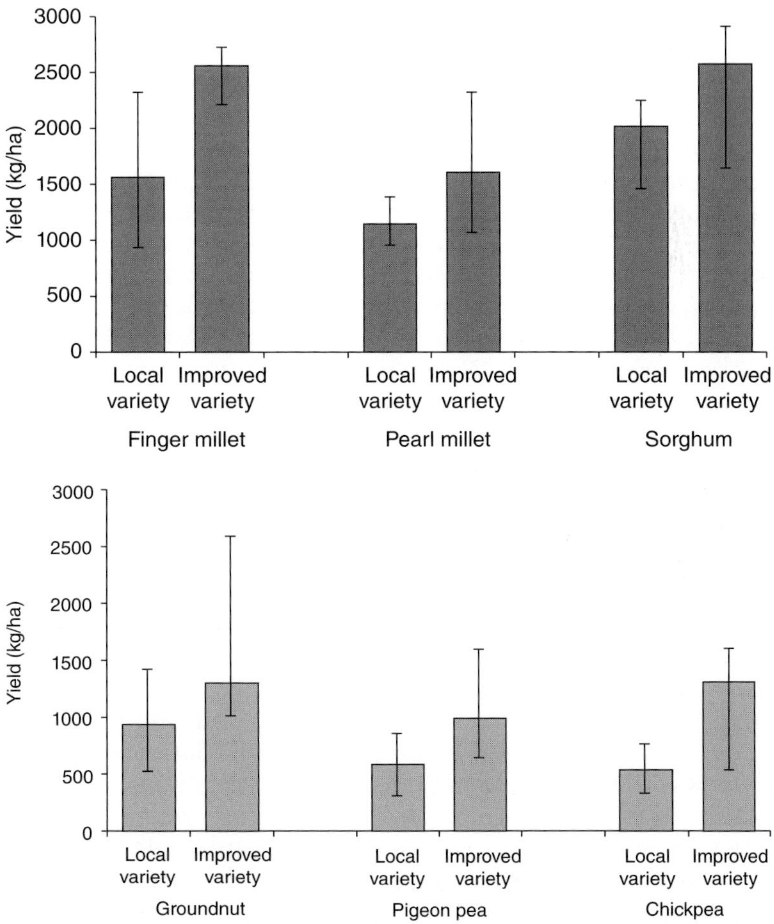

Fig. 10.9. Comparison of crop yield between the local and improved variety of selected cereals (finger millet, pearl millet and sorghum), pulses (pigeon pea and chickpea) and oilseed (groundnut) obtained from farmers' participatory field experiments across the state using data of 2012. Columns show average yields obtained from different crop cutting experiments and error bars show the maximum and minimum range.

10.5 Impact of Bhoochetana on Water Productivity

Crop water productivity (WP) is the amount of grain yield obtained per unit of water used (Tuong and Bouman, 2003; Garg *et al.*, 2012b). Depending on the type of water sources considered, WP is expressed as grain yield per unit water evapotranspired or grain yield per unit of total water input (rainfall under rainfed conditions). In this study, technical WP of IP and FP was calculated using simulated values of ET and yield values obtained for selected cereals, oilseeds and pulses across the entire state. Moreover, economic water productivity (EWP) (₹ per m^3 of water) was also derived using net income obtained against per unit of water use.

Data on average productivity, net income, technical WP and EWP and rainfall are summarized for important cereals, pulses and oilseeds during 2009–2012 in Table 10.2. Average crop yields over the 4-year period were 1810 kg/ha and 2440 kg/ha with FP and IP, respectively. Similarly, the average estimated net income was ₹26,290/ha with FP and ₹35,540/ha with IP, indicating an additional 35% of income by adopting improved management practices. Technical WP with FP and IP was 0.51 kg/m^3 and 0.69 kg/m^3, whereas EWP was ₹5.3/m^3 and ₹7.15/m^3, respectively (Table 10.2).

10.6 Conclusions

Agriculture in India as such, and specifically in Karnataka, assumes much more importance as 60% of people mostly depend on it for their livelihoods. Rainfed agriculture in general is considered as '1 t agriculture' with the perception that not much improvement can be made. Bhoochetana thus became the best example for scaling up in the country and has shown that simple but science-led interventions such as balanced fertilizer application alone can contribute to 30–35% additional yield gain. The total area covered by the programme within the 4-year period between 2009 and 2012 was 3.73 million ha. More than 3 million farmers, including small and marginal farmers from all the 30 districts of Karnataka, benefited from the programme. Field observations and agronomic records also showed that crops were found to be more tolerant to various pests and diseases and yielded better compared with farmers' management practices. The beneficial impact of the Bhoochetana programme is observed in not only wet and normal years but also in dry years. The programme has proven that improved management systems are vital in building the resilience of the farming systems in spite of normal or below normal rainfall in the state. Increase in crop yield and net income by about 30% has contributed to the household budget in rural areas as the benefit–cost ratio ranges from 2 to 20 for different cropping systems and regions.

Table 10.2. Economics of water productivity of selected crops under farmers' practice (FP) and improved practice (IP) for 2009 (wet year), 2010 (normal year), 2011 (dry year) and 2012 (very dry year).

Crop	Data from no. of taluks	Yield (kg/ha)		Net income (₹/ha)		Rainfall (mm)		Technical water productivity (kg/m³)		Economic water productivity (₹/m³)	
		FP	IP	FP	IP	FP	IP	FP	IP	FP	IP
2009											
Paddy	–	–	–	–	–	–	–	–	–	–	–
Maize	5	5,472	7,600	45,965	63,840	691	464	1.18	1.64	7	9.3
Finger millet	3	1,737	2,680	15,891	24,522	581	372	0.47	0.72	8	4.3
Pearl millet	–	–	–	–	–	–	–	–	–	–	–
Sorghum	9	1,286	1,878	11,056	16,149	710	443	0.29	0.42	1.6	2.3
Chickpea	6	1,052	1,407	18,194	24,335	654	415	0.25	0.34	2.8	3.8
Green gram	–	–	–	–	–	–	–	–	–	–	–
Pigeon pea	–	–	–	–	–	–	–	–	–	–	–
Soybean	2	1,775	2,630	24,673	36,557	807	529	0.34	0.5	3.1	4.6
Groundnut	8	1,276	1,761	26,801	36,986	594	377	0.34	0.47	4.4	6.1
Average		2,100	2,993	23,763	33,732	673	433	0.48	0.68	3.57	5.07
2010											
Paddy	–	–	–	–	–	–	–	–	–	–	–
Maize	30	5,435	7,275	47,825	64,021	655	421	1.29	1.73	7.7	10.2
Finger millet	22	1,700	2,317	16,405	22,362	624	405	0.42	0.57	2.8	3.8
Pearl millet	8	1,709	2,394	15,037	21,065	685	426	0.4	0.56	2.3	3.2
Sorghum	6	1,290	1,832	11,610	16,485	614	390	0.33	0.47	1.9	2.7
Chickpea	13	1,352	1,817	23,787	31,978	759	453	0.3	0.4	3.3	4.4
Green gram	14	546	763	17,299	24,183	800	473	0.12	0.16	2.1	2.9
Pigeon pea	14	1,212	1,631	36,364	48,943	836	489	0.25	0.33	4.5	6
Soybean	9	2,008	2,650	28,812	38,022	863	500	0.4	0.53	3.4	4.5
Groundnut	18	1,297	1,814	29,831	41,732	572	371	0.35	0.49	5.2	7.3
Average		1,839	2,499	25,219	34,310	712	436	0.43	0.58	3.69	5

Continued

Table 10.2. Continued.

Crop	Data from no. of taluks	Yield (kg/ha)		Net income (₹/ha)		Rainfall (mm)		Technical water productivity (kg/m³)		Economic water productivity (₹/m³)	
		FP	IP	FP	IP	FP	IP	FP	IP	FP	IP
2011											
Paddy	25	4,477	5,583	48,354	60,294	2,720	776	0.58	0.72	2.4	3.1
Maize	48	3,998	5,348	39,178	52,414	470	320	1.25	1.67	9	12
Finger millet	39	1,552	2,037	16,297	21,393	485	317	0.49	0.64	3.4	4.4
Pearl millet	12	1,714	2,373	16,799	23,259	415	293	0.58	0.81	4.2	5.8
Sorghum	32	1,369	1,833	13,691	18,328	451	304	0.45	0.6	3.2	4.4
Chickpea	31	855	1,151	17,945	24,164	407	278	0.31	0.41	4.8	6.5
Green gram	18	585	826	20,475	28,894	484	329	0.18	0.25	4.3	6.1
Pigeon pea	28	922	1271	29,509	40,663	473	323	0.29	0.39	6.2	8.4
Soybean	11	1,316	1,850	22,247	31,265	520	358	0.37	0.52	4.4	6.3
Groundnut	31	1,381	1,963	37,295	52,998	389	261	0.53	0.75	9.1	13
Average		1,817	2,424	26,179	35,367	681	356	0.5	0.68	5.1	7
2012											
Paddy	48	3,694	4,683	46,180	58,539	1,375	512	0.72	0.91	10.6	13.3
Maize	54	3,939	5,204	46,286	61,148	318	213	1.85	2.44	15.4	20.4
Finger millet	45	1,256	1,676	18,847	25,147	283	173	0.73	0.97	6.8	9
Pearl millet	25	1,546	2,056	18,170	24,153	380	259	0.6	0.79	4.8	6.5
Sorghum	31	1,267	1,686	19,265	25,634	388	259	0.49	0.65	5.5	7.3
Chickpea	30	611	794	17,117	22,223	382	255	0.24	0.31	5	6.5
Green gram	12	819	1,086	36,043	47,777	592	404	0.2	0.27	6.2	8.2
Pigeon pea	41	826	1,066	31,805	41,054	367	243	0.34	0.44	8.8	11.4
Soybean	12	1,167	1,568	25,667	34,485	465	305	0.38	0.51	5.3	7.1
Groundnut	45	792	1,052	29,320	38,932	274	175	0.45	0.6	11.4	15.1
Average		1,815	2,460	26,190	35,457	636	371	0.5	0.68	5.2	7.03

Acknowledgements

We sincerely acknowledge the funding support from the Government of Karnataka and the DoA for implementing the Bhoochetana programme in Karnataka. Our sincere thanks go to the Karnataka State Natural Disaster Management Centre for providing daily rainfall data of different taluks of the 30 districts of Karnataka over a 4-year period between 2009 and 2012. We acknowledge the help and support received from DoA staff, farm facilitators and ICRISAT field staff for conducting crop cutting experiments. Details of total fertilizer consumption (macro and micro) and area coverage under the Bhoochetana programme in each year was collected and provided by the DoA. The authors are also grateful to the National Bureau of Soil Survey and Land Use Planning (NBSS&LUP) and the Bureau of Statistics and Economics for providing data on soil physical parameters and the market price and cost of cultivation for different crops.

References

Boomiraj, K., Wani, S.P., Garg, K.K., Aggarwal, P.K. and Palanisami, K. (2010) Climate change adaptation strategies for agro-ecosystem – a review. *Agrometeorology* 12(2), 145–160.

Central Ground Water Board (CGWB) (2012) *Ground Water Year Book – India*. CGWB, Ministry of Water Resources, Government of India, New Delhi, India.

Central Water Commission (2005) *Handbook of Water Resources Statistics*. Ministry of Water Resources, Government of India, New Delhi, India.

Doorenbos, J. and Kassam, A.H. (1979) *Yield Response to Water*. FAO Irrigation and Drainage Paper No. 33. Food and Agriculture Organization of the United Nations (FAO), Rome, Italy.

Garg, K.K., Bharati, L., Gaur, A., George, B., Acharya, S., Jella, K. and Narasimhan, B. (2012a) Spatial mapping of agricultural water productivity using the SWAT model in Upper Bhima Catchment, India. *Irrigation and Drainage* 61(1), 60–79.

Garg, K.K., Karlberg, L., Barron, J., Wani, S.P. and Rockström, J. (2012b) Assessing impacts of agricultural water interventions in the Kothapally watershed, Southern India. *Hydrological Processes* 26(3), 387–404.

Garg, K.K., Wani, S.P., Barron, J., Karlberg, L. and Rockström, J. (2013) Up-scaling potential impacts on water flows from agricultural water interventions: opportunities and trade-offs in the Osman Sagar catchment, Musi sub-basin, India. *Hydrological Processes* 27(26), 3905–3921.

Government of India (GoI) (2010) *Addressing India's Nutrition Challenges: Report of the Multi-stakeholder Retreat*. Planning Commission, GoI, New Delhi, India.

Government of India (GoI) (2012) *Final Report of Minor Irrigation and Watershed Management for the Twelfth Five Year Plan (2012–2017)*. Planning Commission, GoI, New Delhi, India.

Lovelli, S., Perniola, M., Ferrara, A. and Tommaso, T.D. (2007) Yield response factor to water (*Ky*) and water use efficiency of *Carthamus tinctorius* L. and *Solanum melongena* L. *Agricultural Water Management* 92, 73–80.

National Academy of Agricultural Science (NAAS) (2009) *State of Indian Agriculture*. NAAS, New Delhi, India.

National Rainfed Area Authority (NRAA) (2012) *Prioritization of Rainfed Areas in India*. Study Report 4. NRAA, Ministry of Agriculture and Farmers' Welfare, Government of India, New Delhi.

Rao, A.V.R.K., Wani, S.P., Singh, K.K., Ahmed, M.I., Srinivas, K., Bairagi, S.D. and Ramadevi, O. (2013) Increased arid and semi-arid areas in India with associated shifts during 1971–2004. *Journal of Agrometeorology* 15(1), 11–18.

Rockström, J., Hatibu, N., Oweis, T.Y., Wani, S.P., Barron, J., Bruggeman, A., Farahani, J., Karlberg, L. and Qiang, Z. (2007) Managing water in rainfed agriculture. In: Molden, D. (ed.) *Water for Food, Water for Life: A Comprehensive Assessment of Water Management in Agriculture*. Earthscan, London and International Water Management Institute, Colombo, Sri Lanka, pp. 315–348.

Rockström, J., Karlberg, L., Wani, S.P., Barron, J., Hatibu, N., Oweis, T., Bruggeman, A., Farahani, J. and Qiang, Z. (2010) Managing water in rainfed agriculture – the need for a paradigm shift. *Agricultural Water Management* 97(4), 543–550.

Singh, R., Garg, K.K., Wani, S.P., Tewari, R.K. and Dhyani, S.K. (2014) Impact of water management interventions on hydrology and ecosystem services in Garhkundar-Dabar watershed of Bundelkhand region, Central India. *Journal of Hydrology* 509, 132–149.

Stewart, J.I., Cuenca, R.H., Pruitt, W.O., Hagan, R.M. and Tosso, J. (1977) *Determination and Utilization of Water Production Functions for Principal California Crops*. W-67 CA Contributing Project Report. University of California, Davis, California.

Tuong, T.P. and Bouman, B.A.M. (2003) Rice production in water-scarce environments. In: Kijne, J.W., Barker, R. and Molden, D. (eds) *Water Productivity in Agriculture: Limits and Opportunities for Improvement*. CABI Publishing, Wallingford, UK, pp. 53–67.

Wani, S.P., Sreedevi, T.K., Rockström, J. and Ramakrishna, Y.S. (2009) Rainfed agriculture – past trends and future prospects. In: Wani, S.P., Rockström, J. and Oweis, T. (eds) *Rainfed Agriculture: Unlocking the Potential*. Comprehensive Assessment of Water Management in Agriculture Series. CAB International, Wallingford, UK, pp. 1–35.

Wani, S.P., Rockström, J. and Sahrawat, K.L. (2011a) *Integrated Watershed Management in Rainfed Agriculture*. CRC Press, Leiden, The Netherlands.

Wani, S.P., Rockström, J., Venkateswarlu, B. and Singh, A.K. (2011b) New paradigm to unlock the potential of rainfed agriculture in the semi-arid tropics. In: Lal, R. and Stewart, B.A. (eds) *World Soil Resources and Food Security. Advances in Soil Science*. CRC Press, Boca Raton, Florida, pp. 419–469.

Wani, S.P., Garg, K.K., Singh, A.K. and Rockström, J. (2012a) Sustainable management of scarce water resource in tropical rainfed agriculture. In: Lal, R. and Stewart, B.A. (eds) *Soil Water and Agronomic Productivity. Advances in Soil Science*. CRC Press, Boca Raton, Florida, pp. 347–408.

Wani, S.P., Sarvesh, K.V., Krishnappa, K., Dharmarajan, B.K. and Deepaja, S.M. (2012b) *Bhoochetana: Mission to Boost Productivity of Rainfed Agriculture through Science-led Interventions in Karnataka*. International Crops Research Institute for the Semi-Arid Tropics, Patancheru, Andhra Pradesh, India.

Wani, S.P., Khanwalkar, S., Krishnappa, K., Raju, K.V. and Sarvesh, K.V. (2013) *Bhoochetana: Process Documentation – A Program that Helped Farmers come out of Low Productivity and Poor Economic Status*. Resilient Dryland Systems Report No. 60. International Crops Research Institute for the Semi-Arid Tropics, Patancheru, Andhra Pradesh, India, 134 pp.

11 Social and Economic Benefits

K.H. Anantha,* K.V. Raju, Suhas P. Wani, G. Pardhasaradhi, K. Srinivas, Kaushal K. Garg and K. Krishnappa

International Crops Research Institute for the Semi-Arid Tropics (ICRISAT), Patancheru, India

11.1 Introduction

In order to meet the food demand of the growing population, agriculture has to produce more from the available land area and water through more efficient use of natural resources with minimal impact on the environment (Hobbs *et al.*, 2008). Recent data show a general increase in global food production (FAO, 2010) and this can be attributed to both the expansion of cultivated area and technological progress, leading to increased crop yields (Lal, 2009). This yield gain has been achieved largely due to heavy reliance on fertilizers and pesticides, thereby putting pressure on the environment. It is estimated that by 2025, India's population is expected to reach 1.45 billion (United Nations, 2006) and the cereal grains requirement will be between 257 million t and 296 million t, depending on income growth (Kumar, 1998; Bhalla *et al.*, 1999). Thus, it is clear that current approaches to agriculture and agricultural technology will not be adequate to address food security issues going forward. The future food production must increase by about 5 million t annually to ensure food and nutritional security to the burgeoning population (Kanwar, 2000). Therefore, there is a recognized need to further examine the contextual factors associated with the development of new and environmentally sustainable agricultural technology (Sahrawat *et al.*, 2010; Wani *et al.*, 2011). This requires a truly interdisciplinary

*k.anantha@cgiar.org

holistic approach to deal with the demand for technical knowledge integration along with delivery systems and markets.

The potential of rainfed agriculture to bridge the large yield gaps and enhanced income is rarely contextualized in terms of farm household or share of agriculture in the state's gross domestic product (GDP). Conventionally, the benefits from new technology are measured in terms of higher yields or income per hectare with limited field demonstrations but the diversity of the system is overlooked. This is particularly true of rural areas where crop production is predominantly based on direct rainfall. Thus, the present study focuses on crop yield and income due to adoption of improved technologies in rainfed areas.

This study was conducted in Karnataka, a state with diverse agroecological characteristics which is the second largest rainfed state in India after Rajasthan. About 70% of the cultivated area is rainfed and only 30% is irrigated both by canal and groundwater sources (GoK, 2011). Maize (*Zea mays*), paddy (*Oryza sativa*), sorghum (*Sorghum bicolor*) and finger millet (*Eleusine coracana*) are the major staple crops occupying more than 50% of the land area under cultivation and accounting for more than 60% of the population's caloric intake (GoK, 2011). Until recently, farmers were facing four major constraints (GoK, 2011). First, over the past two decades, average productivity of major rainfed crops in the state has fluctuated around 1 t/ha until 2009. Secondly, degraded and infertile soils as a result of continuous monocropping and insufficient organic matter recycling along with occurrence of rainfall variability marked by frequent dry spells account for low crop yields (Singh *et al.*, 2009; Sahrawat *et al.*, 2010; GoK, 2011; Wani *et al.*, 2012). However, reducing soil degradation only through chemical fertilizer is not adequate. Soil-test-based micronutrient application along with biofertilizers has made a huge impact on the crop yield even if rainfall is very scanty (Wani *et al.*, 2012; Raju *et al.*, 2013). Thirdly, smallholders have limited access to farm inputs such as fertilizers, secondary and micronutrients, and seeds due to low purchasing power and weak market infrastructure. In some instances, the price offered by the state in terms of the minimum support price is low compared with the cost of production, thereby providing little incentive to farmers to produce above the subsistence level. Fourthly, there has been an increase in land fragmentation due to the increase in nuclear families along with labour scarcity for farming operations in the predominantly cereal-based farming systems. This has become a critical issue as a result of the constant migration of people from rural to urban areas. In view of all the above constraints, the Bhoochetana programme was planned and implemented with an overall objective of improving the rural livelihood system by targeting agricultural development in the state. The specific objectives of the programme were: (i) to identify and scale up best-bet options (soil, crop and water management), including improved cultivars, to enhance productivity of the selected crops by 20% in the selected 30 districts; (ii) to train staff of the Department of Agriculture (DoA), Karnataka in stratified soil sampling in villages, analysis of macro- and micronutrients, and preparation of GIS (geographical information system)-based soil maps; (iii) to guide the DoA to establish a high-quality soil analytical laboratory at Bengaluru which can undertake stratified soil sampling and analyses, and share the results in nine districts; and (iv) to build capacity of the stakeholders (farmers and consortium partners) in the sustainable management of natural resources and enhance productivity in dryland areas.

It is reported that the Bhoochetana programme has had impressive effects on the state's overall agricultural scenario (Wani *et al.*, 2012). This has led to the expansion of Bhoochetana from 2009 to include all major rainfed (cereals, legumes, oilseeds) and irrigated crops such as paddy and sugarcane (*Saccharum officinarum*). In the 2009/10 season, the major crop yield in six districts increased by 20–66% depending on the crops, soil types and rainfall situation. There has been a constant increase in crop yield since 2009 and it can be attributed to the implementation of the mission project Bhoochetana.

11.2 Soil Fertility Status

Large tracts of the semi-arid region in India witnessed soil nutrient deficiencies and this resulted in low crop yield (Singh *et al.*, 2009). Soil fertility analysis was an entry point activity under the Bhoochetana initiative, which was useful in determining the deficiency level and taking corrective measures to increase crop yield as well as cut the cost of overdosing of fertilizer. By adopting a statistically proven random stratified sampling method (Sahrawat *et al.*, 2008), about 92,904 soil samples were collected from farmers' fields, covering the rainfed areas in all the 30 districts of Karnataka. The samples collected represented vast spatial variability in rainfall, topography, cropping system, farm size and management. Soil samples were analysed in the state-of-the-art laboratory at the International Crops Research Institute for the Semi-Arid Tropics (ICRISAT), Patancheru, India. Several soil health parameters (i.e. organic carbon (C), available phosphorus (P), potassium (K), zinc (Zn), boron (B), sulfur (S), pH and electrical conductivity (EC)) were analysed and data were used to develop block-level fertilizer recommendations in contrast to conventional blanket recommendations for macronutrients at the state level. The basis used was to recommend the full rate of a nutrient in cases where more than 50% of farmers' fields were deficient in that nutrient, and recommend only half the rate of the nutrient in cases where less than 50% of the farmers' fields were deficient in that particular nutrient. The DoA was empowered to adopt and disseminate soil test results and site-specific fertilizer recommendations through traditional and innovative extension means. Based on region-specific constraints, a package of improved soil–crop–water management practices was designed and shared among the consortium partners and stakeholders.

The soil fertility analysis revealed that there is vast spatial variability in different soil nutrients across different blocks. The results showed that the majority of farmers' fields (52%) were low in C throughout the state. In the whole of Karnataka, 41% farms were deficient in P, indicating that the majority of farms were rather sufficient in P and had the opportunity through site-specific nutrient management to cut the cost on current recommendations. Across the state only 23% of sampled fields tested low in K and a science-led approach calls for a cut on the recommended K. Interestingly, the diagnosis revealed widespread deficiencies of ignored secondary and micronutrients in most farms in Karnataka such that 52% of farms were deficient in S, 55% in Zn and 62% in B. The S, B and Zn deficiencies are even more widespread than deficiencies in the macronutrients P and K, which are mostly focused on, and this is apparently the reason for productivity potential in the semi-arid tropics being held back.

11.3 Impact on Crop Yield and Biomass Production

The new technologies were first demonstrated in farmers' fields to understand the crop responsiveness for further upscaling in the region. All trials were managed by farmers in all the districts with support from extension staff of the DoA and *Krishi Vigyan Kendras* (KVKs), and research technicians of ICRISAT provided recommendations on management of the plots. Scientific knowledge was provided by scientists and research technicians who are located in the districts for liaising with the farmers, farm facilitators and DoA. For trials, fields were selected based on the farmers' willingness to participate and were divided into two parts for demonstrating new technologies (i.e. improved management practice and conventional farmers' practice). A balanced dose of fertilizer was applied along with new varieties under the improved management practice and the farmer's plot was cultivated as per normal conventional practice without improved methods/approaches under the farmers' practice. These trials were regularly monitored by farmers as per the recommendations by DoA staff, KVK staff and research staff of ICRISAT.

11.3.1 Crop yield and biomass estimation

Crop and biomass yields were estimated by adopting standardized crop cutting experiments. At physiological maturity, the entire biomass of two major crops was harvested from ten samples of 5 m × 5 m taken from the farmers' practice, and the same number of samples from the improved management practice treatment in each block. From the demarcated areas, the crop was harvested at ground level from each plot separately and then the pods/heads were separated from the stalks and the total fresh weight of the biomass was recorded. Subsamples were taken and dried in the oven at 65°C and actual yields were calculated. Data from the crop cutting experiment were used to monitor the yield and biomass production of Bhoochetana interventions in the state since 2009.

11.3.2 Impact on crop yield

The Bhoochetana mission project made a huge impact on crop yields of cereals, legumes and oilseeds. The initiative began with an area of 0.2 million ha in six districts covering mostly cereal crops and extended to 3.73 million ha in the subsequent years covering all major cereals, legumes and oilseeds. As the cropping system is diverse, during 2011, the focus was shifted to include even irrigated crops like paddy and sugarcane to realize their potential yield level. Crop cutting experiments data were used to undertake yield analysis by comparing improved management practices with farmers' own practice. To estimate average values for farmers' practice and improved practice, box plots were used to identify extreme cases and outliers for the major crops. The average yields of cereal crops with improved practice were higher than those with farmers' practice. Maize is grown more extensively in almost all the districts, except coastal areas, and yield benefits were impressive during a normal rainfall year. However, poor rainfall and soil fertility status reduced grain yield especially during 2011 and 2012 (Table 11.1). Average maize productivity significantly reduced in 2012 compared with 2009 from 5420 kg/ha to 3700 kg/ha with farmers' practice, and from 7800 kg/ha to 4850 kg/ha with improved practice. As a result, the incremental gain also reduced from 44% to 31% over farmers' practice. The drastic difference is also due to a difference in the type of maize hybrids used, plant spacing and application of supplemental irrigation. However, the major contributing factor for decline in maize yield was rainfall.

Although finger millet is considered to be a drought-tolerant crop, the average productivity declined from 1680 kg/ha to 1380 kg/ha with farmers' practice and 2550 kg/ha to 1830 kg/ha with improved practice during 2009 and 2012, respectively. The improved practice helped farmers to harness a better yield compared with conventional practice despite drought conditions. *Rabi* (post-rainy season) sorghum yield was high around 1370 kg/ha and 1960 kg/ha in a normal rainfall year (2010) and 1450 kg/ha and 1910 kg/ha in a dry year (2011) with farmers' practice and improved practice, respectively. Pearl millet (*Pennisetum glaucum*), one of the staple food grains in northern Karnataka, showed a mixed response to water stress situation during 2010–2013. Wheat

Table 11.1. Grain yield of cereal crops under Bhoochetana during 2009–2013.

Year	Crop	Grain yield				
		Farmers' practice (kg/ha)	Improved practice (kg/ha)	Increase (%)	LSD[a]	*F* value
2009	Maize	5420	7800	44	356**	174.76
	Finger millet	1680	2550	52	157**	120.06
	Rabi sorghum	1230	1780	45	103**	112.73
2010	*Kharif* sorghum	1710	2300	35	387**	9.39
	Maize	5820	7560	30	441**	60.31
	Pearl millet	1690	2250	33	275**	16.03
	Finger millet	1910	2620	37	112**	152.71
	Rabi sorghum	1370	1960	43	368**	10.41
2011	*Kharif* sorghum	2080	2910	40	428**	15.05
	Maize	3900	5110	31	440**	29.32
	Paddy	4340	5120	18	325**	22.20
	Pearl millet	1830	2540	39	459**	9.54
	Finger millet	1500	1960	31	215**	18.12
	Rabi sorghum	1450	1910	32	268**	11.73
2012	*Kharif* sorghum	2050	2670	30	400**	9.95
	Maize	3700	4850	31	437**	26.73
	Paddy	3630	4600	27	310**	38.08
	Pearl millet	1300	1790	38	253**	14.66
	Rabi maize	2190	3100	41	978NS	6.68
	Finger millet	1380	1830	32	204**	18.54
	Rabi sorghum	1210	1590	31	138**	29.87
	Wheat	610	810	32	145**	7.52
2013	*Kharif* sorghum	2110	2770	31	352**	13.72
	Maize	4100	5340	30	220**	121.62
	Paddy	3990	4940	24	164**	128.98
	Pearl millet	1410	1960	39	177**	37.90
	Finger millet	1460	1870	28	116**	47.08
	Rabi sorghum	1330	1690	27	168**	17.94
	Wheat	1000	1380	37	158**	22.67

[a]LSD, Least significant difference; **, significant at 1% level; NS, not significant.

(*Triticum aestivum*) cultivated in a few districts showed a positive response to improved management practices; yield increased from 610 kg/ha and 810 kg/ha in 2012 to 1000 kg/ha and 1380 kg/ha in 2013 with farmers' practice and improved practice, respectively.

Similarly, the average grain yields of legumes and oilseeds presented in Tables 11.2 and 11.3, respectively, reveal that rainfall had a major influence on grain yield. The average yield of black gram (*Vigna mungo*) was 930 kg/ha and 1260 kg/ha in 2010; 670 kg/ha and 890 kg/ha in 2011; 1220 kg/ha and 1640 kg/ha in 2012 with farmers' practice and improved practice, respectively. Chickpea (*Cicer arietinum*), a post-monsoon crop which is generally grown with residual soil moisture, performed better with improved practice compared with farmers' practice. The average chickpea yield was 1140 kg/ha and 1470 kg/ha in 2009; 1380 kg/ha and 1810 kg/ha in 2010; 920 kg/ha and 1210 kg/ha in 2011; 630 kg/ha and

Table 11.2. Grain yield of legume crops under Bhoochetana during 2009–2013.

		Grain yield				
Year	Crop	Farmers' practice (kg/ha)	Improved practice (kg/ha)	Increase (%)	LSD[a]	*F* value
2009	Chickpea	1140	1470	29	178**	13.58
2010	Black gram	930	1260	36	424NS	2.62
	Green gram	470	660	42	122**	10.03
	Pigeon pea	1200	1620	35	135**	36.80
	Chickpea	1380	1810	31	170**	25.12
2011	Black gram	670	890	34	200*	5.41
	Cowpea	270	400	45	34**	57.34
	Field bean	1490	1940	30	607NS	3.25
	Green gram	540	760	41	167*	6.84
	Pigeon pea	915	1260	38	166**	16.77
	Chickpea	920	1210	32	145**	16.38
2012	Black gram	1220	1640	35	128**	49.97
	Field bean	870	1160	33	283*	4.12
	Green gram	840	1110	33	125**	19.28
	Horse gram	100	130	26	90NS	0.56
	Pigeon pea	980	1250	27	274NS	3.69
	Chickpea	630	820	29	128**	8.25
2013	Black gram	570	760	34	145*	6.89
	Cowpea	320	430	34	71**	9.22
	Field bean	640	880	38	189*	6.85
	Green gram	450	590	30	99**	7.15
	Pigeon pea	650	870	34	105**	17.31
	Chickpea	770	1020	32	81**	35.87

[a]LSD, Least significant difference; *, significant at 5% level; **, significant at 1% level; NS, not significant.

Table 11.3. Grain yield of oilseed crops under Bhoochetana during 2009–2013.

Year	Crop	Grain yield: Farmers' practice (kg/ha)	Grain yield: Improved practice (kg/ha)	Grain yield: Increase (%)	LSD[a]	*F* value
2009	Groundnut	1080	1490	39	122**	45.03
	Kharif sunflower	810	1120	38	188**	11.41
	Soybean	1770	2620	48	231**	54.22
2010	Groundnut	1500	2130	42	134**	84.04
	Kharif sunflower	590	720	21	280NS	0.81
	Soybean	1930	2540	32	228**	28.51
2011	Groundnut	1160	1610	39	291**	9.52
	Kharif sunflower	1080	1460	35	530NS	2.08
	Soybean	1480	1990	35	213**	22.98
	Rabi safflower	730	960	31	133**	12.08
	Rabi sunflower	1420	1970	38	737NS	2.37
2012	Groundnut	930	1220	31	140**	16.83
	Kharif sunflower	1010	1320	31	232**	7.13
	Soybean	1060	1400	33	409NS	2.85
	Rabi safflower	510	650	29	171NS	3.05
	Rabi sunflower	720	950	32	250NS	3.48
2013	Groundnut	1310	1700	30	126**	37.99
	Kharif sunflower	1020	1390	36	240**	9.12
	Soybean	1480	1860	26	121**	38.56
	Rabi safflower	730	950	30	109**	18.45
	Rabi sunflower	580	780	34	203NS	3.65

[a]LSD, Least significant difference; **, significant at 1% level; NS, not significant.

820 kg/ha in 2012; 770 kg/ha and 1020 kg/ha in 2013 with farmers' practice and improved practice, respectively. The average groundnut (*Arachis hypogaea*) yield ranged between 1220 kg/ha and 2130 kg/ha during 2009–2013 with improved practice and 930 kg/ha and 1500 kg/ha with farmers' practice during the same period. The productivity difference between farmers' practice and improved practice ranged between 30% and 42% with changing rainfall variability.

The improved practice helped to improve the profitability by 18–52% across cereal crops from 2009 to 2013. The highest yield profitability was observed in finger millet (52%) during 2009. It is significant to note that the soil-test-based fertilizer application along with secondary and micronutrients resulted in enhanced yield of finger millet. Similarly, *rabi* sorghum (45%) and maize (44%) also showed impressive yield benefits during the same period. Although the percentage increase in yield benefits was impressive, the absolute numbers showed a declining trend during dry and very dry years (see Table 11.1).

Box plots were used to capture the yield variability of major crops (Fig. 11.1). There was a declining trend in yield as rainfall decreased. For example, the average yield of maize was 7600 kg/ha in 2009 (wet year) and marginally declined to 7270 kg/ha in 2010 (normal year)

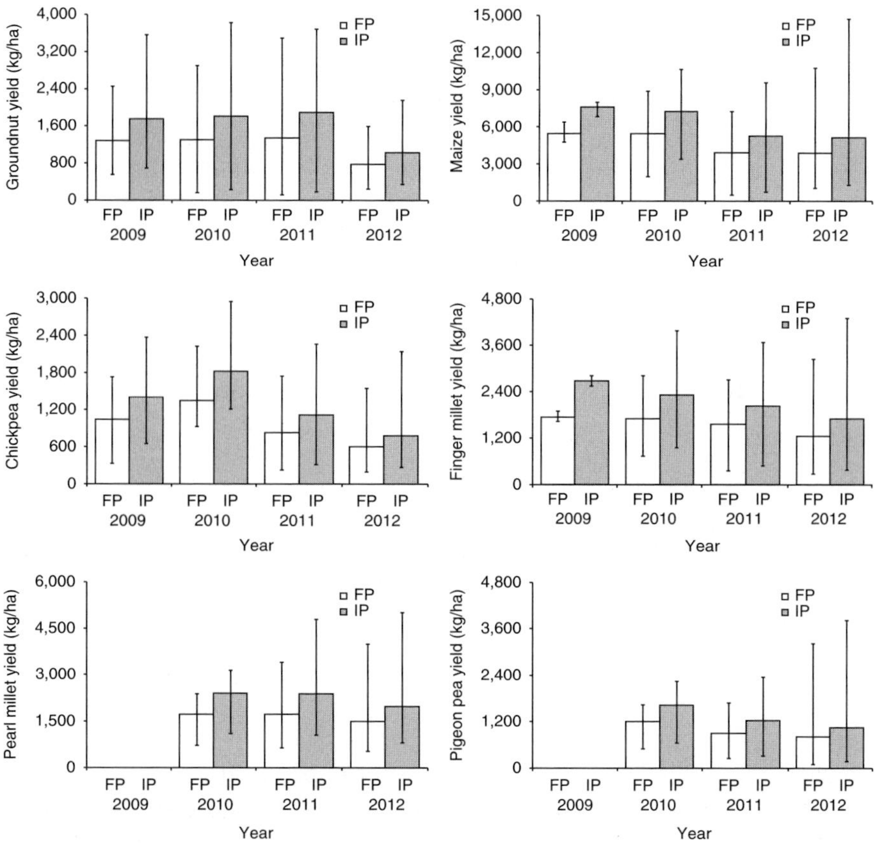

Fig. 11.1. Comparison of crop yield of groundnut, maize, chickpea, finger millet, pearl millet and pigeon pea with improved practice (IP) and farmers' practice (FP) from 2009 to 2012. Columns show average yields obtained from different crop cutting experiments and error bars show the maximum and minimum range.

and further declined to 5350 kg/ha in 2011 (dry year) and 5200 kg/ha in 2012 (very dry year). However, large yield variation during drought year (2012) can be attributed to supplemental irrigation provided by farmers to overcome yield reduction. A similar trend was observed for finger millet also with wide variations between farmers' practice and improved practice. However, the yield reduction was about 37% compared with 2009 yield, which was around 32% for maize during the same period. Similarly, chickpea yield was about 1410 kg/ha in 2009 with improved management practice, which declined to 790 kg/ha in 2012 due to the scarce rainfall situation. Groundnut yield marginally increased in 2009 (1760 kg/ha) and 2011 (1960 kg/ha) but significantly declined in 2012 (1050 kg/ha). The drastic decline in yield was attributed to the poor rainfall situation during 2012, which accounted for about a 44% reduction in chickpea yield and about a 40% yield reduction in groundnut even with improved management practice.

11.3.3 Total vegetative biomass production

Biomass yield was estimated separately for cereals, legumes and oilseeds during the study period and is presented in Tables 11.4, 11.5 and 11.6, respectively. The average biomass yield of maize crop was 8010 kg/ha and 11,140 kg/ha in 2009 and increased to 11,380 kg/ha and 15,500 kg/ha in 2010 with farmers' practice and improved practice, respectively (Table 11.4). Biomass was 8310 kg/ha and 10,990 kg/ha in 2011 and 8100 kg/ha and 10,490 kg/ha in 2012 with farmers' practice and improved practice, respectively. The paddy crop produced average biomass of 9390 kg/ha and 11,300 kg/ha in 2011 and 8340 kg/ha and 10,440 kg/ha in 2013 with farmers' practice and improved practice, respectively. Finger millet, which is one of the important fodder crops in rural areas, produced average biomass of 5570 kg/ha and 7810 kg/ha in 2009, 6450 kg/ha and 8390 kg/ha in 2011 and 5160 kg/ha and 6720 kg/ha in 2012 with farmers' practice and improved practice, respectively.

During a normal rainfall year (2010) the biomass of pigeon pea was 7990 kg/ha, black gram 4160 kg/ha, chickpea 3810 kg/ha and green gram 1920 kg/ha with improved practice compared with 6070 kg/ha, 3000 kg/ha, 2950 kg/ha and 1380 kg/ha with farmers' practice, respectively. As rainfall decreased, biomass production reduced with both farmers' and improved practices. However, the percentage gain is more or less the same as in the normal year with improved management practice over farmers' practice (Table 11.5). Oilseeds produced a more or less equal quantity of biomass irrespective of crops and rainfall situations in all years except in 2012 (Table 11.6). The comparison

Table 11.4. Biomass yield of cereal crops under Bhoochetana during 2009–2013.

Year	Crop	Biomass yield: Farmers' practice (kg/ha)	Biomass yield: Improved practice (kg/ha)	Increase (%)	LSD[a]	*F* value
2009	Maize	8,010	11,140	39	757**	67.12
	Finger millet	5,570	7,810	40	443**	99.04
	Rabi sorghum	4,160	5,620	35	216**	178.48
2010	*Kharif* sorghum	5,080	6,640	31	930**	11.16
	Maize	11,380	15,500	36	1143**	50.15
	Pearl millet	4,880	6,420	31	896**	11.44
	Finger millet	5,660	7,770	37	311**	177.21
	Rabi sorghum	3,940	5,480	39	677**	20.84
2011	*Kharif* sorghum	5,450	7,410	36	1096**	12.70
	Maize	8,310	10,990	32	929**	32.26
	Paddy	9,390	11,300	20	855**	19.23
	Pearl millet	5,070	6,930	37	1710*	4.67
	Finger millet	6,450	8,390	30	937**	16.59
	Rabi sorghum	4,690	6,090	30	883**	9.84
2012	*Kharif* sorghum	6,160	7,970	29	1048**	12.17
	Maize	8,100	10,490	30	958**	24.10
	Pearl millet	3,540	4,750	34	632**	14.38
	Rabi maize	7,660	10,950	43	3200*	8.13
	Finger millet	5,160	6,720	30	826**	13.94
	Rabi sorghum	4,450	5,740	29	486**	27.50
	Wheat	920	1,210	32	288*	4.24
2013	*Kharif* sorghum	6,070	7,860	29	995**	12.61
	Maize	8,390	10,770	28	422**	122.24
	Paddy	8,340	10,440	25	405**	103.63
	Pearl millet	3,810	5,340	40	492**	37.50
	Finger millet	5,720	7,290	28	448**	47.69
	Rabi sorghum	4,050	5,120	26	406**	26.84
	Wheat	2,600	3,440	33	396**	18.44

[a]LSD, Least significant difference; *, significant at 5% level; **, significant at 1% level.

Table 11.5. Biomass yield of major legumes under Bhoochetana during 2009–2013.

Year	Crop	Biomass yield: Farmers' practice (kg/ha)	Improved practice (kg/ha)	Increase (%)	LSD[a]	*F* value
2009	Chickpea	2060	2660	29	251**	22.19
2010	Black gram	3000	4160	39	1188NS	4
	Green gram	1380	1920	40	371**	8.47
	Pigeon pea	6070	7990	32	1095**	11.90
	Chickpea	2950	3810	29	384**	19.65
2011	Black gram	2000	2770	38	607*	6.69
	Cowpea	1280	1910	49	264**	23.43
	Field bean	5490	7230	32	3978NS	1.15
	Green gram	1600	2220	39	385**	10.19
	Pigeon pea	4420	5960	35	1146**	7.12
	Chickpea	1700	2230	32	306**	11.95
2012	Black gram	2760	3670	33	496**	14.64
	Field bean	3450	4510	31	1074NS	3.89
	Green gram	2520	3330	32	308**	28.06
	Horse gram	430	560	30	229NS	1.94
	Pigeon pea	4580	5830	27	1448NS	2.92
	Chickpea	1210	1580	30	223**	10.50
2013	Black gram	1820	2380	31	269**	17.78
	Cowpea	1450	1940	34	272**	12.97
	Field bean	2510	3380	34	887NS	3.86
	Green gram	1420	1880	32	243**	14.06
	Pigeon pea	3140	4220	34	435**	23.60
	Chickpea	1480	1940	32	140**	43.01

[a]LSD, Least significant difference; *, significant at 5% level; **, significant at 1% level; NS, not significant.

between farmers' practice and improved practice also suggests that there is an increase in benefits in the fields where improved practices were adopted with additional benefit in terms of increasing vegetative cover as well as soil moisture content. Total average biomass production was as high as 44% over farmers' practice in 2009 for soybean and 42% for groundnut in 2010. During 2011 and 2012, biomass production decreased by 4–5% compared with 2009 and 2010. Importantly, during normal and dry years, although improved practices showed better performance, the difference was not significant for *rabi* safflower (*Carthamus tinctorius*) during a dry year (2012) (Table 11.6).

Table 11.6. Biomass yield of major oilseed crops under Bhoochetana during 2009–2013.

		Biomass yield				
Year	Crop	Farmers' practice (kg/ha)	Improved practice (kg/ha)	Increase (%)	LSD[a]	*F* value
2009	Groundnut	2410	3170	31	243**	37.29
	Kharif sunflower	2530	3210	27	466**	8.75
	Soybean	3340	4808	44	367**	63.41
2010	Groundnut	3010	4290	42	289**	75.53
	Kharif sunflower	3320	3990	20	1090NS	1.6
	Soybean	3750	5070	35	345**	59
2011	Groundnut	2770	3820	38	691**	9.06
	Kharif sunflower	4500	5860	30	2357NS	1.33
	Soybean	4000	5360	34	690**	15.22
	Rabi safflower	2390	3120	30	432**	11.57
	Rabi sunflower	4910	6320	29	1391*	4.42
2012	Groundnut	2290	2980	30	366**	13.97
	Kharif sunflower	2770	3580	29	579**	7.82
	Soybean	2500	3250	30	689*	4.71
	Rabi safflower	1460	1860	27	417NS	3.57
	Rabi sunflower	2700	3530	31	651*	6.63
2013	Groundnut	3190	4150	30	321**	34.59
	Kharif sunflower	2610	3580	37	578**	11.12
	Soybean	2740	3400	24	181**	52.85
	Rabi safflower	2340	2970	27	246**	29.13
	Rabi sunflower	2360	3130	33	642*	5.68

[a]LSD, Least significant difference; *, significant at 5% level; **, significant at 1% level; NS, not significant.

11.3.4 Spatial variation in grain yield and biomass production

Although Bhoochetana was implemented across 30 districts over a period of 4 years (2009/10–2012/13), the crop yields were not uniform across the regions. Therefore, this provided an opportunity to look into the factors that are contributing to large-scale differences between different agroclimatic zones within the district and state. Karnataka is divided into ten agroclimatic zones based on: (i) the rainfall pattern, amount and distribution; (ii) soil types, texture, depth and physico-chemical properties; (iii) elevation; (iv) topography; (v) major crops; and (vi) type of vegetation.

The crop and agroclimatic zone-wise analysis revealed that legumes had significant yield gains. Although the yield of field bean (*Vicia faba*; broad bean) was quite impressive in all the agroclimatic zones, the difference was not significant when compared to farmers' practice. The comparison of chickpea grain yield between different agroclimatic zones revealed that the North Eastern Transition Zone performed better during 2010, 2012 and 2013 compared with other agroclimatic zones under improved practice. During this period, the average grain production was 1660 kg/ha and 2310 kg/ha, 800 kg/ha and 980 kg/ha and 1220 kg/ha and 1590 kg/ha under farmers' practice and improved practice, respectively.

A close look at regional variation revealed that during the first year (2009) the Northern Dry Zone performed better in terms of groundnut yield (1720 kg/ha and 2330 kg/ha, respectively), whereas in other years the yield fluctuated as there was rainfall variability. Among the six targeted agroclimatic zones, the Northern Dry Zone performed better in 2010 for maize yield followed by the Hilly Zone, the Central Dry Zone and the Southern Transition Zone. Similarly, the maize crop responded better to improved management practices during 2011 in the Northern Dry Zone, the Southern Transition Zone and the Hilly Zone, compared with farmers' practice. However, we observed wide variation among these regions due to variation in rainfall, soil type and climatic conditions. Groundnut samples obtained from five agroclimatic zones showed that the Southern Transition Zone responded better followed by the Southern Dry Zone and the Northern Dry Zone.

The spatial presentation of biomass production revealed that the average biomass of maize with improved practice was 19,450 kg/ha compared with 12,317 kg/ha with farmers' practice in 2010 in the Southern Transition Zone. However, with decreasing rainfall, biomass production also decreased significantly in the Eastern Dry Zone compared with other agroclimatic zones. The highest average biomass yield of paddy was 17,990 kg/ha with improved practice during 2013 in the North Eastern Dry Zone compared with 13,706 kg/ha with farmers' practice. In fact, paddy stalk is used as a major dry fodder for livestock. In this context, the increased biomass production under improved practice is very significant to overcome fodder scarcity. Therefore, crop–livestock integration assumes greater importance in the context of rainfed agriculture.

The variation in biomass production across the years was captured using box plots (Fig. 11.2). Groundnut showed more biomass variability during 2010 and 2012 as the highest biomass produced was about 8000 kg/ha with improved practice, whereas the lowest was < 2000 kg/ha. In all the situations, yield was less with farmers' practice than with improved practice. It is important to note that millets showed great variability across the years, irrespective of rainfall variability in the state. This can be attributed to sowing date, varieties and quantity of micronutrient application.

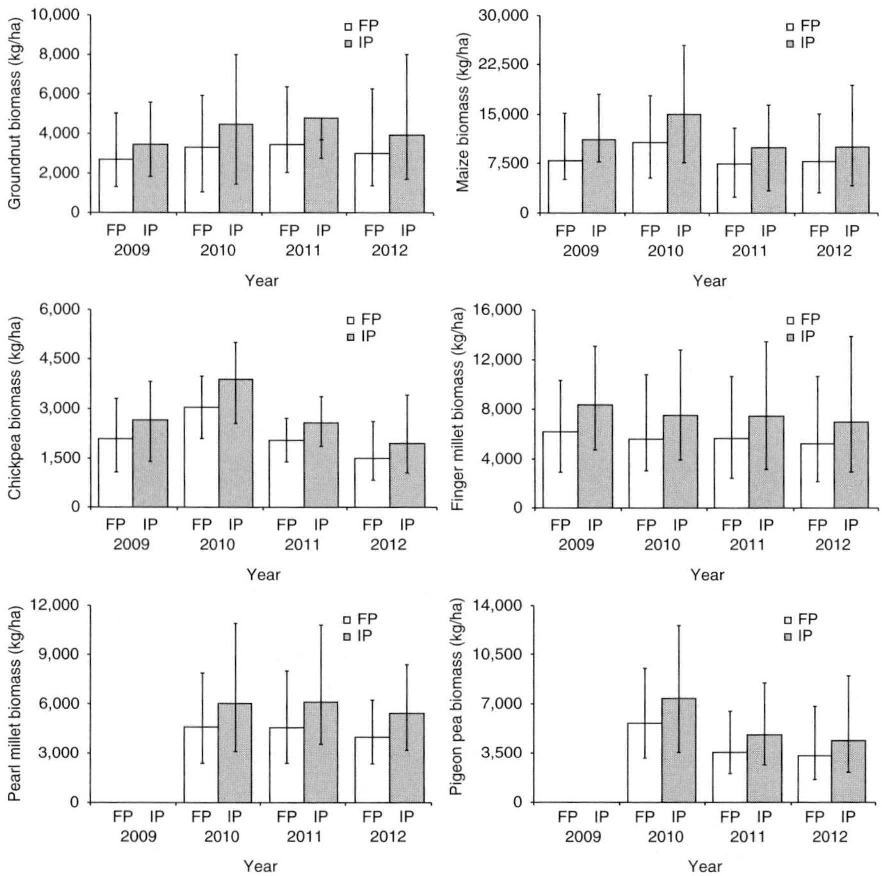

Fig. 11.2. Comparison of aboveground biomass production of groundnut, maize, chickpea, finger millet, pearl millet and pigeon pea with improved practice (IP) and farmers' practice (FP) from 2009 to 2012.

11.4 Profitability Analysis

The economic performance of the system was assessed using standard benefit–cost (BC) ratio analysis to determine production costs and profitability. Additional income and BC ratios were estimated separately for cereals, pulses and oilseeds for 5 years (2009–2013). We considered additional direct benefits – in terms of increased agricultural income – due to project interventions compared with farmers' practice. Gross income generated from crop yield was estimated using the market price (minimum support price). Subsequently, net economic returns were calculated by subtracting the cost of cultivation from the gross income. Additional income due to Bhoochetana interventions was derived as the difference in net income between improved practice and farmers' practice. Further additional BC was estimated from additional net income and additional investment on secondary and micronutrients.

To estimate average values for farmers' practice and improved practice, box plots were used to identify extreme cases and outliers for major crops under these practices. Net income per hectare from crop production was defined as gross returns minus variable cost (i.e. the cost of micronutrients). Finally, we compared the BC ratio of other improved practices followed in different agroecological regions of India with that of Bhoochetana to show the comparative advantage of improved crop management practices to bridge the large yield gaps in rainfed agriculture.

11.4.1 Crop yield and net income

Tables 11.7 and 11.8 compare the crop yield, net income and BC ratio between normal, dry and very dry rainfall years in Karnataka. To compare the economic benefit from different crops, the net income was calculated after subtracting the cost of cultivation from gross income both for farmers' practice and for improved practice. The results are most revealing. Maize, pigeon pea and groundnut were more remunerative during a normal rainfall year (2010) as the yield of these crops were 7250 kg/ha and 5410 kg/ha (maize), 1630 kg/ha and 1210 kg/ha (pigeon pea), and 1810 kg/ha and 1300 kg/ha (groundnut), under improved practice and farmers' practice, respectively. Increased yield resulted in enhanced net income of ₹16,220/ha, ₹14,670/ha and ₹11,900/ha, respectively, due to improved practice. Other dryland crops such as chickpea, soybean and green gram also performed better as the median yield increment was around 35% between farmers' practice and improved practice, resulting in increasing net income of ₹9770/ha, ₹9250/ha and ₹7970/ha, respectively, under moderate to good rainfall conditions. The improved technologies were very effective and improved profitability by 32–40%.

During 2011, although the yield of maize, pigeon pea, groundnut and paddy with improved practice was less than in 2010, it increased significantly over farmers' practice. Net income obtained from improved practice was ₹12,930/ha for maize, ₹12,450/ha for pigeon pea, ₹15,210/ha for groundnut and ₹12,160/ha for paddy. A similar trend was observed during 2012 where the net income was lower compared with 2011 in groundnut (₹9400/ha), and pigeon pea (₹9030/ha) but marginally increased for maize (₹14,450/ha) and paddy (₹12,950/ha). The effect of declining rainfall was more evident in all the years. The yield of all crops except green gram and soybean was reduced by 21–133% with improved practice during a very dry year (2012) when compared with a normal year (2010). However, improved practice helped to withstand the shock and subsequently increased net income.

Table 11.7. Average crop yield, net income and benefit–cost (BC) ratio of selected crops with improved practice (IP) and farmers' practice (FP) in normal (2010), dry (2011) and very dry (2012) rainfall years in Karnataka.

Crop	No. of taluks	Mean rainfall (mm)	Yield IP (kg/ha)	Yield FP (kg/ha)	Increase (%)	Net income (₹/ha)	BC ratio
2010 (Normal year)							
Chickpea	13	759	1,820	1,350	34	9,770	5.3
Green gram	14	800	760	550	40	7,970	4.4
Groundnut	18	572	1,810	1,300	40	11,900	6.7
Maize	31	655	7,250	5,410	34	16,220	9.1
Pearl millet	8	685	2,390	1,710	40	6,030	3.3
Pigeon pea	14	836	1,630	1,210	35	14,670	7.9
Finger millet	22	624	2,320	1,700	36	5,960	3.5
Sorghum	8	614	2,410	1,780	36	5,710	3.1
Soybean	9	863	2,650	2,010	32	9,250	5.1
Mean		712	2,560	1,890	36	9,720	5.4
2011 (Dry year)							
Chickpea	32	407	1,110	830	35	8,030	4.4
Green gram	18	484	830	580	41	9,620	5.4
Groundnut	32	389	1,900	1,340	42	15,210	9.0
Maize	55	470	5,270	3,950	33	12,930	8.5
Paddy	25	2720	5,580	4,480	25	12,160	9.7
Pearl millet	12	415	2,370	1,710	38	6,460	3.6
Pigeon pea	29	473	1,230	890	38	12,450	7.5
Finger millet	39	485	2,040	1,550	31	5,100	3.8
Sorghum	13	451	3,010	2,150	40	8,640	5.1
Soybean	11	520	1,850	1,320	41	9,020	5.0
Mean		681	2,520	1,880	36	9,960	6.2
2012 (Very dry year)							
Chickpea	33	382	780	600	30	5,330	3.1
Green gram	12	592	1,090	820	33	11,730	6.3
Groundnut	46	274	1,030	780	33	9,400	6.0
Maize	61	318	5,080	3,850	32	14,450	9.2
Paddy	50	1375	4,770	3,760	27	12,950	8.9
Pearl millet	26	380	1,980	1,490	33	5,750	3.4
Pigeon pea	42	367	1,040	810	29	9,030	5.2
Finger millet	45	283	1,680	1,260	33	6,300	4.4
Sorghum	11	388	2,510	1,940	29	8,650	4.7
Soybean	12	465	1,570	1,170	34	8,980	5.1
Mean		482	2,150	1,650	31	9,260	5.6

11.4.2 Benefit–cost (BC) ratio

Improved practices gave an acceptable and impressive return on investment during all the years. The mean BC ratios for 2010, 2011 and 2012 were 5.4, 6.2 and 5.6, respectively. Improved management practice contributed to enhance the BC ratio above the mean level for maize

Table 11.8. Benefit–cost (BC) ratio on a full cost basis in Bhoochetana in Karnataka.

Crops	No.of taluks	Average rainfall (mm)	Average yield (kg/ha)		Net income (₹/ha)		BC ratio	
			Farmers' practice	Improved practice	Farmers' practice	Improved practice	Farmers' practice	Improved practice
2009								
Finger millet	3	581.2	1,740	2,680	6,900	13,780	1.8	2.3
Maize	5	691.1	5,470	7,600	26,250	42,050	2.3	2.9
Chickpea	6	653.5	1,050	1,410	4,460	8,540	1.3	1.5
Soybean	2	807.4	1,770	2,630	10,730	21,140	1.8	2.4
Groundnut	8	593.6	1,280	1,760	8,050	16,610	1.4	1.8
Rabi sorghum	9	710.2	1,280	1,880	310	4,010	1.0	1.3
2010								
Pearl millet	8	685	1,710	2,390	6,040	10,340	1.7	2.0
Finger millet	22	624	1,700	2,320	7,410	11,520	1.8	2.1
Maize	30	655	5,440	7,270	28,450	42,920	2.5	3.0
Pigeon pea	14	836	1,210	1,630	19,240	29,870	2.1	2.6
Chickpea	13	759	1,350	1,820	10,440	17,040	1.8	2.2
Soybean	9	863	2,010	2,650	14,470	21,810	2.0	2.3
Green gram	14	800	550	760	7,110	12,300	1.7	2.0
Groundnut	18	572	1,300	1,810	10,970	21,130	1.6	2.0
Black gram	5	1012	930	1,260	13,820	21,600	2.1	2.5
Rabi sorghum	6	614	1,290	1,830	590	3,800	1.1	1.3

2011								
Pearl millet	12	415	1,710	2,370	7,810	12,660	1.9	2.2
Finger millet	39	485	1,550	2,040	7,300	11,000	1.8	2.1
Maize	48	470	4,000	5,350	19,770	31,600	2.1	2.6
Pigeon pea	28	473	920	1,270	12,190	21,750	1.8	2.2
Chickpea	31	407	850	1,150	4,100	8,610	1.3	1.6
Soybean	11	520	1,320	1,850	8,460	16,150	1.6	2.1
Green gram	18	484	590	830	10,350	17,130	2.1	2.5
Groundnut	31	389	1,380	1,960	18,090	32,250	2.0	2.7
Black gram	7	644	550	750	5,070	9,580	1.4	1.7
Rabi sorghum	32	451	1,370	1,830	2,430	5,460	1.3	1.5
Paddy	25	2720	4,480	5,580	13,640	24,240	1.4	1.7
2012								
Pearl millet	26	377	1,490	1,980	8,480	12,810	1.9	2.2
Finger millet	45	283	1,260	1,680	9,850	14,700	2.1	2.4
Maize	54	318	3,940	5,200	26,960	40,470	2.4	3.0
Pigeon pea	41	367	830	1,070	14,750	22,500	1.9	2.3
Chickpea	30	382	610	790	3,490	7,110	1.3	1.5
Soybean	12	465	1,170	1,570	11,950	19,510	1.9	2.3
Green gram	12	592	820	1,090	25,980	36,140	3.6	4.2
Groundnut	45	274	790	1,050	10,380	18,540	1.6	2.0
Black gram	5	654	1,210	1,630	39,930	56,590	4.3	5.2
Rabi sorghum	31	388	1,270	1,690	8,120	12,990	1.8	2.1
Paddy	48	1375	3,700	4,680	11,500	22,540	1.3	1.6

Table 11.9. Comparison of net returns for improved versus traditional base practices reported since 2000 for different crop and natural resource management interventions in rainfed crop production in India. (Adapted from Harris and Orr, 2014.)

Intervention[a]	Crop	Net returns (US$/ha/season)		Increase (%)	Benefit–cost (BC) ratio		Reference
		Base	Improved		Base	Improved	
Tillage							
Tillage + fertilizer (Inceptisol), 9 years	Pearl millet	207	254	23	1.93	2.33	Maruthi Sankar *et al.* (2012)
Tillage + fertilizer (Vertisol), 9 years	Pearl millet	153	285	86	1.89	3.52	Maruthi Sankar *et al.* (2012)
Tillage + fertilizer (Aridsol), 9 years	Pearl millet	44	86	95	1.12	1.26	Maruthi Sankar *et al.* (2012)
Rotations, fallows, intercropping							
Alley cropping	Soybean, safflower, tree products	117	156	33	1.88	2.27	Mutanal *et al.* (2009)
Alley cropping, discounted at 12%	Soybean, safflower, tree products	39	58	49	1.88	2.27	Mutanal *et al.* (2009)
Leucaena-based agroforestry	Cowpea, timber	145	542	274	1.86	3.17	Prasad *et al.* (2010)
Biomass retention, double cropping	Rice-vegetable sequences	84	752	795	0.46	1.82	Das *et al.* (2008)
Crop mixtures, inter-cropping	Wheat, lentil, toria	101	437	333	1.79	2.1	Kumar *et al.* (2008)
Intercropping	Maize, black gram	89	194	118	1.45	1.78	Sheoran *et al.* (2010)
Intercropping	Pigeon pea, maize	123	346	181	2.61	2.75	Marer *et al.* (2007)

Fertilizers and soil amendments							
Phosphorus and biofertilizers	Pigeon pea	224	444	98	2.51	4.09	Singh and Yadav (2008)
Fertilizer + FYM	Rice, niger	175	303	73	2.07	2.21	Gogoi *et al*. (2010)
Fertilizer + organic inputs	Sesame	54	248	359	1.39	2.43	Deshmukh and Duhoon (2008)
Foliar spray with calcium nitrate	Paddy	194	327	69	0.86	1.38	Kundu and Sarkar (2009)
Foliar spray with potassium chloride	Hybrid cotton	317	454	43	1.87	2.24	Aladakatti *et al*. (2011)
Phosphorus + VAM	Wheat	159	268	68	1.55	1.86	Singh and Singh (2008)
Pest and disease control							
Improved weed control	Wheat	208	398	91	0.60	1.37	Singh *et al*. (2010)
Improved varieties							
Improved versus local varieties, farmers' fields	Chickpea	196	360	84	4.28	5.6	Shiyani *et al*. (2001)
Improved versus local varieties	Chickpea	142	199	40	1.34	1.58	Kiresur *et al*. (2010)
Improved versus local varieties (mean overall crops in farmers' fields)	Pearl millet, sorghum, mung bean (green gram), groundnut, wheat, barley, mustard and chickpea	208	283	36	2	2.58	Mann *et al*. (2009)

[a]FYM, Farmyard manure; VAM, vesicular arbuscular mycorrhiza.

(9.1), pigeon pea (7.9) and groundnut (6.7) during 2010 while the same crops performed poorly during 2012 due to poor rainfall. However, all the above crops and paddy performed well with a BC ratio above the mean level (see Table 11.7).

Table 11.9 presents the results of the literature survey, including outliers, showing levels of net income from rainfed cropping with and without improved technologies in different regions of India. The net income obtained by adopting improved management practices increased in the range of 23–795% in different crops. The BC ratio, which was calculated by taking full cost and returns of the system in different regions, showed a significant difference between farmers' practice and improved practice. It varied from as low as 1.37:1 in wheat with improved weed control measures to 4.09:1 in pigeon pea with application of phosphorus and biofertilizers. Although it is not comparable as the measures are different in both the cases, it is important to understand the return on investment. In Bhoochetana, the overall BC ratio (which was calculated by taking into account additional cost and additional income) of the cropping system in Karnataka is above 5:1. This suggests that the integrated approach, including soil-test-based fertilizer application, improved varieties of seed and integrated pest management measures, has the potential of producing a higher BC ratio as compared with any single management approach.

11.4.3 Net profits and additional value of crop production

At the state level, the improved crop yield contributed to enhanced net income and additional value of the product. Net profits have been arrived at by subtracting the cost of micronutrients. At the state level, by the end of the fifth year, the net profit accrued from the programme was about ₹9727 million from all the 30 districts. During the first year, about ₹155 million was generated from six districts covering 0.2 million ha and during the second year the programme extended to another ten districts with an additional net profit of ₹1077 million. Similarly, during the third and fourth years the programme was implemented in all 30 districts and the net profit increased to ₹3886 million in 2011 and ₹4609 million in 2012 (Plate 6). The result of crop cutting experiments revealed that the state has potential in bridging the large yield gap by adopting holistic science-led crop management interventions with farmers' participation.

The value of additional production due to improved management practices is presented in Plate 7. The results showed that during the first phase of Bhoochetana, the additional value generated from crop production is about ₹11,190 million. It is interesting to note that the additional value of crop production during the first year is about ₹182 million, which covers only six districts and the second year generated about ₹1275 million. Similarly, during 2012 (a very dry year) the total additional value of the product is about ₹5267 million which is significantly higher than the dry year (₹4466 million in 2011). In fact, 2011 and 2012 were drought years in terms of rainfall quantity and distribution.

11.5 Measuring Social Benefits

Apart from yield and economic benefits, we also have attempted to analyse social benefits of the programme as the yield and economic benefits have immensely contributed to improving the social status of the participating farmers. A household survey was carried out in eight districts representing four revenue divisions of the state. In each selected district, a representative taluk (block) was selected and within the taluk two villages were chosen for household survey. Twenty households were randomly selected in each village and a total of 320 households were surveyed. The simple social benefit measures such as increased investment on assets formation, gender equity and enhanced knowledge were applied to estimate the benefits.

11.5.1 Reinvestment of additional income on asset formation

About 40% of farm households have reinvested the additional income obtained from Bhoochetana on agriculture and agriculture-related infrastructure. A proportion (13%) of households had also invested in white goods (luxury goods) such as a fridge, ceiling fan, mixer grinder, mobile and vehicles. It is important to note that about 10% of the households have invested income obtained from Bhoochetana on loan repayment, house infrastructure and education. The additional income was useful for the many households (7%) to overcome health-related expenditure, which is significant as it ensures better working conditions for the family members. The Bhoochetana programme also helped to take care of domestic expenditures as it facilitates small savings due to crop improvement (Plate 8).

11.5.2 Knowledge improvement

Bhoochetana is a holistic process-based mission project which was intended to not only increase crop productivity but also enhance stakeholders' knowledge regarding agricultural operations. The analysis covered major activities which are part of the Bhoochetana mission project and periodic training or capacity building programmes were organized to disseminate the knowledge. The results are more revealing. First, the knowledge dissemination process initiated by ICRISAT through master trainers from the University of Agricultural Sciences and the DoA, Karnataka had a most positive impact on farmers, as more than 50% of the households have acknowledged improved knowledge on major aspects of agricultural development in the state. Secondly, the knowledge about soil health status, micronutrient application and seed varieties, which are critical components of agricultural development, improved significantly. More than 85% of rural households reported that their knowledge enhanced on these critical components. Thirdly, nearly 80% of households have learnt new methods to control pests and diseases to enhance their crop yield in rainfed agriculture (Fig. 11.3).

11.5.3 Addressing gender equity

The gender equity issue was addressed by analysing the decision-making process by men and women farmers who are following Bhoochetana practices. The analysis revealed that women exclusively have a very meagre role in decision making with regard to selection of crop, variety, land preparation, fertilizer and manure application, irrigation, harvesting, threshing and marketing (Fig. 11.4). However, most of the critical decisions related to the above-mentioned activities were taken jointly, which shows that there is a consensus among men and women to carry out agricultural activities in the dryland areas. It is evident that women are mostly involved in laborious activities on which they have decision-making control, namely, transplanting (23%), hand weeding (19%) and interculture (12%). It is worth mentioning that men have greater control over decision making in almost all the activities and among which marketing is the critical aspect. More than 70% of the men and women farmers jointly made decisions regarding harvesting, threshing and seed selection and storage. This reflects that certain activities essentially benefited with women's decisions, which are critical in agricultural operations.

11.6 The Way Forward

11.6.1 Crop yield, biomass production and benefit of micronutrients

The increase in crop yield with improved practice was 18–52% in cereals, 26–45% in legumes and 21–48% in oilseeds. Among the cereals, finger millet recorded the highest yield increase of (52%) over farmers' practice during 2009, whereas in other years the yield difference was around 37%, 31%, 32% and 28% during 2010, 2011, 2012 and 2013, respectively. Given the perception that finger millet responded to improved management practices very meagrely, the increased rate of yield in Bhoochetana is significant. It was observed that rainfall variation

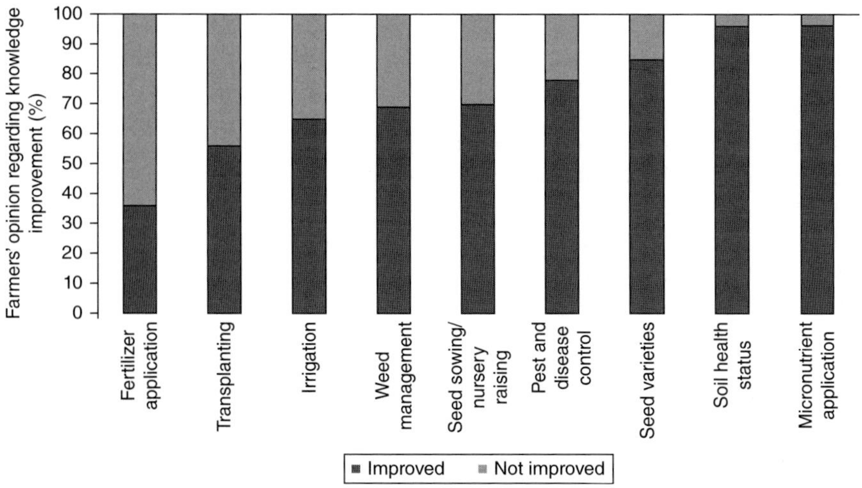

Fig. 11.3. Contribution of Bhoochetana to increased farmers' knowledge regarding agricultural development in Karnataka.

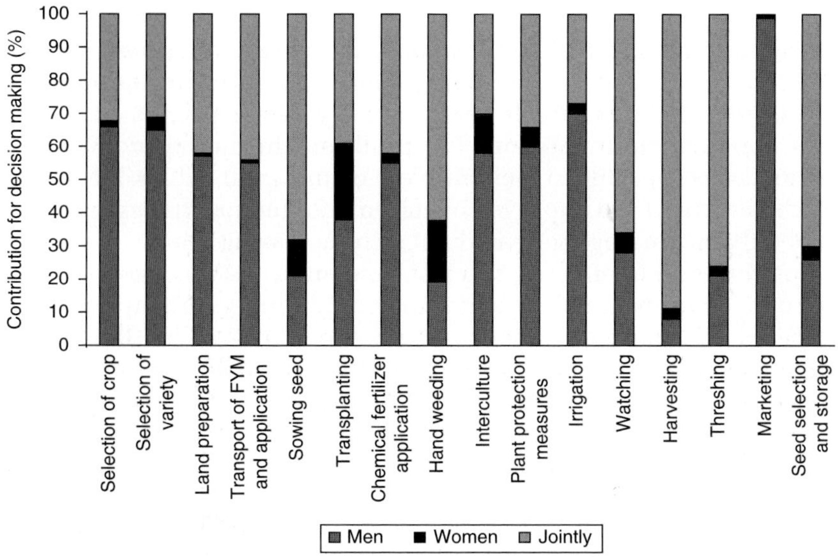

Fig. 11.4. Bhoochetana implementation helped to minimize gender inequality and enhanced decision making of women and men in agriculture-related activities. FYM, Farmyard manure.

had a negative impact on crop yield and as a result, the yield gain was reduced during low rainfall years. However, crop yield was considerably higher when compared with farmers' practice. It showed that the use of micronutrients, new varieties and other improved practices had greater effect on the crop yield. Similarly, legumes and oilseed crops also performed well throughout the project period. Since legumes form a major part of staple food crops in the state, efforts are needed to revive the legume-based cropping system with new knowledge, practices, methods and approaches.

The total vegetative biomass from various cropping systems was highest with improved practice compared with farmers' practice. These amounts are sufficient to have a positive impact on crop and water balance when the biomass is incorporated into the soil, and/or can improve animal health when used as fodder. This has implications on the fertilizer usage as well as fodder quality where livestock is more integrated into the farming system. Biomass retention in the field also increases soil permeability and water infiltration as a result of attracting termites to the plots that make many subsurface tunnels. Most communities burn the crop residues without knowing the value of the crop residues for enriching the soil. Therefore, communities need to be sensitized to retain crop residues (Ngwira *et al.*, 2012).

11.6.2 Maximizing the profit due to an integrated approach

The cumulative effect of an integrated management approach on crop yield resulted in impressive BC ratios for major crops in the state. The analysis revealed that cereals, legumes and oilseeds performed better in terms of return on investment. The profit maximization proposition is high for cash crops due to their inherent commercial nature. However, it was observed that the improved management practices have contributed to not only enhancing crop yield but also increasing the income and return on the investment. The soil-test-based nutrient management along with other improved technologies increased crop yields up to 66% as compared with farmers' practice (Wani *et al.*, 2012). The Bhoochetana programme initially started with 0.2 million ha in six districts and expanded to 3.73 million ha in all 30 districts of Karnataka and in the subsequent years has contributed immensely to the state's GDP. Overall, a net profit of about ₹9726 million was generated by adopting improved practices. About 3.6 million farmers benefited from Bhoochetana. The net income obtained by adopting improved technologies varied from ₹5000/ha to ₹16,000/ha depending on the crop, soil type and rainfall situation. The return on investment ranged between 3 and 10 depending on crop, soil type and rainfall condition and was an acceptable proposition for future investment in dryland agriculture with improved management practices. Our analysis also suggests that crops were found to be tolerant to an increasing drought situation and pests and diseases, resulting in increased yield compared with conventional farmers' practice. This is evidenced not only during normal years but also during dry and very dry years, thereby indicating that improved management practices are critical in building the resilience of the farming system with increased climate variability. The increased yield as well as the increase in income (up by about 30%) have contributed to households' economy in rural areas. The additional net income obtained from the project was invested on asset creation and building social capital. This exemplary strategy has contributed to enhance not only the yield and income but also the knowledge and decision-making ability of men and women in agriculture-related activities.

11.6.3 Future scope for the integrated crop management approach

The integrated approach often offers to minimize risks related to production, maximizes water use efficiency and minimizes production costs (Wani *et al.*, 2003, 2012). In the case of Bhoochetana, science-led innovative approaches were implemented to realize higher yields with modifications in soil, water and crop management technologies. These modifications in crop management often require significant changes in technological and economic support to the farmers, especially in regions where farmers are not accustomed to using micronutrients to rejuvenate the soil and enhance the yield. Thus, rainfed areas of semi-arid regions could be more favourable for adoption of this integrated approach (Bhoochetana) because farmers are more receptive to new interventions and familiar with integrated technologies to enhance the crop yield.

The present study demonstrated the technical performance of science-led interventions at field level through action research. There is a need to understand major drivers and hindrances beyond the field level. For example, an understanding of market conditions, interactions among stakeholders and other institutional and political dimensions become important to achieve profitability. The availability of inputs such as seeds, fertilizers, micronutrients and crop protection chemicals ensures timely sowing and planting and also that other agricultural operations are carried out easily. The same applies to farmers' access to credit and to markets for agricultural inputs and produce. The innovative institutional arrangements helped to reach out to millions of farmers, bringing improved knowledge to their doorstep.

11.7 Conclusion

The preceding discussion reveals that the yield and economic benefits of the Bhoochetana programme are contributing towards improving rural livelihoods in Karnataka. The improved practices demonstrated with farmers' participation gave improved yield of cereals, legumes and oilseeds. It was possible to increase total biomass production in farmers' fields using integrated crop management practices without reduction in grain yield of the crops. Similarly, the improved crop management system resulted in higher income per hectare than with farmers' practice. The integrated approach adopted with the great willingness of the consortium partners, including line departments, helped to bridge the large yield gaps that existed in the state. The net additional value generated by Bhoochetana has contributed significantly to the state's economy. The high return on investment suggests that there is a great hope for future investment in dryland agriculture. The discussion suggests that there are two situations where crop production can provide a viable and achievable solution for food security and poverty alleviation in rainfed regions. First, smallholders can adopt improved management practices to increase their farm productivity and income. This is possible when the government is willing to introduce science-led innovations to bridge the large yield gaps. Secondly, it is not possible for smallholders to increase their farm size with increased land fragmentation. Thus, intensification is one of the viable options to enhance the efficiency of the land with improved knowledge and technologies. The above discussion has produced enough evidence to take this proposition forward to address the land quality and food security issues together and to build resilience in the system. However, even when the BC analysis at farm level indicated economic benefits, farmers may lack the opportunity to purchase agricultural inputs to undertake the sowing operations or lack credit. In this context, the incentive mechanism proposed by the Government of Karnataka to handhold smallholders is also another option to promote sustainable intensification in dryland agriculture.

Acknowledgements

The authors sincerely acknowledge funding support from the Government of Karnataka for implementing Bhoochetana. Thanks are also due to DoA staff, farm facilitators and ICRISAT field staff for conducting crop cutting experiments.

References

Aladakatti, Y.R., Hallikeri, S.S., Nandagavi, R.A., Naveen, N.E., Hugar, A.Y. and Blaise, D. (2011) Yield and fiber qualities of hybrid cotton (*Gossypium hirsutum*) as influenced by soil and foliar application of potassium. *Karnataka Journal of Agricultural Sciences* 24(2), 133–136.

Bhalla, G.S., Hazell, P. and Err, J. (1999) Prospects for India's cereal supply and demand for 2020. *Food, Agriculture and the Environment*. Discussion Paper 29. International Food Policy Research Institute, Washington, DC.

Das, A., Patel, D.P., Munda, G.C., Hazarika, U.K. and Bordoloi, J. (2008) Nutrient recycling potential in rice–vegetable cropping sequences under *in situ* residue management at mid-altitude subtropical Meghalaya. *Nutrient Cycling in Agroecosystems* 82, 251–258.

Deshmukh, M.R. and Duhoon, S.S. (2008) Effect of organic inputs on sesame (*Sesamum indicum* L.) under rainfed situation in Kymore Plateau Zone of Madhya Pradesh (India). *Journal of Maharashtra Agricultural Universities* 33(3), 323–324.

Food and Agriculture Organization of the United Nations (FAO) (2010) *The State of Food Insecurity in the World 2010: Addressing Food Insecurity in Protracted Crises*. FAO, Rome.

Gogoi, B., Barua, N.G. and Baruah, T.C. (2010) Effect of integrated nutrient management on growth, yield of crops and availability of nutrients in Inceptisol under rainfed rice (*Oryza sativa*)–niger (*Guizotia abyssinica*) sequence of Asom. *Indian Journal of Agricultural Sciences* 80(9), 824–828.

Government of Karnataka (GoK) (2011) *Karnataka Agriculture Budget – 2011–12*. GoK, Bangalore, India.

Harris, D. and Orr, A. (2014) Is rainfed agriculture really a pathway from poverty? *Agricultural Systems* 123, 84–96.

Hobbs, P.R., Sayre, K. and Gupta, R. (2008) The role of conservation agriculture in sustainable agriculture. *Philosophical Transactions of the Royal Society B: Biological Sciences* 363(1491), 543–555.

Kanwar, J.S. (2000) Soil and water resource management for sustainable agriculture imperatives for India. In: *Managing Natural Resources for Sustainable Agricultural Production in the 21st Century*. International Conference 14–18 February 2000, New Delhi. Water Technology Centre and Indian Agricultural Research Institute, New Delhi, India, pp. 4–6.

Kiresur, V.R., Bantilan, M.C.S., Parthasarathy, R.P., Rao, G.D.N., Padmaja, R., Anupama, K.V., Suhasini, K. and Kulkarni, G.N. (2010) *Chickpea Breeding and Seed Delivery Efforts to Enhance the Impact on the Livelihoods of the Poor in Drought Prone Areas of South Asia – Insights from Baseline Studies*. Summary Report. International Crops Research Institute for the Semi-Arid Tropics, Patancheru, Andhra Pradesh, India.

Kumar, P. (1998) Food demand and supply projections for India. *Agricultural Economics Policy Paper* 98-01. Indian Agricultural Research Institute, New Delhi, India.

Kumar, N., Prakash, V., Mina, B.L., Gopinath, K.A. and Srivastva A.K. (2008) Evaluation of toria (*Brassica campestris*) and lentil (*Lens culinaris*) varieties in intercropping system with wheat *(Triticum aestivum*) under rainfed conditions. *Indian Journal of Agronomy* 53(1), 47–50.

Kundu, C. and Sarkar, R.K. (2009) Effect of foliar application of potassium nitrate and calcium nitrate on performance of rainfed lowland rice (*Oryza sativa*). *Indian Journal of Agronomy* 54(4), 428–432.

Lal, R. (2009) Soil degradation as a reason for inadequate human nutrition. *Food Security* 1, 45–57.

Mann, J.S., Jat, H.S., Kumar, V., Kumar, S., Chaturvedi, O.H., Saha, S., Kumar, D. and Bhaka, B.L. (2009) A participatory approach to improve the economic condition of resource poor farmers in semi-arid regions of Rajasthan. *The South Asian Association for Regional Cooperation (SAARC) Journal of Agriculture* 7(1), 1–10.

Marer, S.B., Lingaraju, B.S. and Shashidhara, G.B. (2007) Productivity and economics of maize and pigeonpea intercropping under rainfed condition in Northern Transitional Zone of Karnataka. *Karnataka Journal of Agricultural Sciences* 20(1), 1–3.

Maruthi Sankar, G.R., Mishra, P.K., Sharma, K.L., Singh, S.P., Nema, A.K., Kathmale, D.K., Upadhye, S.K., Sidhpuria, M.S., Osman, M. and Ravindra Chary, G. (2012) Efficient tillage and nutrient practices for sustainable pearl millet productivity in different soil and agro-climatic conditions. *Experimental Agriculture* 48(1), 1–20.

Mutanal, S.M., Patil, S.J., Patil, H.Y., Shahapurmath, G. and Maheshwarappa, V. (2009) Performance of soyabean–safflower under different tree species in black soils. *Karnataka Journal of Agricultural Sciences* 22(2), 377–381.

Ngwira, A.R., Aune, J.B. and Mkwinda, S. (2012) On-farm evaluation of yield and economic benefit of short term maize legume intercropping systems under conservation agriculture in Malawi. *Field Crops Research* 132,149–157.

Prasad, J.V.N.S., Korwar, G.R., Rao, K.V., Srinivas, K., Rama Rao, C.A., Srinivasarao, C., Venkateswarlu, B., Rao, S.N. and Kulkarni, D. (2010) Effect of modification of tree density and geometry on intercrop yields and economic returns in *Leucaena* based agro-forestry systems for wood production in Andhra Pradesh. *Southern India Journal of Experimental Agriculture* 46(2), 155–172.

Raju, K.V., Wani, S.P. and Anantha, K.H. (2013) *Bhoochetana: Innovative Institutional Partnerships to Boost Productivity of Rainfed Agriculture in Karnataka, India.* Resilient Dryland Systems Report No. 59. International Crops Research Institute for the Semi-Arid Tropics, Patancheru, Andhra Pradesh, India.

Sahrawat, K.L., Rego, T.J., Wani, S.P. and Pardhasaradhi, G. (2008) Sulfur, boron and zinc fertilization effects on grain and straw quality of maize and sorghum grown on farmers' fields in the semi-arid tropical region of India. *Journal of Plant Nutrition* 31, 1578–1584.

Sahrawat, K.L., Wani, S.P., Pardhasaradhi, G. and Murthy, K.V.S. (2010) Diagnosis of secondary and micronutrient deficiencies and their management in rainfed agroecosystems: case study from Indian semi-arid tropics. *Communications in Soil Science and Plant Analysis* 41, 346–360.

Sheoran, P., Sardana, V., Singh, S. and Bhushan, B. (2010) Bio-economic evaluation of rainfed maize (*Zea mays*)-based intercropping systems with blackgram (*Vigna mungo*) under different spatial arrangements. *Indian Journal of Agricultural Sciences* 80(3), 244–247.

Shiyani, R.L., Joshi, P.K. and Bantilan, M.C.S. (2001) *Impact of Chickpea Research in Gujarat.* Impact Series No. 9. International Crops Research Institute for the Semi-Arid Tropics, Patancheru, Andhra Pradesh India.

Singh, S.R. and Singh, U. (2008) Efficacy of vesicular arbuscular mycorrhizae as influenced by phosphorus application in wheat (*Triticum aestivum*) under rainfed conditions of Kashmir. *Indian Journal of Agricultural Sciences* 78(9), 771–776.

Singh, R.S. and Yadav, M.K. (2008) Effect of phosphorus and biofertilizers on growth, yield and nutrient uptake of long duration pigeonpea under rainfed condition. *Journal of Food Legumes* 21(1), 46–48.

Singh, P., Aggarwal, P.K., Bhatia, V.S., Murty, M.V.R., Pala, M., Oweis, T., Benli, B., Rao, K.P.C. and Wani, S.P. (2009) Yield gap analysis: modelling of achievable yields at farm level. In: Wani, S.P., Rockström, J. and Oweis, T. (eds) *Rainfed Agriculture: Unlocking the Potential.* CAB International, Wallingford, UK, pp. 81–123.

Singh, G., Singh, O.P., Singh, S. and Prasad, K. (2010) Weed management in late sown wheat (*Triticum aestivum*) after rice (*Oryza sativa*) in rice–wheat system in rainfed lowland. *Indian Journal of Agricultural Sciences* 55(2), 83–88.

United Nations (2006) World Population Prospects: the 2006 Revision and World Urbanization Prospects: the 2005 Revision. Population Division of the Department of Economic and Social Affairs of the United Nations Secretariat. Available at: http://www.un.org/esa/population/publications/wpp2006/English.pdf (accessed 24 August 2014).

Wani, S.P., Maglinao, A.R., Ramakrishna, A. and Rego, T.J. (eds) (2003) *Integrated Watershed Management for Land and Water Conservation and Sustainable Agricultural Production in Asia. Proceedings of the ADB-ICRISAT-IWMI Annual Project Review and Planning Meeting, 10–14 December 2001, Hanoi, Vietnam.* International Crops Research Institute for the Semi-Arid Tropics, Patancheru, Andhra Pradesh, India.

Wani, S.P., Dixin, Y., Li, Z., Dar, W.D. and Chander, G. (2011) Enhancing agricultural productivity and rural incomes through sustainable use of natural resources in the semi-arid tropics. *Journal of the Science of Food and* Agriculture 92, 1054–1063.

Wani, S.P., Sarvesh, K.V., Krishnappa, K., Dharmarajan, B.K. and Deepaja, S.M. (2012) *Bhoochetana: Mission to Boost Productivity of Rainfed Agriculture through Science-led Interventions in Karnataka.* International Crops Research Institute for the Semi-Arid Tropics, Patancheru, Andhra Pradesh, India.

12 Lessons Learnt and a Way Forward

Suhas P. Wani* and K.V. Raju

International Crops Research Institute for the Semi-Arid Tropics (ICRISAT), Patancheru, India

12.1 Introduction

The Bhoochetana project is one of a kind – a scaling-up initiative implemented during 2009–2012 by the Department of Agriculture (DoA), Government of Karnataka, India, with technical support from the consortium led by the International Crops Research Institute for the Semi-Arid Tropics (ICRISAT). This unique project has had a good impact as discussed in the earlier chapters (Chapters 1–11) of this book and provides an excellent opportunity for synthesizing the lessons learnt from this novel initiative by a government agency in India and ICRISAT, an international research organization. The Bhoochetana project has achieved the impact on a large scale, spread over 5.1 million ha in 2013/14 and improved the livelihoods of 4.75 million farmers directly and another 16–18 million people indirectly in 30 districts of the state of Karnataka over 4 years.

This large-scale initiative was based on agricultural productivity enhancement, a pilot programme under the Sujala–ICRISAT watershed project supported by the World Bank in Karnataka during 2002–2005 (more details in Chapters 2 and 6 of this book). The unique mechanism in terms of institutions, policies, capacity building, monitoring, planning and implementation are responsible for the success of this project

*s.wani@cgiar.org

on a larger scale in Karnataka. Across the world, this is a rare project on rainfed agriculture, wherein science-led development helped millions of farmers to increase crop productivity, profitability and sustainability for achieving food and nutritional security as well as to improve the livelihoods of millions of smallholder farmers in rainfed areas. The process adopted from planning to implementation and monitoring was evolutionary and the business approach evolved through iterations during the project period. It was a continuous learning process for all the stakeholders as well as the consortium members, and lessons learnt from the fields helped to refine the policies and institutions for implementation of the project during the 4-year period in the state. The key processes and the quick major decisions along with the outputs are listed in Table 12.1.

12.2 A Demand-driven Project

In Karnataka, the agricultural growth rate during 2001–2008 was stagnant at around 1% or even lower. The state was in dire need of increasing agricultural productivity as the livelihoods of 60% of the population (in 2008) were dependent on agriculture. As there was a demand for increasing the productivity, government officials and policy makers were looking for technological solutions/innovations that could help scaling up for the benefit of farmers. During the brainstorming session in 2008, the Watershed Commissioner, who had piloted the productivity enhancement initiative with the help of ICRISAT in 3700 ha in different micro-watersheds, pursued rainfed agriculture – an area that was already of particular interest to ICRISAT scientists. As Karnataka has 65% of the 12.31 million ha (cultivable area) under rainfed agriculture, the state government quickly picked up the possibility of scaling up the interventions which were proven earlier by ICRISAT to be successful for increasing the productivity of dryland agriculture under the Sujala–ICRISAT watershed initiative.

12.3 Realistic Planning and Implementing a Strategy

Once it was decided that rainfed agriculture potential can be harnessed with technical support from ICRISAT, the DoA in the Government of Karnataka wanted to design a long-term technical proposal to boost agricultural growth in the state. It was important that a technical proposal needed to have a clear understanding of: (i) the available technologies; (ii) the possible adoption rates considering the situation on the ground; (iii) existing bottlenecks for scaling up in the current setup; and (iv) the possible increase based on actual experiences. The government was seeking for clear deliverable targets from the project in a time-bound

Table 12.1. Key processes, major decisions and output.

Key processes	Major decisions	Output
Selection of resource agencies	• Invited ICRISAT as technical support organization • Invited three state agricultural universities to provide region-wise handholding to field team, through training and technologies	• Timely availability of information and knowledge • Close follow-up on adoption of technology
Strong monitoring system and quick decision-making process	• High-powered committee review • Weekly progress review • Weekly video conference to address the issues discussed in the weekly review meeting • Field visit by district nodal officers. At the initial stage, there was a nodal senior officer for each district, which made a big impact in strengthening work at ground level	• Better planning • Timely execution • Increased reach at field level • Proper information sharing at all levels
Appointing farm facilitators (FFs) and lead farmers	• One FF from the village for every 500 ha; two to three lead farmers to implement concepts at field level	• A village link person is available for timely sharing of information and proper follow-up • Easy demonstration of new and improved technologies
Input procurement	• A well-established procurement system to ensure quality procurement of inputs • Properly monitored at higher level	• Quality input available at cluster and *hobli* (group of villages) level • Easy access for farmers • Timely application
Storage facility at cluster level and village level	• Increase reach of farmers to inputs stored at cluster and village level	• Inputs are stored at cluster level and available for farmers as per their requirement
Special budget for FF, lead farmers, storage and transportation of input	• Provision in the state budget to ensure timely activities in the field	• Inputs are stored well in time and close to villages • Transportation at farmers' doorstep made application easy and timely

Continued

Table 12.1. Continued.

Key processes	Major decisions	Output
		• FF able to link more farmers to adopt better practices • Lead farmers could take up demonstrations on their plots
Efficient communication	• Wall writings • Pocket booklets • Community meetings • Soil health cards • Web-based soil health information, access to identify soil health issues and recommendations • Slogan writings • Soil maps on village walls	• Farmers are aware about the importance of micronutrients in keeping crops healthy • Increased awareness • Farmer awareness regarding outcome of each micronutrient, after application in the field

manner and the project could arrive at an achievable target of increased productivity of 20% in 4 years. To work out the possible and achievable impact on a larger scale, science-based quantitative data for increased yields, adoption rates and possible constraints needed to be identified. This was essential owing to existing institutional mechanisms in the state and required refinements. The tangible benefits from the project were possible to achieve on an annual basis, where enabling institutions and policies are firmed up by the government in a timely manner; accordingly, an agreement between the government agency and ICRISAT was signed to achieve the success of the Bhoochetana programme.

12.4 Identification of Constraints

The priority for the project team was to make a scientific assessment of the yield gaps for the major crops grown and the constraints for potential yield. The yield gap between the current farmers' yield and the achievable potential yield for different crops was two- to fourfold (Wani *et al.*, 2012) and the knowledge about the constraints (technical, social, institutional, policies, infrastructural, etc.) were identified and incorporated into the project design. The project team identified the possible interventions and strategized the implementation in a phased manner for enhancing the agricultural productivity. A realistic assessment of constraints on the farm level and detailed knowledge of the systems in the state are critical for highlighting the intervention.

12.5 Knowledge-based Entry Point Activity

The initial interventions were made for a large-scale project and these were expected to provide tangible economic and visible benefits to small farmers. For this purpose, the entry point activity had to be taken up very carefully, ensuring that activity carried out in farmers' fields would succeed with 90–95% probability and everyone would be able to see the benefits by adopting these interventions. For this purpose, the project team undertook soil health mapping as the major intervention for piloting productivity enhancement in six districts covering 0.23 million ha in 1440 villages during 2009/10. The agricultural soils in the whole state were mapped for their nutrient content by adopting a stratified soil sampling method on different toposequences and socio-economic situations in the villages (Wani *et al.*, 2011).

For efficient use of scarce water availability in dryland agriculture, soil health/fertility is a major driver; but unfortunately, farmers are not aware about their soil nutrient status. As a result, farmers apply the fertilizer indiscriminately based on the suggestions/recommendations by the fertilizer dealers, which resulted in low use efficiency of fertilizers and water, thereby increasing the cost of cultivation but not the productivity (Wani *et al.*, 2013). Based on earlier experience, widespread deficiencies of secondary and micronutrients along with macronutrients were holding back the potential of agriculture in different districts of Karnataka. To obtain the correct and quantitative knowledge about the soil nutrient status, soil health mapping for the whole state was undertaken in a phased manner. During the first year, samples from only 15 districts were collected by adopting farmer participatory sample collection. This may be a unique example in which 92,904 soil samples from farmers' fields were collected by the farmers and these were analysed in a state-of-the-art laboratory in ICRISAT (more details in Chapter 3 of this book). All 92,904 soil samples were analysed for all the plant nutrients as well as soil pH and electrical conductivity and these results were shared with the community, policy makers and the public by adopting information and communication technology (ICT) tools available for such dissemination (Wani *et al.*, 2012). It is clear from large-scale mapping of soil nutrient status in the state covering all taluks that widespread deficiency of multiple secondary and micronutrients was holding back the agricultural potential and the current recommendations of generally applying NPK (nitrogen–phosphorus–potassium) fertilizers to different crops at state/eco-regional level did not hold good for developing recommendations for the farmers. Taluk-wise recommendations based on soil nutrient mapping were prepared to demonstrate the benefits to farmers.

12.6 A Community Participatory Approach is Essential for Impact

Since a large number of smallholder farmers reside in villages, a community participatory approach was adopted while undertaking technology evaluations in the villages. For popularizing the approach, soil samples were collected by the farmers. Awareness among farmers was increased through village meetings, and thus, farmers' curiosity to understand the nutritional status of their soils played an important role in popularizing the new taluk-wise recommendations based on soil nutrient analysis.

12.7 Soil Health Information Dissemination to the Community and Policy Makers

After undertaking soil nutrient mapping, it was important to disseminate the information in a simple language that could be understood by the farmers as well as the policy makers. In Bhoochetana, as farmers collected the soil samples, they were very curious to know the results and this curiosity helped to build awareness among the farmers about balanced fertilizer application for enhancing the efficiency and protecting the environment. Soil nutrient content status was discussed in the village through village meetings and this was also written on the walls in the local language, so that farmers could understand, along with the recommendations using different sources of fertilizers. In addition, soil health cards for individual farmers along with recommendations for major crops grown in the village were prepared and disseminated to the farmers. For policy makers and other interested persons, soil nutrient information was put in the public domain through the Internet, a soil health atlas (Plate 10) and small pocket books in the local language (Wani *et al.*, 2011). All the information was made available to the farmers before they sowed their crops, which helped the farmers to make a decision about adoption of improved management practices (more details in Chapter 3 of this book).

12.8 Enabling Policies and Proper Incentives

Available knowledge or information would not benefit farmers automatically and availability of recommended materials was important to achieve the impact. Through appropriate discussions, evidence-based suggestions for policy change were made to the government and policy makers, which in turn ensured dissemination of the knowledge and based on that information, farmers could rectify the deficiencies. This played an important role in the success of the project. Farmers were ensured the availability of the recommended fertilizers at the village/cluster level so that they need not travel a long distance at the time of sowing, which could discourage them for adoption of the recommendations. The Government of Karnataka, through convergence of different schemes, ensured the availability of incentives such as covering 50% of the cost of micronutrients to all farmers who wanted to use the recommendation and also ensured the availability of inputs by proper planning at district and taluk levels (more details in Chapter 6 of this book). To ensure the availability of fertilizers, during the season godowns (warehouses) were hired in villages. With such a single window system, farmers could access the information as well as use the recommended fertilizers and other improved practices for increasing agricultural productivity.

12.9 Creation of an Alternate Extension Mechanism through Farm Facilitators (FFs) and Lead Farmers

The existing extension system was not able to reach the farmers, and farmers had no access to scientific information about improved management practices except only what was provided through their peer groups and/or input dealers. Bhoochetana created a new cadre of para-extension workers known as farm facilitators (FFs) and lead farmers to act as a link between farmers in the villages and the DoA, Karnataka. It was proposed that for every 500 ha, one FF should be employed as a para-extension worker and paid an honorarium for providing services to help farmers during the crop season (more details in Chapters 9 and 10 of this book). During the peak activities of the crop season such as sowing and harvesting, three to four lead farmers were generally assigned to each FF. These FFs were trained properly by master trainers who in turn were trained particularly for adopting the new paradigm of taluk-wise fertilizer recommendations, based on soil nutrient mapping in the taluk.

12.10 Seeing is Believing

It is important for farmers to see the process and effect of new technologies in their own villages. With the help of FFs, the improved technologies such as seeds of improved cultivars, soil-test-based fertilizer application, including secondary and micronutrients, integrated water management practices, biocontrol agents, and soil and water conservation measures were demonstrated to the farmers by the farmers. We have observed that farmer-to-farmer knowledge dissemination about improved management practices is more effective than extension workers/scientists disseminating the knowledge to farmers in the villages (more details in Chapter 10 of this book). For this purpose, good demonstrations on large plots with appropriate controls were laid out on farmers' fields and conducted by adopting farmer participatory strategies. Farmers were involved from the beginning in the decision-making process to decide what to do and how to do it and also continuously monitored the changes. During field days, they explained to other farmers their methodology and observations.

12.10.1 Field days for dissemination of results

To list the results from farmer participatory demonstrations with improved management practices, farmers' field days were organized during the crop season, with appropriate publicity, for farmers from the surrounding villages. A large number of farmers were invited and exposed to the overall strategy of the whole programme as well as interventions that were made under the programme and the process adopted (more details in Chapter 10 of this book). Through field days, technology dissemination was achieved quickly as confirmed by farmers; also clarifications were provided by FFs or lead farmers on the spot.

12.10.2 Farmer-to-farmer videos for effective dissemination

FFs were trained to document through short video recordings various improved interventions undertaken by farmers. These videos presented the whole process, as well as the output achieved, in the local language by using battery-operated Pico projectors. Edited short videos helped FFs to demonstrate the improved technologies, besides concerns of farmers (more details in Chapter 5 of this book). This method of farmer-to-farmer videos has been proved to be more effective than other methods of knowledge dissemination in the project.

12.10.3 Crop cutting experiments (CCEs)

For quantitative evaluations of impact, scientifically proven crop cutting experiments (CCEs), were undertaken by adopting the stratified random sampling method for each taluk. This process involved farmers, officials from the DoA, the Department of Statistics and Economics, and state agricultural universities (SAUs); thereby, this process has provided good authenticity, transparency and publicity as the CCEs were done in the presence of farmers in farmers' fields. By weighing the samples in the field to obtain the fresh weight, the farmers themselves, as well as other farmers, could see the impact and they started discussing the results even before the results were finalized. This also resulted in proper dissemination of the project activities and their associated impacts (more details in Chapter 8 of this book).

12.11 A Result-based Framework for Monitoring and Evaluation

It is important that the implementing agency is clearly aware at the beginning of the project about the performance indicators which will be monitored for assessing the progress at different levels, and also whether these indicators are quantitative and/or qualitative. Quantitative performance indicators are always preferred to avoid ambiguity and to help all stakeholders to assess the progress and identify good and weak performers who need to be supported appropriately in order to achieve the best result. Weekly online monitoring and evaluation at different levels has helped to provide complete appreciation by all actors involved in the implementation as well as the monitoring process to check the needed quantitative indicators regularly. It is a well-established fact that 'what gets measured gets delivered'. The success of Bhoochetana is largely because of objective monitoring and evaluation at all levels, as well as close on-site visits by the authority at all levels, which also resulted in enhancing awareness among the farmers as well as officials at different levels (more details in Chapter 6 of this book). Using this specified format and quantitative indicators from FF level upward, results are reported at weekly intervals to be collated at taluk, district and state levels for further discussion and evaluation by the higher authorities.

12.11.1 Weekly video conference

A weekly video conference was conducted by higher authorities at state level for the officials of all the 30 districts of Karnataka who were involved in implementation of Bhoochetana. During the weekly review through video conference meetings, the overall situation of the rainfall and the project interventions were discussed and summaries of the results across the districts for different parameters were presented and discussed. Good and weak performers were identified for the specific indicators and the difficulties faced by the weak performers as well as the innovations adopted by the good performers were discussed and shared with all participants. Through this process of learning from each other, good agricultural practices were adopted in other districts also.

12.11.2 Wider publicity

It is important to widely publicize the project strategies, the objectives and the progress and this is a pressure on all the stakeholders as public commitments are made and all are aware about the targets to be achieved. The results of field days, CCEs as well as farmer-to-farmer videos were disseminated widely using audio, video and print media to different stakeholders at different levels starting from taluk, national and international levels for dissemination. Such a strategy helped in putting the achievements of the project in the limelight and also for further dissemination for scaling up in different states and countries (more details in Chapter 9 of this book). This also served as a motivator for stakeholders in the state, who are associated with the implementation as it was a personal recognition at different levels starting from individual district officials up to the state policy makers which strengthened the strategy and implementation further and rectified the weak links to enhance the impact of the various initiatives.

12.12 Key Lessons Learnt

Key lessons learnt that emerged from the earlier chapters as well as from the processes described above are provided in this section.

12.12.1 Realistic planning

- For careful planning, it is essential to undertake scientific assessment of all constraints existing in the state and based on that knowledge and the potential which can be achieved realistically, the strategy needs to be developed.
- One needs to be realistic in expecting the impact on the ground. Considering the variables in socio-economic conditions, implementing arrangements in different districts as well as shortcomings and constraints (institutional policies, operational procedures, human resources, etc.) at local level, using earlier experience, realistic targets need to be fixed.
- Holistic and science-led development strategies which are well proven need to be identified and demonstrated under different socio-economic and agroecological conditions in the state for establishing the sites of learning (i.e. the 'seeing is believing' principle plays an important role in scaling up technologies).
- For each intervention, the alternative approach available needs to be highlighted. For example in the case of secondary and micronutrients, if application as a basal dose is delayed or missed, farmers

should be advised to undertake spraying of micronutrients at the appropriate stage to get the benefits accordingly.

- Farmers are very responsive to knowledge-based improved agricultural practices and come forward willingly to undertake evaluations through participatory research, as is evident from the participation of farmers with the use of micronutrients across 40% of the area in different districts of Karnataka.
- Smallholder and marginal farmers need to be integrated and provided with the necessary technical handholding support to implement improved interventions. Then farmers become good spokespersons for the project to reach the large number of smallholder farmers and marginal farmers.
- There could be some unknown good examples which are not expected in the programme and these need to be considered carefully. For example in Bhoochetana where a neglected crop such as finger millet (*Eleusine coracana*), which was considered as a non-responsive crop, showed a far better response and benefited smallholder farmers making the programme popular in the state.
- Farmer-to-farmer dissemination of improved technologies played a very important role along with print media in the state. Such dissemination was achieved by adopting the principles of 'seeing is believing' through demonstrations by the farmers, farmer-to-farmer videos and dissemination during field days by the farmers themselves.
- Capacity building of all the stakeholders played an important role. There is a need to overcome the deficiencies in the existing extension system. Labour arrangements of FFs and lead farmers trained by the master trainers from the SAUs played a very important role in achieving the impact of the Bhoochetana programme.
- Publicity using the print/audio/video media in the field and sensitizing the media people about the achievements played a very important role and finally, the programme became farmer driven rather than scientist driven. Thus inputs and knowledge played a very important role in the success of the project.
- Changing the mindset of different actors in order to adopt the proposed changes takes time and concerted efforts need to be made with humility. On many occasions the leader as well as team members need to keep their professional expertise, experiences and personal feelings in check in the interest of the success of the project while dealing with government officials and policy makers.

12.12.2 Convergence

- Resources and knowledge played an important role in scaling up the benefits of the holistic and consortium approach, involving a consortium of different actors.

- The integrated holistic approach adopted to unlock the potential of rainfed agriculture not only increased productivity and profits during normal rainfall years but also showed good performance in terms of increased crop yields over farmers' practice during the deficit rainfall years, which helped in building the trust of the farmers in the DoA.
- The convergence of various actors involved in knowledge dissemination was very much needed as conflicting messages/information provided to the farmers creates confusion on the ground and adversely affects the programme. In such scenarios, the higher authorities need to handle the situation carefully and avoid any such conflicts across various players in the consortium to ensure benefits for the farmers.

12.12.3 Effective implementation and monitoring

- For effective implementation and monitoring it was essential at the start of the project to identify constraints and put together a package of possible solutions to promote the improved management practices. Farmers as well as other stakeholders need appropriate capacity building measures to internalize the new interventions.
- Regular review meetings at all levels and use of quantitative indicators enhanced the transparency and accountability in the programme, which helped in building the trust and performance of all the actors.
- Timely availability of resources (financial, human, materials, infrastructure, etc.) at district and taluk levels was very crucial. Thus timely communication enhanced the implementation and impact of the programme.
- To overcome the operational bottlenecks at different levels, at the start of the project it was essential that the State Level Coordination Committee for Bhoochetana met frequently. Internalization and institutionalization of the programme through appropriate capacity building initiatives at all the levels played a very important role for achieving the targets of the mission project.
- Timely planning, availability of inputs, and human resources were ensured through appropriate institutional and policy interventions.

12.12.4 Scaling up

- In order to scale up any initiative, it should be need-based and demand driven rather than supply driven for the science-led development in the state/country.

- For success of the scaling-up programme, a 'champion' at a high level in the Karnataka Government played a very crucial role to address the issues at different levels in the department. The champion had to monitor and evaluate as well as understand the constraints and appreciate the good performers who contributed significantly to the success of the project.
- For scaling up large programmes, an appropriate operational unit needs to be identified, and for dealing with natural resource management, it is beneficial to embark on existing area-based approach programmes like integrated watershed management.
- Execution of large programmes like Bhoochetana should be done in a phased manner and only well-proven interventions/technologies which give tangible economic benefits (e.g. soil-analysis-based fertilizer applications and/or use of improved seed) should be used as an entry point demonstration; any technologies where there is a chance of failure should be avoided for demonstration plots.
- For scaling-up initiatives like Bhoochetana, where convergence, consortium and collective action are important, interventions needed clarity in communication at different levels, as internalization of strategy is crucial not only for the farmers but also for policy makers, researchers and extension agents to avoid confusion among the farmers as well as other stakeholders.
- The strategy should be flexible, having an evolutionary model, as the process of learning continually and mid-term correction are very much needed.
- Adoption of a science-led farmer participatory development approach helped in organizing communities for collective action and wide dissemination for scaling up the programmes in the state.
- Transaction costs are high as many stakeholders are involved. Moreover, frequent transfers of government officials associated with project implementation, as well as monitoring, call for continued investment of time and resources.
- A dedicated team of scientists with attitude and aptitude to technically support such a scaling-up initiative is a must.
- Identifying a suitable champion to lead the scaling-up initiatives is important. The person should be well experienced, polite and skilled to interact with government officials and policy makers and also have a good reputation to deliver the outputs and work closely with the team.
- A holistic approach enabled solutions from start to finish, for example from diagnosis of nutrient deficiencies in the soil and other constraints that hold back the potential of rainfed agriculture in the state through to providing workable solutions. Critical components for the success of the programme were: (i) diagnosing the constraints; (ii) following diagnosis, taking the necessary steps for awareness

building through dissemination of the information among different stakeholders (farmers, FFs, DoA officials, policy makers and scientists); (iii) after dissemination of information, ensuring availability of the needed inputs in a timely manner with enabling policies and institutions; and (iv) monitoring and evaluation.

- Transforming scientific knowledge into information which can be understood by the farmers was a critical step and also ensuring information and input delivery to the farmers through a suitable channel is important.

12.13 Way Forward

The Bhoochetana scaling-up programme has clearly established a 'proof of concept' for the science-led development to make a difference through increasing agricultural productivity and improving the livelihoods of millions of smallholder farmers in the country. The importance of soil science, which has been known for a century, could have an impact of benefiting millions of farmers; however, only by understanding the impact pathway (from making a discovery in a laboratory or field plot, through the stages of piloting and proof of concept, to scaling up to the final impact on a large scale, where millions of farmers can benefit) can gaps in that pathway be identified. Generally, scientists develop a technology or knowledge or product and take it from the discovery phase to the pilot stage or proof of concept stage and leave it there for someone to take it forward to achieve the outcomes and impacts. However, in reality, the gap between proof of concept and the impact/outcomes stage is a missing link for achieving the impacts, as a number of players are involved in the process of taking it up to the millions of farmers. The Bhoochetana mission initiative has demonstrated that this gap between the proof of concept and the impact/outcomes stage can be bridged successfully through crossing the 'death valley' of impact as shown in Fig. 12.1.

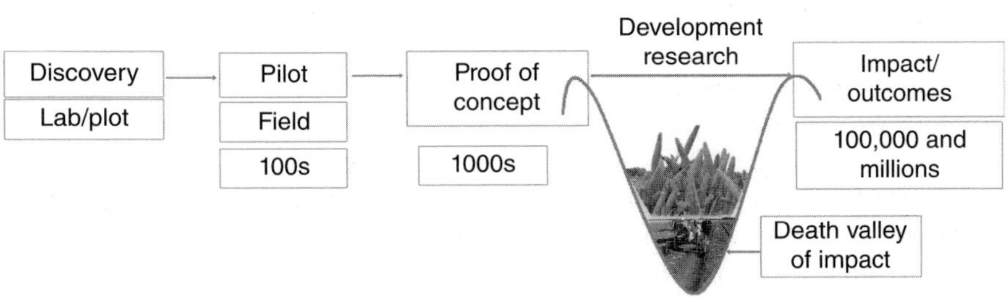

Fig. 12.1. Improving livelihoods through crossing the 'death valley' of impact.

Based on the lessons learnt, this initiative has demonstrated a pathway to scale up an increase in agricultural productivity by two- to threefold to make agriculture productive and profitable by adopting a holistic approach and integrating the science with practitioners, like farmers in this case, as observed from existing yield gaps. These lessons learnt in the state of Karnataka may not apply universally and may vary from case to case for different projects as well as different provinces, countries, etc. However, by bridging the yield gaps, a second green revolution in rainfed areas in other states of India and other countries may be feasible. The following steps can guide the scaling-up initiatives in other states/countries:

- A systematic and planned initiative for a given agroecology or a state needs to be undertaken to understand the major constraints withholding the potential of agriculture in the country/region.
- As evident from the experiences in Bhoochetana in Karnataka, understanding the soil nutrient content in other states or the country as a whole is very critical, which would increase the productivity of the crops with soil-test-based fertilizer recommendations. Different states in the country, and the country as a whole, need to adopt stratified soil sampling for assessing the nutrient deficiencies in different parts of the country.
- Based on the soil analysis, a strategy specific for the different states needs to be prepared taking into account the existing delivery systems (knowledge, inputs, etc.).
- For undertaking reliable and quality soil analysis, specific laboratories in the region need to be identified and accredited for undertaking soil analysis, by adopting uniform methods in the country to enable comparison and cross learning.
- Knowledge delivery (public extension) systems are quite weak in India and alternative knowledge delivery systems need to be developed and adopted for reaching the unreached by using ICT.
- Enabling policies to promote balanced fertilization need to be ensured and appropriate incentives to promote integrated nutrient management should be in place.
- Farmers' livelihoods can be improved with increased production, provided the market prices are stable and remunerative. Appropriate marketing support is essential for smallholder farmers to increase productivity and profitability.
- A value chain approach for enabling more productivity for the farmers needs to be ensured through public–private partnerships, and quality assurance for the services as well as inputs provided to the farmers need to be ensured.
- Different states in the country can use the results-based framework adopted in Bhoochetana as well as the scaling-up approach.

Imbibing the lessons learnt from Bhoochetana can also help the states to break the ceiling that has existed in agricultural growth for a long time.
- New technologies such as remote sensing, geographical information system (GIS), ICT and new partnerships between researchers and development agencies, government departments and corporates, as well as community groups, need to be harnessed for mutual benefits in the country.
- At the country level, soil health maps need to be prepared to benefit farmers and the knowledge delivered from the soil analysis needs to be communicated as simple information to the farmers for adoption.
- A new mechanism for science-led development in the country needs to be adopted and refined further, in order to achieve the impacts that could benefit millions of smallholder farmers by making agriculture productive and profitable through mechanization and use of new science tools and technologies.
- There is an urgent need to change the paradigm of development in the country by undertaking demand-driven research by the researchers and at the same time ensuring the impact of their research by adopting the approach of 'science with a human face'. This is possible by addressing the newly emerging area of delivery of science for which appropriate skill development initiatives are needed to build partnerships. Also, it is important to establish the sites of learning while making sure that policy makers put in place the requisite enabling policies and institutions to ensure the benefits/impacts are delivered to those that need them.

References

Wani, S.P., Sahrawat, K.L., Sarvesh, K.V., Baburao Mudbi and Krishnappa, K. (2011) *Soil Fertility Atlas for Karnataka, India*. International Crops Research Institute for the Semi-Arid Tropics, Patancheru, Andhra Pradesh, India.

Wani, S.P., Sarvesh, K.V., Krishnappa, K., Dharmarajan, B.K. and Deepaja, S.M. (2012) *Bhoochetana: Mission to Boost Productivity of Rainfed Agriculture through Science-led Interventions in Karnataka*. International Crops Research Institute for the Semi-Arid Tropics, Patancheru, Andhra Pradesh, India.

Wani, S.P., Khanwalkar, S., Krishnappa, K., Raju, K.V. and Sarvesh, K.V. (2013) *Bhoochetana: Process Documentation – A Program that Helped Farmers come out of Low Productivity and Poor Economic Status*. Resilient Dryland Systems Report No. 60. International Crops Research Institute for the Semi-Arid Tropics, Patancheru, Andhra Pradesh, India.

Index

Page numbers in **bold** type refer to figures, tables and boxed text.